Validation and Verification of Automated Systems

Andrea Leitner • Daniel Watzenig •
Javier Ibanez-Guzman
Editors

Validation and Verification of Automated Systems

Results of the ENABLE-S3 Project

Editors
Andrea Leitner
AVL LIST GMBH
Graz, Steiermark, Austria

Daniel Watzenig
Virtual Vehicle Research Center
Graz University of Technology
Graz, Steiermark, Austria

Javier Ibanez-Guzman
API : TCR RUC 0 52
Renault S.A.
Guyancourt, France

ISBN 978-3-030-14630-6 ISBN 978-3-030-14628-3 (eBook)
https://doi.org/10.1007/978-3-030-14628-3

This Springer imprint is published by the registered company Springer Nature Switzerland AG.
The registered company address is: Gewerbestrasse 11, 6330 Cham, Switzerland

Preface

ENABLE-S3 has been built as pan-European project that brings together both value chains and ecosystems from six different domains in the field of validation and testing of automated cyber-physical systems. The consortium, consisting of 68 partners from 16 European countries, led to a unique network of experts working on the verification and validation of automated cyber-physical systems (ACPS). ENABLE-S3 has emerged as a European cornerstone project that contributes significantly to the advancement of the state of the art in terms of new methods, tools, and validation processes covering "Design-for-Safety," "Design-for-Reliability," and "Design-for-Security" to impact end-user acceptance. The technology resulting from this work strongly supports the market introduction of automated systems by:

- Enabling safe, secure, and functional systems across six application domains (automotive, aerospace, rail, maritime, farming, and health care)
- Establishing a modular comprehensive test and validation framework for scenario-based testing
- Driving open standards for higher interoperability.

The diversity and the major challenges to validation and testing that remain to be resolved to advance automation closer to reality have motivated the edition of this book. All the contributions have been carefully selected to cover validation and testing aspects of the three building blocks of automation—sense, reason, and act—from an academic and industrial perspective. As an industry-driven project, there is also a section highlighting the achievements by the industrial use cases. All the contributions review the current state of practice and state of the art in their respective domains. The articles envision exciting results of ENABLE-S3, future trends, along with the specific challenges in the validation and testing of automated cyber-physical systems, sensors (i.e. radar, LiDAR, and camera), the concepts for improved availability, as well as quality assurance and reproducible test setups for automation.

ENABLE-S3 can be considered as one of the first industry-driven initiatives to make autonomy safe, reliable, and acceptable by society. However, we need several more collaborative projects to progress beyond where ENABLE-S3 left off.

We strongly believe that this book on validation and testing of automated cyber-physical systems greatly summarizes the key achievements of this remarkable project. ENABLE-S3 provides an overview of current and emerging technical challenges in the validation of highly automated cyber-physical systems. It provides in-depth insights into industrial demands. We hope that the reader will be inspired by the different technical contributions, selected project results, and major outcomes from the cooperation work across the consortium over the past 3 years.

It is a pleasure to introduce you to the fascinating world of validation of ACPS; the various contributions should create awareness of the complexities and solutions available in this domain.

Finally, we would like to express our sincere appreciation and gratitude to all authors and coauthors, who made the publication of this book possible. We are grateful to Silvia Schilgerius at Springer for her professionalism and support during the preparation of this book.

Enjoy your reading.

Graz, Austria — Andrea Leitner
Guyancourt, France — Javier Ibanez-Guzman
Graz, Austria — Daniel Watzenig

Acknowledgement

This work has received funding from the Ecsel Joint Undertaking Under Grant Agreement No 692455. This joint undertaking receives support from the European Union's Horizon 2020 Research and Innovation Programme and Austria, Denmark, Germany, Finland, Czech Republic, Italy, Spain, Portugal, Poland, Ireland, Belgium, France, Netherlands, United Kingdom, Slovakia, and Norway.

Contents

Part I
Introduction and Motivation

Challenges for Scenario-Based V&V of Automated Systems from an OEM Perspective

Javier Ibanez-Guzman

The beginning of the twenty-first century is marking a major turn in the technological development by mankind. The combination of powerful computing platforms, networking, advanced algorithms, the ability to manipulate large quantities of data and machine learning methods has resulted into their tight integration with physical processes. Cyber-Physical Systems (CPS) represent this integration enabling the monitoring and control of the physical processes, with feedback loops resulting in strong interactions between the controlling systems and physical processes.

The deployment of highly automated CPS has moved from laboratories to public spaces, different levels of automation are being experimented with. The most known operational trials are represented by robo-taxis which aim to deploy driverless passenger vehicles as the main form of transport in our cities. Navigating autonomously in such environments represents perhaps one of the major challenges within the CPS realm. Multiple applications beyond land, air and sea transportation have rapidly emerged, these include medical, agricultural and factory applications all with their own levels of complexity. CPS represent a paradigm change, they introduce a very strong symbiosis between highly automated systems and the environment, whilst their control is shifted from humans to computers. It opens multiple alternatives as shown by the lasting interest in autonomous vehicles beyond vehicle OEMs and the substantial investments made over the past 5 years.

For managers, to decide whether CPS are to be deployed, substantial factual assurances are needed. Investments are high due to the different technologies embedded in CPS; acceptability by society is a key issue. That is, any major malfunctions leading to hazardous situations could result in the rejection of CPS. As systems become under more computer control for safety related applications, the need to know how these would react (if at all) in all types of operating situations

J. Ibanez-Guzman (✉)
Research Division, Renault S.A., Guyancourt, France
e-mail: javier.ibanez-guzman@renault.com

A. Leitner et al. (eds.), *Validation and Verification of Automated Systems*,
https://doi.org/10.1007/978-3-030-14628-3_1

is a major requirement. Edge-cases (the long tail) represent the largest challenges, it is difficult to design for all of them. To solve such issues, it is important to validate the operational component of CPS by running tests that are statistically credible. However, this brings out a major challenge, how can physical tests be done for an almost infinite number of situations that CPS will encounter. For example, an autonomous vehicle crossing a simple intersection, will face thousands of likely combinations by simply considering the presence within its evolving environment of several entities e.g. other passenger vehicles, trucks, emergency service vehicles, pedestrians (adult, elderly, children), motorcycles, bicycles, etc. All converge at different speeds and different time intervals. If one considers, road, weather and light conditions, the combinations are too many. Could it be possible to make such tests physically? Culture influences the way people react. Do we need to test under different geographical situations? What about the age distributions of the actors involved? There are many other similar examples. New techniques are necessary, novel testing frameworks are needed, entire system models are required (e.g. a digital twin).

ENABLE-S3 represents a major contribution to the emergent field of validation and testing of CPS. Its strength resides in the combination of domain knowledge Europe-wide. Different industry led use cases have been studied, different solutions proposed, from formal methods, simulation and real experimentation as well as theoretical fundamentals. From an automotive perspective, it has enabled us to understand the complexities of the problem, to gain knowledge from efforts done in different domains and on the interaction with test & measurement specialists, suppliers, academics and our peers at large. Different use cases are examined at various levels of detail. Some results are very successful others opened a pandora-box. From the testing of autonomous vehicles perspective, it has allowed us to appropriate for ourselves of the ENABLE-S3 framework, to define a clear partition between the system under test (SUT) from the testing systems (TS), to find about the difficulties of integrating different simulation software into a coherent SUT, to examine how simulation models can be credible. The different chapters included in this publication should provide an overview of state-of-the-art solutions, approaches whose scope originated through work on specific industrial use cases.

ENABLE-S3 was formulated more than 4 years ago, within this period we have witnessed the renaissance of Artificial Intelligence, the use of deep machine learning which currently is being applied not only for classification but is also part of the decision-making process. Progress in this domain is exponential. From a validation point of view, this raises major difficulties, that of understanding how machine learning based methods operate and to comprehend why they have failed. Whilst with conventional algorithmic methods it is possible to understand the whole process, in the ML approach this is not explicit. It is still an emergent area of research; methods that facilitate this testing are in progress. Another challenge is to demonstrate that the testing space covers all the likely scenarios and on how the experimental space can be minimized to validate the simulations via sampled experimentation. The whole is known as explicability and robustness of Artificial

Intelligence based methods. Currently this is the subject of extensive exploratory research.

The validation of CPS requires common efforts, however, at some point this should be considered highly-competitive, those companies able to validate complex CPS will be the first to market, potentially scalable products.

Challenges for Scenario-Based V&V of Automated Systems from an Academic Perspective

Daniel Watzenig

Automation is seen as one of the key technologies that considerably will shape our society and will sustainably influence future mobility, transportation modes, and our quality of life. Many benefits are expected ranging from reduced accidents, efficient people/freight/goods movement, effective multi-modal transport (on-road, off-road, sea, air), better road/vehicle utilization, or social inclusion. Beyond mobility, automation creates tremendous impact in production, health-care, and eventually our everyday life.

In automotive, it is envisioned that automated driving technology will lead to a paradigm shift in transportation systems in terms of user experience, mode choices, and business models. Over the last couple of years, significant progress has been made in vehicle automation. Given the current momentum, automated driving can be expected continuously to progress, and a variety of products will become commercially available within a decade. Not only technological advancements have been demonstrated in many venues, several states and federal governments in Europe, the US, and in Asia have moved forward to set up regulations and guidelines preparing for the introduction of self-driving cars.

However, despite tremendous improvements in sensor technology, high performance computing, machine learning, computer vision, data fusion techniques, and other system technology areas, market introduction of a fully automated vehicle that is capable of unsupervised driving in an unstructured environment remains a long-term goal. In order to be accepted by drivers and other stakeholders, automated vehicles must be reliable and significantly safer than today's driving baseline. The ultimate safety test for automated vehicles will have to point out how well they

D. Watzenig (✉)
Graz University of Technology, Graz, Austria

Virtual Vehicle Research Center, Graz, Austria
e-mail: Daniel.Watzenig@v2c2.at

A. Leitner et al. (eds.), *Validation and Verification of Automated Systems*,
https://doi.org/10.1007/978-3-030-14628-3_2

can replicate the crash-free performance of human drivers especially at the level of conditional and high automation within mixed traffic.

Beyond the technological issues, several regulatory action items for faster introduction of automated vehicles still have to be resolved by the governments in order to ensure full compatibility with the public expectations regarding legal responsibility, safety, and privacy.

Compared to typical functions (SAE level 2, available on the market) which are designed for specific situations, SAE level 3+ functions need to be designed and tested for a significantly higher number of situations. The complexity of the real-world situations (including edge/corner cases and rare but safety-relevant events) does not allow a complete specification to be available in the requirement phase when developing along the V-cycle process. Thus, the traditional static specification sheet as it is used in the automotive development and testing process is no longer applicable. With increasing the level of automation, the complexity of the systems and therewith the need for agile development and scenario-based/situation-dependent testing increases exponentially. Conventional validation methods, tools, and processes are not sufficient anymore to test highly-interacting vehicle functions which have to be robust in complex traffic situations, evolving scenarios, and adverse weather conditions. The transition towards a higher level of artificial and swarm intelligence poses even greater demands.

As the technology for automated driving becomes more and more advanced, the research focus is shifting towards the emerging issues for their type approval, certification, and widely agreed homologation procedures affecting the entire value chain from semiconductor industry via component suppliers and software vendors to integrators. Within this complex chain of sensing-processing-controlling-actuating with a driver-in/off-the-loop many types of failures might occur and have to be avoided by design and during operation. These include failures of components and hardware deficiencies, software failures, degraded system performance due to inoperable sensors or the vehicle is completely inoperable, deficiencies in sensing (road, traffic, and environmental conditions), or faulty driver and vehicle interaction (mode confusion and false commanding). Additionally, the question on how to periodically re-qualify and re-certificate suchlike systems is currently open and needs to be addressed. This is of particular interest when software updates and/or hardware upgrades come into play.

Consequently, there is a strong need for an independent and reproducible validation of automated cyber-physical systems. Without a traceable demonstration of the maturity (technological readiness), reliability, and safety, the societal acceptance will lack. Reliable and safe enough means rare enough, i.e. the failure probability rate should be less than 10-n per hour. If somehow an adequate sample size n can be argued—which is currently not the case—then appropriate safety and reliability can be demonstrated by driving (testing) in the order of 10n hours. For $n > 5$ this becomes effectively infeasible since effort and related costs will increase tremendously. In fact, we must test even longer, potentially repeating tests multiple times to achieve statistical significance. Yes, there is a strong need for methods and tools to supplement real-world testing in order to assess automated vehicle safety and shape appropriate policies and regulations. These methods may include but are not limited to accelerated testing, residual risk quantification, virtual testing and simulations, scenario and behavior testing as well as extensive focused testing of hardware and software components and systems. And yet, even with these methods, uncertainty will remain. This poses significant liability and regulatory challenges for policymakers, insurers, and developers of the technology.

Challenges for Scenario-Based V&V of Automated Systems from a Tool Provider Perspective

Kai Voigt and Andrea Leitner

AVL has long experience as solution provider for various automotive testing applications with a focus on powertrain development and vehicle integration. The advent of autonomous vehicles poses several new challenges for tool and service providers in the automotive field. The number of assistant systems and conditionally automated functions in vehicles will continue to grow quickly and will be deployed in all kinds of vehicles. New testing and validation methodologies must not only be applied to autonomous vehicles, but also to conventional vehicles. Therefore, it is vital for tool providers to take up the challenge of testing autonomous systems.

First, autonomous systems need to take their environment into consideration and take decisions based on what is happening around them. This means that new simulation environments representing the surroundings and the behavior of other traffic participants are required. Autonomous vehicles are further equipped with different kinds of environment sensors to perceive the environment. Hence, new sensor technologies need to be considered in the testing process and thus also in simulation and a lot of effort needs to be spend on the development of accurate and realistic sensor models. This is not only a requirement for pure simulation, but also for vehicle-in-the loop environments where these sensors need to be stimulated appropriately.

The important role of sensors for autonomous vehicles further introduces new data types that need to be considered throughout the testing tool chain. Besides traditional time-series data, object data as well as sensor raw data need to be handled. This does not only mean a paradigm shift in the type of data, but also for the amount of data that needs to be processed. Autonomous systems are equipped with various sensors, each perceiving their environment with high resolution. To be assessed, this data needs to be recorded and processed. As a result, new data

K. Voigt (✉) · A. Leitner
AVL List GmbH, Graz, Austria
e-mail: Kai.Voigt@avl.com; andrea.leitner@avl.com

A. Leitner et al. (eds.), *Validation and Verification of Automated Systems*,
https://doi.org/10.1007/978-3-030-14628-3_3

logging technologies and big data solutions for evaluation and data management are required. Apart from that, autonomous cyber-physical systems are very complex software systems—usually also including certain parts of Artificial Intelligence. This is a further huge step towards software-driven systems in a traditionally mechanics driven industry. Ensuring the reliability of systems based on AI is still a major research challenge, since it is highly dependent on the amount and quality of data used to train the system. Development and testing tools need to take this new development paradigms of train instead of coding into consideration.

Driven by venture capital, there are a lot of software startups popping up every day promising to provide dedicated solutions for generating training data or ways to accelerate the autonomous system development process. Most of them don't have an automotive background and thus are not completely aware of requirements for safety-critical systems. For an automotive development tool provider this means that there are many potential competitors or, depending on the perspective, potential technology partners in this dynamic business environment. Keeping an overview of the players is a challenge but necessary to succeed in the market.

Currently there is still no commonly agreed testing methodology and there are no regulations available. It is still unclear, which testing environments (model-in-the-loop, hardware-in-the-loop, vehicle-in-the-loop, or proving ground or road testing) will be required in the long term and to which extend they will be efficiently used. Nevertheless, a complete system validation can most likely only be done with a combination of different test environments.

We as AVL believe that we must be capable to provide integrated testing solutions, covering the whole process from test preparation (e.g. test case generation), test execution in different environments, to test evaluation and reporting. We also fully understand that not all aspects can ever be covered by one tool or even by tools from a single provider. Therefore, we are committed to openness and we support standardization activities facilitating the composition of tools into an integrated and open development platform, enabling efficient and reliable validation of autonomous systems.

ENABLE-S3: Project Introduction

Andrea Leitner

1 Introduction

The ECSEL JU funded project ENABLE-S3 [4] (May 2016–May 2019) was launched due to the need for verification and validation (V&V) solutions in the field of automated cyber-physical systems (ACPS). ACPS are expected to improve safety as well as efficiency tremendously. There are already technology demonstrators in various application domains available (cars, ships, medical equipment, etc.)—nevertheless, productive systems are not yet available on the market. A main obstacle for commercialization is the systems' interaction with the environment. Not only the automated cyber-physical system itself has to be tested, but also if it reacts to the behavior and specifics of its surroundings in a correct way. This leads to a huge amount of potential scenarios the system has to cope with.

For automated driving, Winner et al. [1] as well as Wachenfeld et al. [2] predict that more than 100 million km of road driving would be required to statistically prove that an automated vehicle is as safe as a manually driven vehicle. A proven-in-use certification is simply not feasible by physical tests.

This challenge is not constrained to a specific application domain. Therefore, industry and research partners from different application domains (automotive, aerospace, rail, maritime, health and farming) have joined forces to develop the required technology bricks for the V&V of automated cyber-physical systems.

The EU-research project ENABLE-S3 aspires to develop the missing technology solutions for the verification & validation of automated cyber-physical system, proposing a scenario-based virtual V&V approach. The main testing effort should be shifted to a virtual environment represented in terms of models. This has

A. Leitner (✉)
AVL List GmbH, Graz, Austria
e-mail: andrea.leitner@avl.com

A. Leitner et al. (eds.), *Validation and Verification of Automated Systems*,
https://doi.org/10.1007/978-3-030-14628-3_4

several advantages: tests can be conducted much earlier, cheaper, safer, and in a reproducible way. The input for the testing process are scenarios. Within the project, the following multi-domain definition was defined:

> A scenario class is a formalized description of the multi-actor process, including its static environment, its dynamic environment and environmental conditions. In a scenario class, the parameters are described and may have parameter ranges or distributions. A scenario class may include activities, events, goals of the activity and decisions of actors.

Scenarios can be collected in several ways. They can be engineered (e.g. based on safety or security analysis), collected from the real-world measurement data, extracted from accident databases, and so on.

Due to the diverse background of the project partners and the application domains, the project does not aim for a common, generic software solution. The idea is to have a common methodology and a set of reusable technology bricks (tools, methods, models, etc.), which can be used to build up a testing environment for a certain use case.

Figure 1 shows the main scope of the project: the development of a modular framework for validation and verification of automated cyber-physical systems.

Because of the large scope and complexity of the problem, it has been split into two parts. The validation methodologies on the left side describe the necessary steps and research on data acquisition and storage, scenario and metrics selection, as well as test generation methods. Since the project aims at a scenario-based validation approach, scenarios are an integral aspect. Countless variations for these scenarios exist (i.e. for different environmental conditions, different persons/traffic

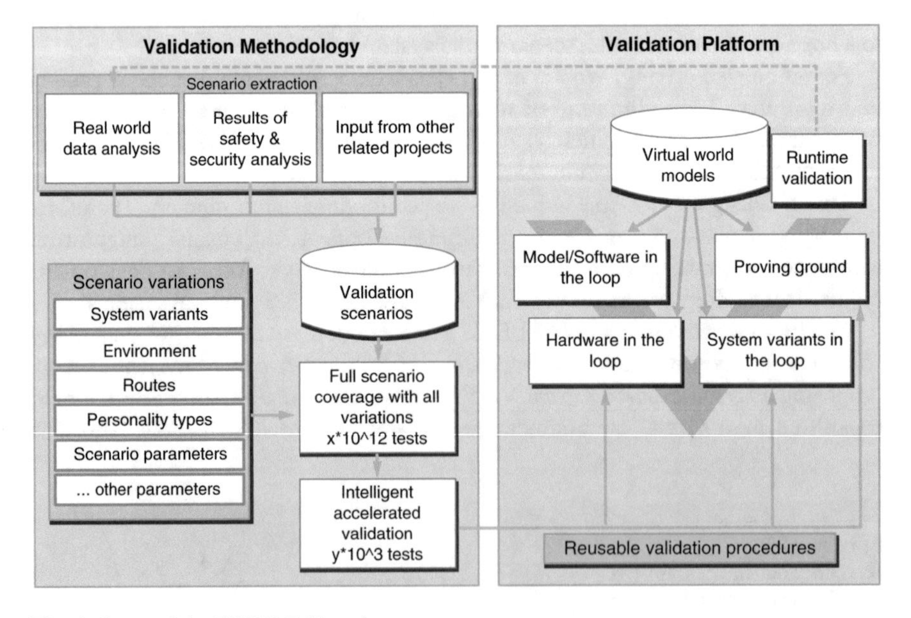

Fig. 1 Scope of the ENABLE-S3 project

participants involved, etc.) leading to an enormous number of potential test cases. The goal is to provide intelligent methods to select the required test scenarios in a way that ensures sufficient test coverage.

The validation platform on the right side focuses on reusable validation technology bricks (tools and models), which can seamlessly support various development and testing environments (model-in-the-loop, hardware-in-the-loop, system-in-the-loop, e.g. vehicle-in-the-loop, as well as real-world testing). By combining both parts and the respective technology bricks, the project works towards a significant reduction of the required test effort or even to enable testing of highly automated cyber-physical systems at all, respectively.

2 Generic Test Architecture

A major goal of the ENABLE-S3 project is to firstly deliver reusable technology bricks, promoting the development of models and tools that are easily reusable in different contexts, and secondly seamless development environments. This can be realised by setting up a testing environment where virtual representations can easily be exchanged by physical components. For both, the use of a modular structure with well-defined interfaces is essential. As a result, a generic ENABLE-S3 test architecture as shown in Fig. 2 was defined to support the integration of different technology bricks in a test system instance. It consists of three main layers and includes the most essential parts for testing automated cyber-physical systems across the six ENABLE-S3 application domains (automotive, aerospace, rail, maritime, health care and farming). The blocks' characteristics depend on the specific use cases. For some use cases, the blocks might be interpreted slightly different or are not required at all.

On high level, we differentiate between the Test Framework and the Test Data Management. The Test Data Management covers all aspects, which are valid across test phases and are reusable for testing different products. The test framework summarizes all aspects required for the planning (Test Management) and execution of tests (Test Execution Platform) for a specific product.

2.1 Test Framework

The test framework is divided into two parts: Test Management and Test Execution Platform, which are described in more detail in the following.

2.1.1 Test Execution Platform

The Test Execution Platform, as illustrated in more detail in Fig. 2, covers all relevant aspects for testing an ACPS including the SuT (either as a model, as

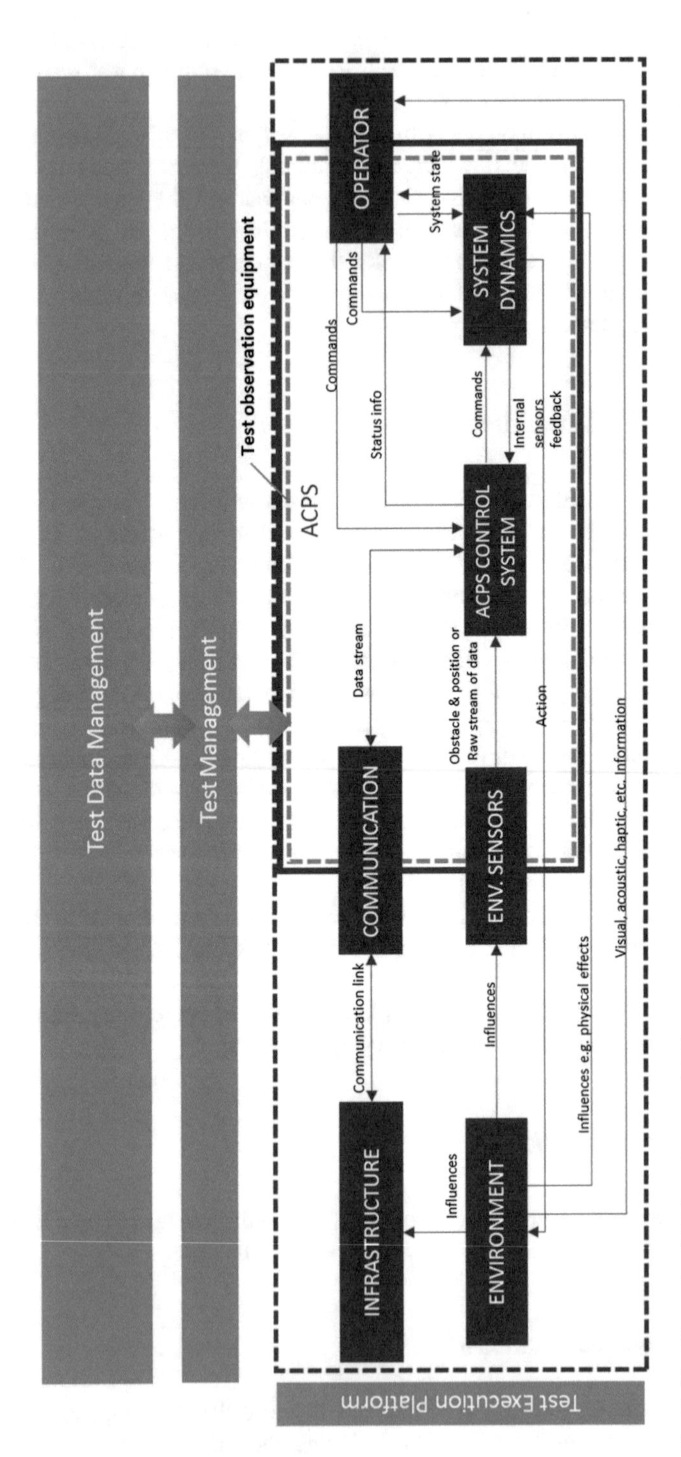

Fig. 2 Test Execution Platform in more detail

software component, as subsystem or as complete system). The ACPS control system interacts with its environment (e.g. driving on a road, which is shared with other traffic participants, etc.), perceiving its environment either via sensors or the communication to the infrastructure or both. The system itself is described by its physical dynamics, which again need to be fed back to the environment and so on. The arrows show the basic interactions of these testing architecture blocks. The description of the interface depends on the application domain as well as on the respective use case. For certain aspects, standardization of the interfaces is proposed as stated in Section "Open Simulation Interface".

Depending on the development stage, there are different instances of the test platform/architecture. For example, in a MiL environment, all components will be available as simulation models. Later simulated components will be successively substituted by real physical components, resulting in a mixed environment of real-time and non-real-time components. In this case, the ACPS control system describes the main system under test (SUT). In later development stages, more aspects are integrated in the SUT (e.g. real sensors).

2.1.2 Test Management

Figure 3 shows the different aspects of the Test Management part in more detail covering the generation of a representative set of test cases from scenarios. An interface to a scenario database is required to query the required information (relevant scenario classes and parameters, e.g. weather, type of operator, type of route, equipment, etc.). Depending on the testing purpose, this module has to include intelligent methods to select and instantiate the required scenario instances and prepare test cases. Then, the test cases need to be handed over to and executed in the test execution platform (or more concretely in the Environment block of the Test Execution Platform). The results need to be recorded, processed and potentially also visualized for inspection. In the last step, the overall safety of the system needs to be justified.

2.2 Test Data Management

Test Data Management focuses on all aspects that are valid across different test environments and includes the management of different types of data (such as measurement results, scenarios, etc.) as highlighted in Fig. 4. It includes the establishment of a managed tool chain (validated/qualified tool chain, configuration management, etc.), which is an important aspect in the long term—especially if virtual validation environments will be used for homologation and certification.

The ENABLE-S3 project aims for a scenario-based verification and validation approach. A major prerequisite is the existence of a set of scenario classes, which can be instantiated. These scenario classes and their potential variation parameters can either be extracted from real-world data as well as generated synthetically. The

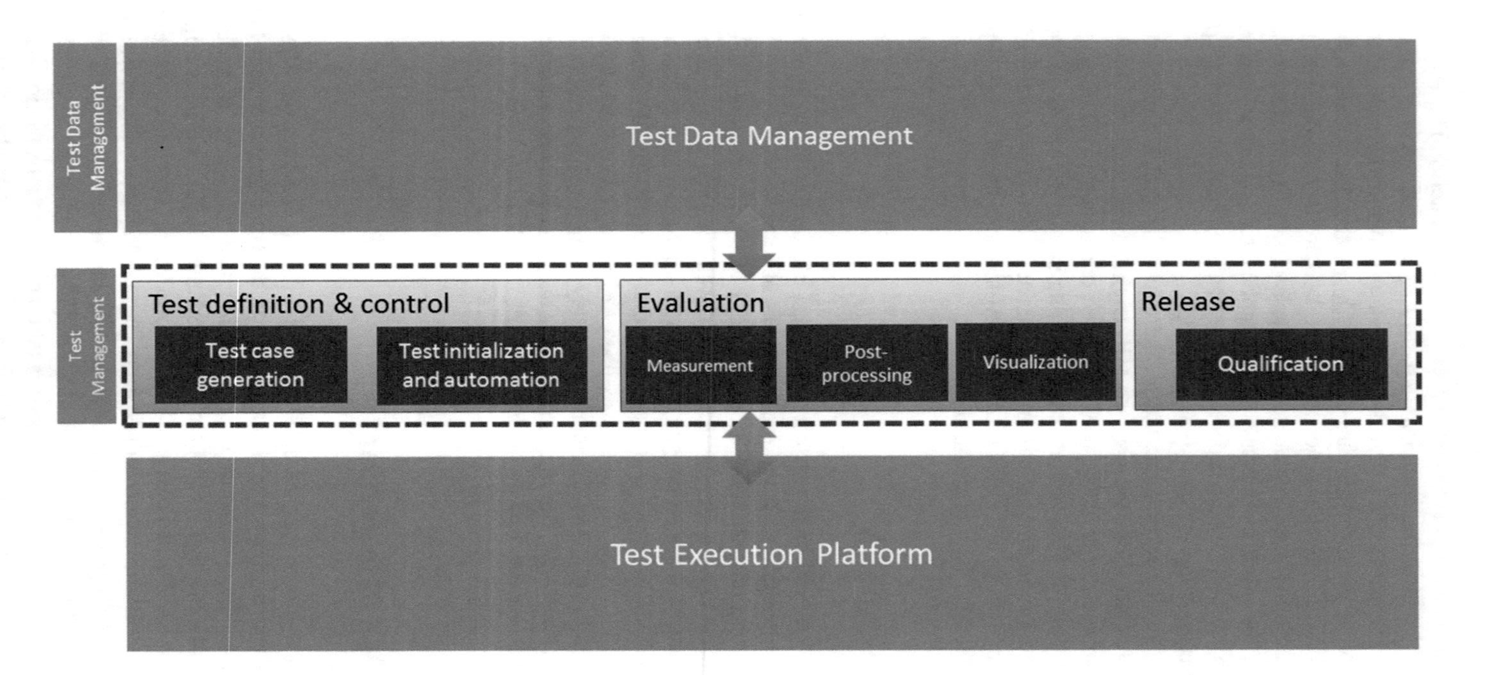

Fig. 3 Test Management in more detail

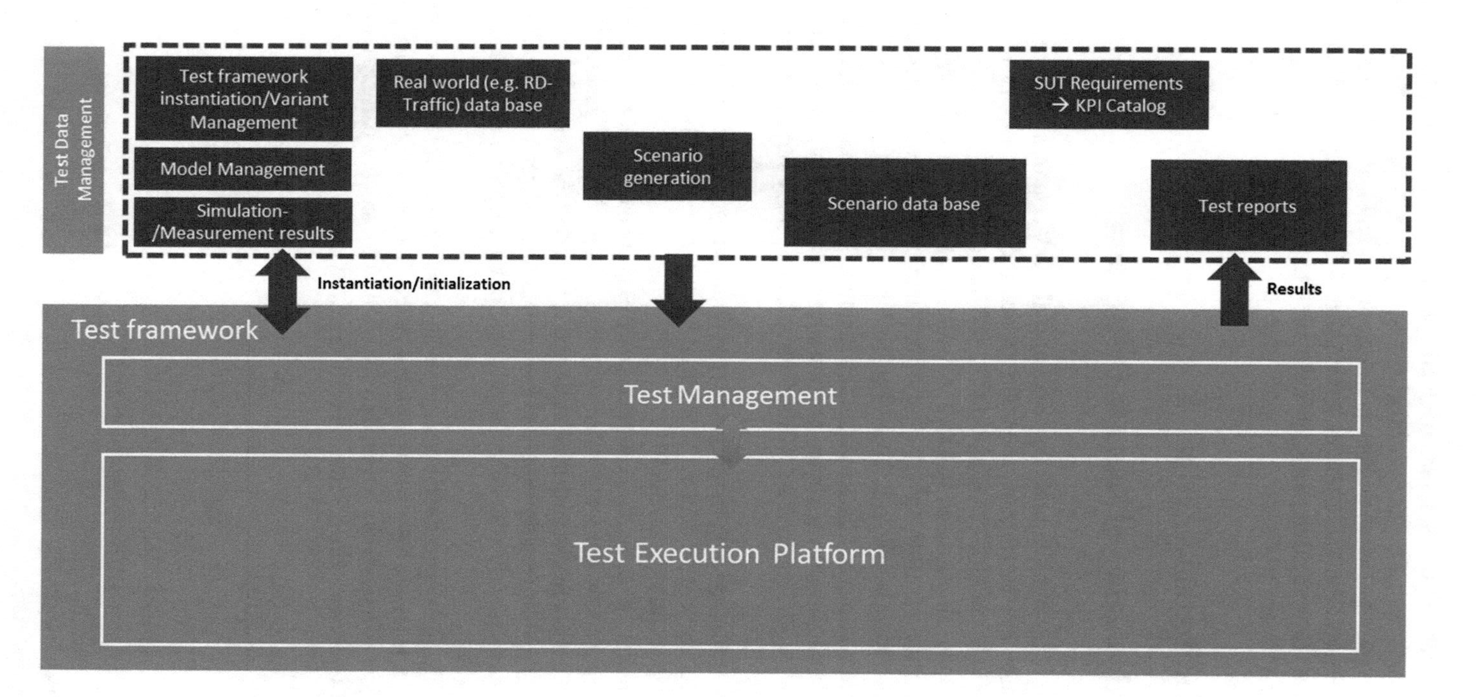

Fig. 4 Test Data Management in more detail

"scenario generation" block summarizes all activities, methods and tools which are required to extract scenario information (e.g. by identifying and transforming critical real-world situations), which can be transformed into something executable by an environment simulation engine. For traceability and reproducibility, it is further required to store all test artefacts and their interrelations.

3 Standardization Activities

3.1 Standardization of Scenario Descriptions

In the previous section, the scenario-based virtual V&V approach has been introduced. To make this practicable, scenarios (scenario classes as well as instances) need to be reusable in different environments and be shareable.

They need to be represented in a format, which is understood and interpreted by different simulation tools in the same way. For automotive, there are currently two open formats (OpenDrive®[1] and OpenScenario®[2]).

OpenDrive is an already quite established specification for describing the logical view on the road networks (i.e. road curvature, lane information, speed limits and directions for single lanes). This specification is supported by various environment simulation tools.

Currently, the specification is restricted to automotive applications. Nevertheless, certain aspects and design decisions might be reused in other application domains (e.g. to describe routes for vessels).

OpenScenario® is an open file format describing dynamic contents in driving simulation applications. Currently in its early stage, OpenScenario® just starts to be supported by environment simulation tools and is targeting the dynamic aspects of the scenario (i.e. traffic participants and their interaction). Again, the specification is currently developed for the automotive domain, but might be adapted to other domains as well.

The specifications have been used in the different automotive use cases of the project, the experiences have been fed back in the OpenScenario working group. Several project partners have committed themselves to actively participate in the evolution of OpenScenario, continuing the work even after the project has ended.

[1] http://www.opendrive.org

[2] http://www.openscenario.org

3.2 *Standardization of Sensor Model Interfaces*

The Open Simulation Interface (OSI) [3] is an upcoming standard to describe the data structure (message-based) of virtual perception sensors. It has been introduced by BMW and the Technical University of Munich and has been published as an Opensource project. The specification covers ground truth (as output of the environment simulation) as well as sensor information like lidar point clouds or object lists, which are relevant as possible output of simulated perception sensors and sensor systems. Regarding the generic test architecture, OSI thus provides a standardized data interface between the environment simulation and the perception sensor as well as for perception sensor data, which is used by automated driving functions. Hence, this interface enables the connection between function development frameworks and the simulation environment. A standardized interface for the description of environment data is helpful to provide compatibility between different frameworks.

3.3 *Sustainability of Results*

All specification described above have already existed at the beginning of the project. Within the ENABLE-S3 project, we identified these specifications as essential for the automotive domain. Therefore, we applied them and identified further requirements from our project use cases.

The specifications have been handed over to the ASAM e.V. standardization organization and are now managed there in dedicated working groups. As a result, the findings of the project are sustained, accessible and further developed after the project has ended.

4 Conclusions and Future Work

The project has tackled existing challenges for the verification and validation of automated cyber-physical systems. Nevertheless, there are still open issues, which have either been out of scope for this project or have been identified as important future work as a result of the project.

There is a trend to employ machine learning and other Artificial Intelligence (AI) techniques for ACPS product differentiation. The use of AI for object recognition in automated vehicles, first prototypes of AI-based Air Traffic Management or AI-based identification of health issues based on medical images are just three out of many examples how products can be improved by the means of AI and how AI technologies create the basis for intelligent decision making in future systems. Higher levels of automation combined with the use of AI pose significant new verification and validation challenges to ensure safety and security of advanced ACPS. Moreover, future ACPS will require frequent, fast updates and upgrades after product release in order to deal with changing environmental conditions, to leverage fast technology innovations and to cope with safety and security issues experienced in field operation. ACPS are no longer immutable. Instead, learning systems based on AI combined with online updates and upgrades are employed in such a way that a systems' implementation evolves during its lifetime. This trend of customization and updates will increase the already high number of existing ACPS product variants by an additional order of magnitude. Due to this system evolution, the initial verification of a systems' safety and security is no longer valid across the full product lifecycle. This requires to ensure that changes, such as over-the-air updates of (safety-critical) functions and upgraded hardware-components, do not compromise the overall safety and security of the systems.

All three ACPS-trends—autonomy, use of AI, and continuous system evolution—pose new verification, validation and certification challenges.

A certification solely based on virtual verification & validation approaches (without any physical testing) would impose high demands to model fidelity or might be even impossible—especially because of the required in-depth qualification of sensor and environment models, used simulation frameworks, V&V tools, artefact management, etc. An intelligent combination of in-depth virtual V&V plus economically feasible physical testing in the field is required—and new certification procedures and standards that reflect this approach. The combination of virtual and physical certification need to ensure economically feasible effort for the simulation-based V&V and a significantly reduced number of physical testing—meeting the required safety and security levels of ACPS in line with European standards and faster time-to-market for ACPS products.

Due to the complexity of ACPS, the "Open World" problem, the use of AI and the potentially changing environment conditions after product release, no approach can ensure error-free operation of ACPS along their lifecycle. Thus, systematic self-monitoring and self-diagnosis mechanisms need to be built into safety-relevant ACPS together with suitable fail-operational mechanisms to avoid accidents in unexpected situations. Examples of these situations could be e.g. component failures or unanticipated rare environment conditions. The self-monitoring and self-diagnosis mechanisms are also necessary to provide the in-field information for corrective system updates and upgrades. These self-monitoring, self-diagnosis and fail-operational mechanisms also need to be considered in the overall certification processes to ensure that they indeed implement safely degrading functionality of the automation (e.g. limited speed).

Acknowledgements This work has received funding from the Ecsel Joint Undertaking Under Grant Agreement No. 692455. This Joint Undertaking receives support from The European Union's Horizon 2020 Research And Innovation Programme and Austria, Denmark, Germany, Finland, Czech Republic, Italy, Spain, Portugal, Poland, Ireland, Belgium, France, Netherlands, United Kingdom, Slovakia, Norway.

References

1. Winner, H., Wachenfeld, W.: Absicherung automatischen Fahrens, vol. 6. FAS-Tagung München, Munich (2013)
2. Wachenfeld, W., Winner, H.: Die Freigabe des autonomen Fahrens. In: Autonomes Fahren. Springer, Berlin (2015)
3. Hanke, T., Hirsenkorn, N., Van-Driesten, C., Gracia-Ramos, P., Schiementz, M., Schneider, S.: Open simulation interface: a generic interface for the environment perception of automated driving functions in virtual scenarios. Research Report. http://www.hot.ei.tum.de/forschung/automotive-veroeffentlichungen/ (2017). Accessed 28 Aug 2017
4. European Initiative to Enable Validation for Highly Automated Safe and Secure Systems. www.enable-s3.eu

Part II
V&V Technology Bricks for Scenario-Based Verification & Validation

Using Scenarios in Safety Validation of Automated Systems

Sytze Kalisvaart, Zora Slavik, and Olaf Op den Camp

1 Introduction: Why Scenarios?

Automated systems are tested against requirements through test runs. Many systems operate in an open world environment that continuously changes: not like a robot operating in a fixed fenced factory location but like an automated ship in full sea or an automated vehicle in an urban environment. Since the automation takes over the role of the operator, the automation needs to be flexible enough to handle 'all possible situations'. Tests are required in all different development stages of the system to check if the system under development meets the requirements and to determine if the system performs according to expectations once it is deployed. Each set of tests provides information to possibly steer the development or to make choices in adapting or extending the design. For this reason, the test cases that are used as input to the tests are required to be a realistic representation of all possible and relevant real-world situations as perceived by the system.

With higher levels of automation, the functionality cannot be tested fully as isolated components or functions with explicit input/output relationships—it must be tested at full system level with integral and realistic inputs to the system under test.

Scenarios are introduced to represent 'all possible situations'. That means that observed or generalized situations are formalized into exchangeable scenarios. In other words, the collection of all relevant situations that the system under test is expected to possibly encounter when deployed in the real world, is described by a

S. Kalisvaart (✉) · O. Op den Camp
TNO Integrated Vehicle Safety, Helmond, The Netherlands
e-mail: sytze.kalisvaart@tno.nl; olaf.opdencamp@tno.nl

Z. Slavik
Forschungszentrum Informatik, Karlsruhe Institut für Technologie, Karlsruhe, Germany
e-mail: slavik@fzi.de

A. Leitner et al. (eds.), *Validation and Verification of Automated Systems*,
https://doi.org/10.1007/978-3-030-14628-3_5

compilation of well-described scenarios, that are grouped in a limited number of scenario classes. Such a formalization has several advantages:

- A significant cost benefit can be achieved by reusing scenarios across system types, versions and variants as long as the functionality is comparable. The alternative would be to perform endless costly test runs until 'all situations' have been encountered.
- A selection of scenarios can be used to define dedicated test cases that meet the objectives of specific tests.
- Formalized scenario descriptions enable the analysis of the completeness of testing. A metric can be set up to compute how well the compilation of scenarios describes the possible situations that the system-under-test can be subjected to in the real world.

Scenarios have been used in an informal or more formal way for decades in accidentology, type approval and consumer testing. For automated systems, there is a strong need to use scenarios as the key element for verification, validation and certification. Consequently, scenarios need to be formalized and quantified. Moreover, there is a need for harmonization of definitions and terminology to come to a common understanding in the large community of researchers, engineers and other stakeholders that make use of scenarios for verification, validation and certification.

The work presented is the result of the ENABLE-S3 scenario working group where many project partners convened to create consensus on how to use scenarios in domains as diverse as maritime, healthcare, automotive, agriculture, aerospace and rail.

This chapter first introduces applications of scenarios. Then it discusses various definitions of scenarios and concludes with the ENABLE-S3 definition of *scenario class*. The usage of scenarios in various application domains is discussed and conclusions are drawn.

2 Scenarios in the Development Cycle

Scenarios can be used in many stages of system development, verification and market surveillance. The various ENABLE-S3 partners clearly represent this diversity. Historically, the V-model developed in the US [1] and Germany [2] is frequently used even while under significant critique. The author introduces an extended V-model called the Triangle model (Fig. 1) where the possible uses of scenarios are shown [3]. It adds the application domain (green) to the V-model (blue). A set of scenarios represents a collection of knowledge about how a system is used during operation, thus providing a model of the application domain. Over various generations of a system, more knowledge can be built up and release cycles can become shorter and shorter.

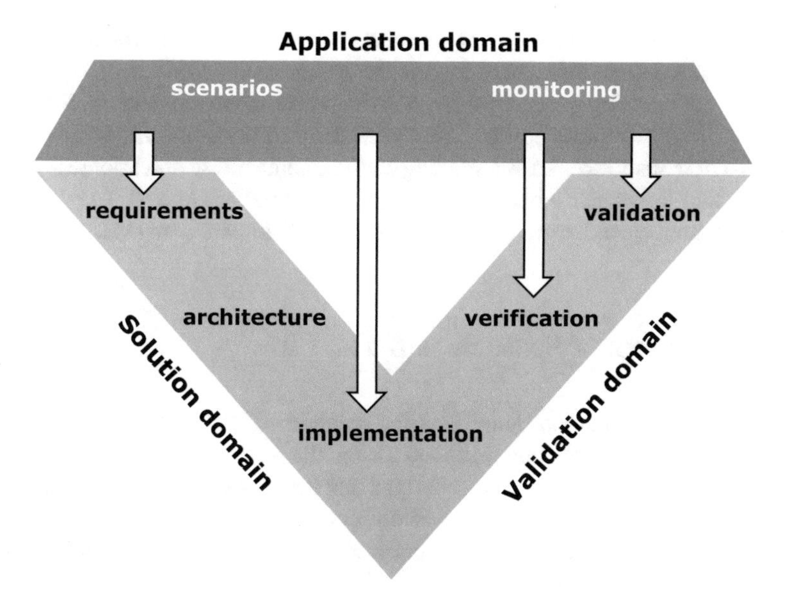

Fig. 1 Triangle model showing uses for scenarios [3]

A number of general statements about the purpose of a scenario can be made:

- A scenario deals with multiple actors and their interaction. This can involve the interaction between multiple machines, systems, and humans.
- A scenario involves the evolution of the interaction over a relevant period of time.
- A scenario is set in a context which has a role in the interaction. A ship on full sea may behave different than a ship in a canal.
- The application of the scenario determines the perspective from which the scenario is defined. If traffic management is the purpose, the scenario will look at the full street, where vehicle testing could be limited to only involve a couple of surrounding traffic participants.
- Scenarios can be manually described scenarios from real-world instances ('observed scenarios'), can be detected automatically ('mined scenarios'), can be designed by engineers ('desktop scenarios' or 'designed scenarios') and we can synthesize scenarios based on a model ('synthetic scenarios'). All sources of scenarios are relevant.

The purpose of a scenario may vary in nature and detail with the application. Open loop testing is implemented for monitoring and testing scenarios with detached systems-under-test (SuT) or functions-under-test (FuT). Contrary, closed loop testing aims at determining interactions between a scenario and a SuT, thus influencing the preferable definition. Usability testing requires a scenario description that contains a unique series of events within a full workflow. Similar to the verification and validation of a sensor model against an annotated ground truth, it belongs to the category of open loop test setups. Performance tests regarding particular elements of realistic environments such as noise and artefacts in sensor testing, do not require a feedback loop to the scenario.

Closed loop testing is required for ADAS and sensor model testing. Simulation goals can be for example performance tests of sensor systems, ADAS and ADAS functions in realistic scenarios including environmental noise and artefacts. Finally, a feedback to the scenario is essential for performance tests of a SUT or VUT with partial components in hardware (HiL) or full vehicle in hardware (ViL).

3 The Concept of Scenario and Test Case

Before providing common definitions for scenarios, first the concept of scenario and test case is explained, and it is shown what the difference is between scenarios and test cases. It is generally acknowledged that test cases for the safety assessment of automated systems should be based on real-world scenarios. Nevertheless, the terms *scenario* and *test case* are often confused; even the combination *test scenario* is often heard in discussions.

In ENABLE-S3, the term *scenario* describes a situation that can happen or has happened in the real-world. In other words, scenarios are used to describe any type of situation that a system in operation can encounter during its lifetime. Since scenarios are collected in the real-world (or mined from data), the set of scenarios in the scenario database will not fully cover all possible situations that can occur in real-world. In the Fig. 2 this is represented by the fact that the available scenario set (light blue set) does not cover all relevant situations in the real-world (dark blue set).

It is expected that scenarios differ from one use case or application field to the other. In that case sets of scenarios identified in a different area of the world might also cover different parts of the real-world. Note that also a large overlap in scenarios for different application areas is expected.

Apart from scenarios, it is important to know a system's Operational Design Domain, to determine the set of relevant test cases. SAE automotive standard J3016 [4] defines an Operational Design Domain (ODD) for the automotive

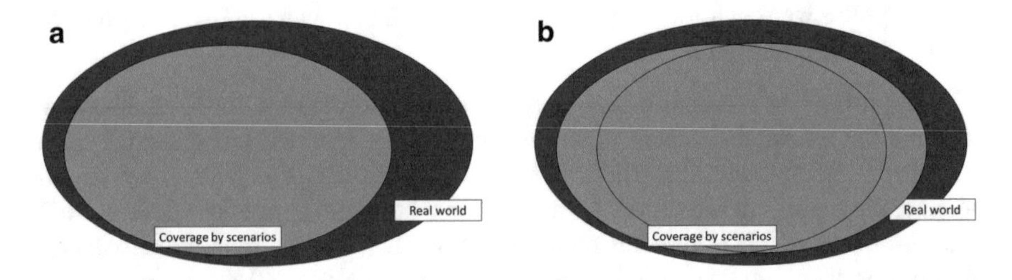

Fig. 2 Set of scenarios covering part of the real-world (**a**). Scenarios mined in a different area might cover a different part of the real world (**b**)

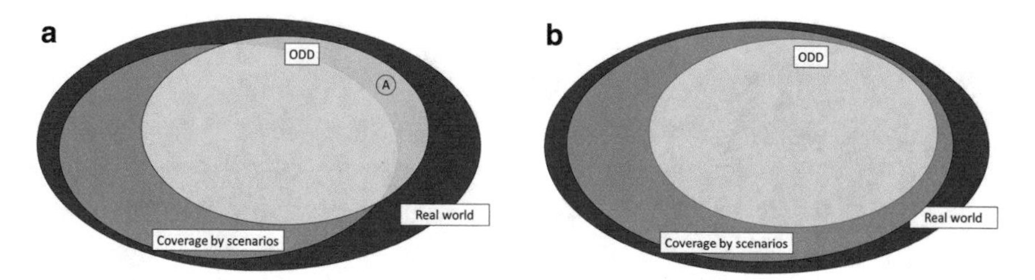

Fig. 3 Representation of an ODD (**a**). Extending the scenario set to be able to describe all possible scenarios in the complete ODD (**b**)

sector as "operating conditions under which a given driving automation system or feature thereof is specifically designed to function, including, but not limited to, environmental, geographical, and time-of-day restrictions, and/or the requisite presence or absence of certain traffic or roadway characteristics." The ODD very much depends on the application of an Automated Vehicle (AV), and usually is the result of the design of the AV in relation to the requirements of the application. An ODD covers a dedicated and limited area of the real-world as indicated in the following figure by the yellow set.

The ODD is very likely to differ for different types of vehicles, or even for a different application of a vehicle. Figure 3 considers the ODD of one vehicle type for a given application. Moreover, the figure shows that part of the ODD is covered by the scenario set, but possibly not all scenarios will ever be identified to cover all of the ODD. The objective of continuous scenario mining is to increase the coverage by scenarios so that the complete ODD can be covered by scenarios, reducing the area (A) to zero. A scenario set is considered *complete* in case it covers all relevant ODDs.

Once the ODD is known, and there are scenarios that cover (part of) the ODD, *test cases* can be generated. Test cases are generally based on a subset of scenarios, as not all scenarios are relevant for each type of system or each type of application. However, test cases are always generated from scenarios; therefore, it is assumed here that no test case is generated in areas not covered by a scenario set.

In this example (Fig. 4), test cases are provided for an area (B) outside the ODD, and the set of test cases does not cover the complete ODD leaving an area (C) of the ODD that is not covered by test cases. It seems unnecessary to provide test cases outside the ODD, as outside the ODD, the system is not expected to respond. It might provide for valuable test cases though in case the test cases just outside the ODD are intended to check the sensitivity of the system-under-test for false positive responses. Such test cases usually are selected close to the 'boundaries' of the ODD, leaving a small area (B) at the boundaries of the ODD.

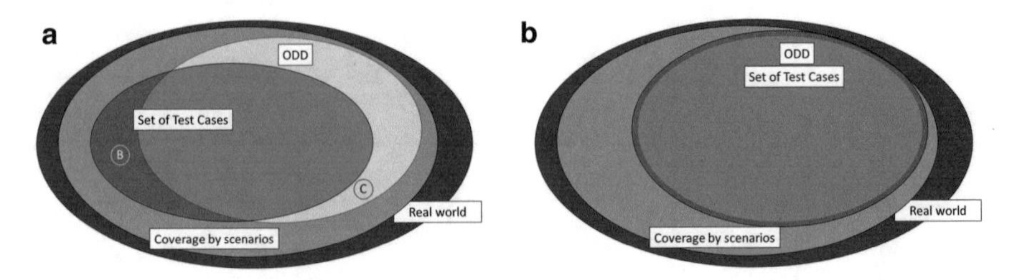

Fig. 4 Set of test cases not well aligned with the ODD (**a**). The ideal situation in which the test cases nicely cover the ODD (**b**)

For a complete assessment, the full ODD is required to be sufficiently covered by test cases, reducing the area (C) to zero. This results in the ideal figure in which the ODD is fully covered by test cases, and test cases are available just outside the ODD. In the test case generation process, the ODD can be used to filter the relevant scenarios from which the relevant set of test cases needs to be determined.

It should be noted that test cases are discrete by nature: one test case is represented as a single point in the set of test cases. Although also concrete scenarios exist, scenarios have a continuous character, covering a certain area limited by the ranges of the scenario specific parameters. Following this reasoning, the set of test cases consists of a finite number of distinct test cases.

4 A Definition for Scenarios

If scenarios are intended to be used for validation and type approval of automated driving, it is crucial to speak the same language and hence define the terminology first. In the end, scenarios will be used for training a system, testing and certification. They will become important for communication within industry, authorities and the public. They will also play a role in ethical and legal considerations.

In the following, existing scenario definitions will be introduced in order to develop a cross-domain definition of scenario. Each definition is then discussed shortly to develop a common general definition based on comparison of definitions.

The publication by Ulbrich, Menzel, Reschka, Schuldt and Maurer [5] is widely used in the automotive industry. The authors define scenarios as follows:

> A scenario describes the temporal development between several scenes (in a sequence of scenes). Every scenario starts with an initial scene. Actions & events as well as goals & values are (may be) specified to characterize this temporal development in a scenario. Other than a scene, a scenario spans a certain amount of time.

A scenario in this definition connects a series of scenes (time snapshots containing perception information) over time (Fig. 5). The relationship between the

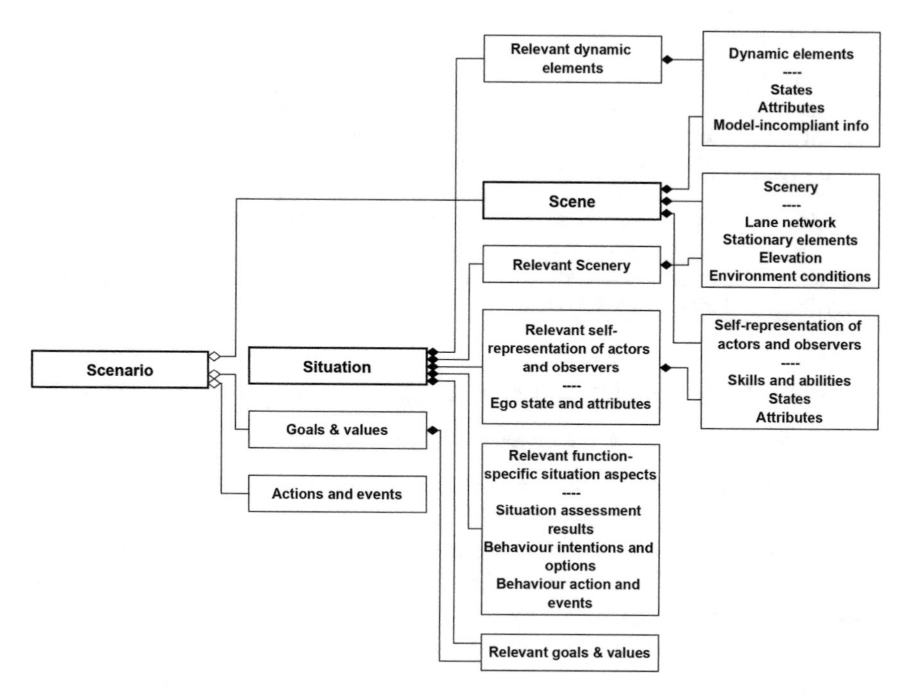

Fig. 5 Schematic overview of terms used in Ulbrich et al. (diagram by authors)

actors is considered part of scene. Tracking of actors over time is not captured in the definition of scenario or scene, though the actions and events will result in such relationships over time. The actions of the actors (like road users) causes the situation to change to a new status. For an outside observer, such an action produces an event. A situation overlaps with a scene and focuses on parts of the scene relevant to an automated driving function.

In an important paper of Geyer et al. [6], a unified ontology for scenarios and its elements scenes and situations is discussed. It introduces the theatre metaphor for scenarios and scenes. The elements of a scenario are discussed and described:

> a scenario includes at least one situation within a scene including the scenery and dynamic elements. However, scenario further includes the ongoing activity of one or both actors. According to the movie and theatre metaphor previously introduced, the term scenario can be understood as some kind of storyline—including the expected action of the driver—but does not specify every action in detail.

The draft standard for Safety of the Intended Functionality ISO PAS SOTIF 21448 is very relevant in this context as it extends functional safety standard ISO 26262 to the area of perception and control algorithms in highly variable environmental conditions originating from outside the system under test. This is an important step towards operational safety.

Table 1 Pegasus levels of scenarios

Scenario type	Description
Functional scenario	A functional scenario specifies a scenario on a high-level basis. This is often done in a human-readable mostly textual (prose)/graphical (sketches). It can be quite broad, e.g. "any cut-in manoeuvre"
Logical scenario	A logical scenario is a concretization of a functional scenario. It has a more formal character and is a description for a set of concrete scenarios (e.g. a cut-in manoeuvre coming from the left lane).
Concrete scenario	A concrete scenario is a fully defined sequence. It describes a single instance from a logical scenario.

The SOTIF standards [7] using as its definition an abridged version of Ulbrich:

> A scenario describes the temporal development between several scenes.

Depending on the purpose, the required level of detail that should be provided by a scenario is different. The later in the V&V process, the more information is required to be included in the scenario description. Within the Pegasus project based on the work of Bagschik [8] three levels of detail were defined (Table 1) based on the widely used scenario description by Ulbrich.

Alternatively, TNO [9] introduces the concept of "scenario class" as:

> A scenario class refers to multiple scenarios with a common characteristic. A scenario is an instance of a scenario class.

A scenario class can be considered an equivalent of the Pegasus logical scenario and a test case the equivalent of the Pegasus concrete scenario. In the TNO definition, scenario classes are a category, whereas a scenario is an observed unique instance of such a scenario class.

TNO defines scenarios as follows [9]:

> A scenario is a quantitative description of the ego vehicle, its activities and/or goals, its dynamic environment (consisting of traffic environment and conditions) and its static environment. From the perspective of the ego vehicle, a scenario contains all relevant events.

Philips defines scenarios [10] as follows:

> A scenario is a description of the X-ray system process, its activities and/or goals, its static environment, and its dynamic environment.

Other scenario components are defined in Table 2. As can be seen, few of the sources define all elements. The result is a scattered image of various scenario components and definitions.

Table 2 Scenario components and definitions of terms

Scenario component	Definition
Scene	A scene describes a snapshot of the environment including the scenery and dynamic elements, as well as all actors' and observers' self-representations, and the relationships among those entities (Ulbrich).
Scenery	The scenery is initially a structured collection of single static elements that enable the stage director to construct—by various combinations—a suitable surrounding for the scene (Geyer).
Activity	An activity is a time evolution of state variables such as speed and heading to describe for instance a lane change, or a braking-to-standstill. The end of an activity marks the start of the next activity. The moment in time that marks the beginning or ending of an activity is called event (TNO). An activity refers to the way a state evolves over time. (Philips)
Trace	All events associated with one case instance.
Artefact	The state of the system as a vector of multiple states, such that we can abstract from the "unknown" or irrelevant parts of the system behaviour, for each scenario (Philips).
Event	An event marks the time instant at which a transition of state occurs, such that before and after an event, the state corresponds to two different activities (TNO, Philips).
Situation	A situation is the entirety of circumstances, which are to be considered for the selection of an appropriate behaviour pattern at a particular point of time. It entails all relevant conditions, options and determinants for behaviour (Ulbrich).
Test case	A scenario selected from a broader scenario class with all values specified to be used for testing (TNO).
Stage	Stage is a moment in time, when decisions are taken (Kaut [11]).
Period	Period is a time interval between two stages (Kaut).

5 Multi-domain Usage of Scenarios

For scenario definitions, the challenge is not only to find a common terminology, but also to keep in mind application and context. As it becomes obvious from the following chapters, scenarios are used very differently in various domains such as automotive, health care or farming. The reasons lie partially in the testing goals and partially on the previous history of each domain. For example, in automotive applications there often exists a device or a system of devices that should be tested in various scenarios. In health care, the focus might be located rather on clinical workflows, while automated devices are rather part of the scenario. Another example regarding different histories comes from comparing automotive with commercial domains such as farming, where the underlying sensor technology is the same such as camera or radar. Accordingly, the aim is rather to evaluate how existing automotive ADAS sensors or functionalities can be applied in commercial applications and how ADAS need to be modified implying different requirements on scenarios. Thus, implies additionally, that despite the differences a cross-domain scenario definition is useful to reduce effort and mitigate misconceptions in cross-domain applications. In the following, aspects of several domains are illuminated together with their implications.

6 Automotive Applications

Scenarios cover a wide range of applications in the automotive domain, with the main purpose of creating virtual tests on system or sensor level. Scenarios incorporating communication channels also become important.

The testing purpose defines the level of detail that has to be fulfilled by the simulation components, e.g. functional or physical level (Fig. 6). Probably the most common test system is an environment that addresses the perception of optical and other sensors mounted on an ego-vehicle that drives within an environment simulation like VIRES VTD or TASS PreScan. Another main objective within the automotive domain is traffic flow simulation that can also include driver behavior models [12]. System level simulation that models ECU behavior plays an important role, especially with fault injection.

For scenario generation, on the one hand, relevant scenarios should be used for testing. On the other hand, exceptional or even critical situations have to be included in the test coverage.

When real world scenarios are used for test case generation, a database is required as well as a methodology for scenario extraction such as TNO's StreetWise [9]. It provides a description of what any vehicle might encounter on the road, independent of the functions that have been implemented in the vehicle. Figure 7 shows the

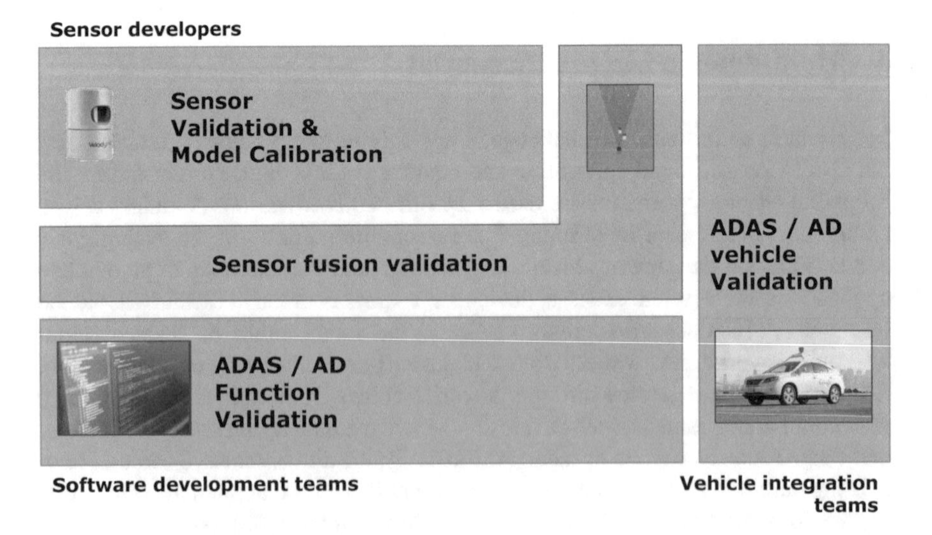

Fig. 6 Automotive applications for scenarios [13]

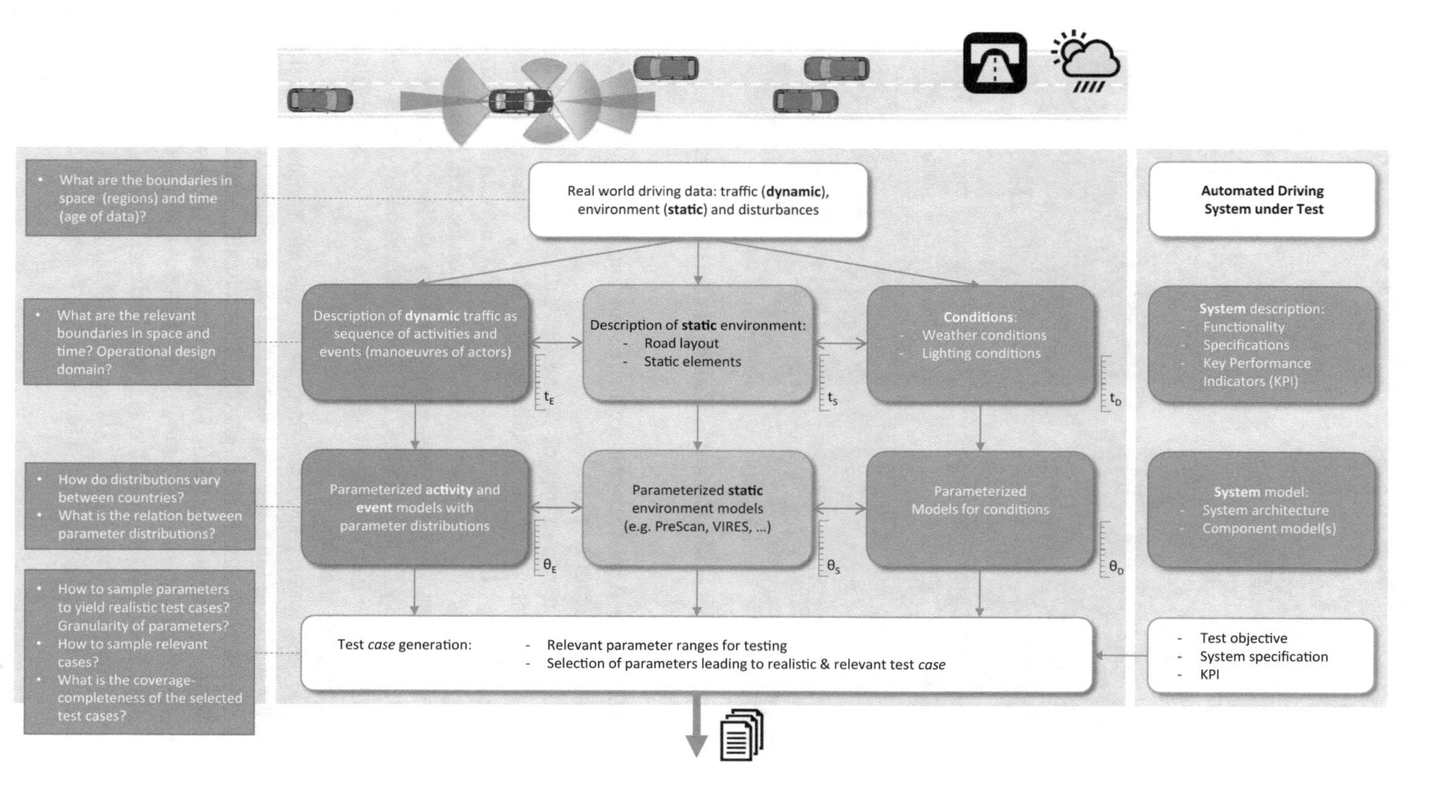

Fig. 7 Schematic overview of the information flow from real-world data collection to test case generation [9]

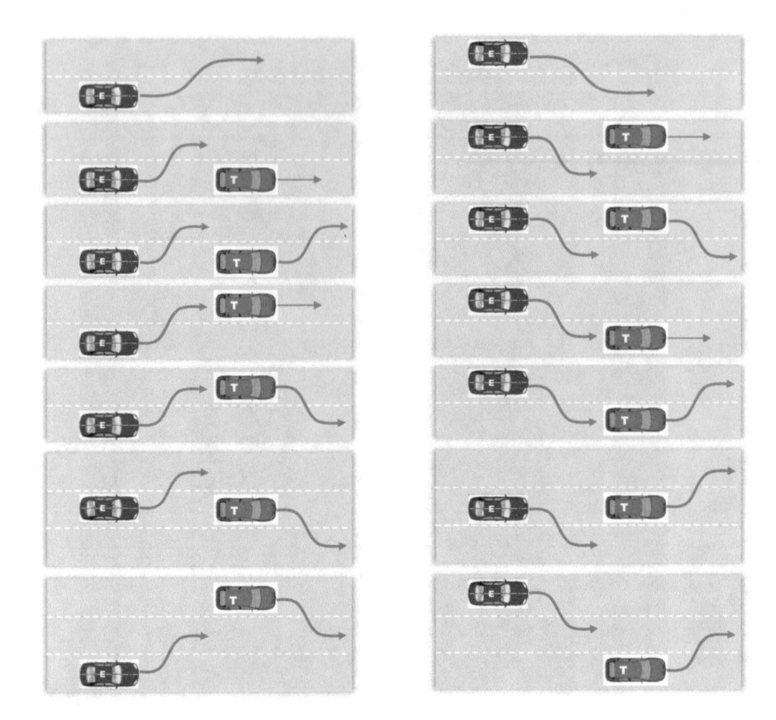

Fig. 8 Sample combinations of activities to describe scenarios for two road users

relation (and differences) between real-world driving data, scenarios, test cases and the system-under-test according to TNO.

In StreetWise, the dynamic part of scenarios is decomposed into underlying longitudinal and lateral activities (longitudinal: braking, cruising, accelerating; lateral: changing lane left/right, stay in lane) for each road user. In Fig. 8, examples of possible scenarios for two road users are shown based on the underlying activities.

Euro NCAP defines tests for cars equipped with AEB Systems (AEB = Autonomous Emergency Braking) in [14, 15]. These tests consider typical scenarios for possible collisions with other traffic participants like pedestrians, bicyclists and cars based on accident analysis.

The scenario description parameters describe the test environment (e.g. day, dry and flat test track, size of the test track), describing the movement (trajectories and velocity profiles) of ego vehicle and target objects and finally parameters describing the appearance (size and colors) of the target objects used. In this way, the Euro NCAP test procedure permits to compare and evaluate for example the performance of different AEB equipped cars, which was performed by HAGL.

Using the simulation tool Vires VTD and the definitions in [14, 15] HAGL has created templates for generating simulated AEB scenarios as defined by Euro

Fig. 9 Application of CBNA scenario to Intersection Crossing

NCAP. In a second step, HAGL has created templates for generating simulated AEB scenarios in intersections and parking related environments based on the scenarios defined by Euro NCAP. These can be exported as OpenSCENARIO files [16] for exchange with simulators. The generated simulated scenarios may be used as test cases in HIL and SIL environments for various use cases. HELLA Aglaia elaborated AEB scenarios based on Euro NCAP [14].

The scenario selected in the following example is: Car to Bicyclist Nearside Adult 50% (CBNA-50)—a collision in which a vehicle travels forwards towards a bicyclist crossing its path, cycling from the nearside. The frontal structure of the vehicle strikes the bicyclist when no braking action is applied.

Figure 9 shows the application of the CBNA scenario to the use case Intersection Crossing: The traffic light is green for the ego vehicle crossing the intersection and a distracted bicyclist is crossing the ego vehicle path at the same time coming from the near (right) side.

Possible variations in the simulation to generate new test cases are for example to consider other traffic signs giving priority of way to the ego vehicle, night or twilight, rain or snow, wet lane surface (reduced braking coefficient), different appearance of bicyclist (e.g. colors of clothing), velocity and trajectory of the bicyclist, velocity and trajectory of the ego vehicle.

7 Medical Applications

In the medical domain [10], the interaction between surgeon (or radiologist) and the imaging device is crucial for both safety and quality of the intervention or diagnosis. Even though procedures exist and a preferred work flow is suggested in training, variation in the actual user behavior is large. Scenarios are used to describe the observed behavior of the user with the device.

Philips distinguishes between positive and negative scenarios.

A positive scenario is a scenario where under a certain condition, a goal is met. Example: when enabling X-ray and then pressing the exposure pedal (=condition), the radiation will be on (=goal).

In a negative scenario, the goal is not met, and the reason might be the order of the activities or missing/new activities. E.g. when pressing the exposure pedal and then enabling X-ray, the radiation will not be on.

Each scenario (positive and negative) will be formalized:

1. Description: the definition of the scenario in text.
2. Condition: expressed as a state vector, which can contain time constraints. If met, the goal can be checked.
3. Goal: expressed as a state vector, which can contain time constraints. If met, it means an instance of the scenario was found.

Scenario instance: one unique trace for which the scenario condition and goal were met (Fig. 10).

Description: Looking for instances of "correct" behaviour, where no pedal is pressed and radiation is off, and enabled. The expected consequence is that pressing fluoroscopy pedal will lead to a state change of the artifact RadiationStatus, meaning that the radiation will be turned on.

Condition: S = (Enabled,ExamStarted,X,Y,Off) where X = None ∨ FluoroReleased ∨ ExposureReleased and $Y \in S_4$

Time constraints:
- none

Goal: S = (Enabled,ExamStarted,FluoroPressed,Y,On) where $Y \in S_4$

Time constraints:

- $0 \leqq \mathit{timeIn(On)} - \mathit{timeIn(FluoroPressed)} \leqq 5$ seconds
- $\mathit{timeIn(FluoroPressed)} - \mathit{timeIn(ExamStarted)} \geqq 0$ seconds

Fig. 10 Example of positive scenario

8 Farming Applications

Similar to other domains whereas the systems under test have a highly specialized purpose such as commercial vehicles, the main objective for according test systems is to test their functionality [17]. Although these vehicles also need to drive on regular roads such as regular vehicles, driving behavior on public streets is not the focus of testing. Therefore, farming scenarios are strongly connected to the targeted application and purpose of the system under test. In addition, the aim is to identify the potential and limitation of existing automotive ADAS sensors and functionalities for farming applications that can differ a lot from automotive scenarios. The application and key scenarios for testing can include even more than one ego-vehicle, especially when the collaboration between several vehicles are the focus of testing. Within ENABLE-S3, a sequence of scenarios was defined for demonstrating virtual validation and verification of sensors and communication. The sequence of scenarios illustrates a drone-supported harvesting procedure. Each sequence is related to a harvesting operation step. Accordingly, the three defined sequences depicted in Fig. 11 are:

- Pre-flight of a drone: Drone scouts the field for mapping the field
- Harvesting: Drone supported harvesting, until the harvester's grain cart is full
- Grain cart offloading: during an autonomous parallel driving maneuver, a tractor collects the grain from the harvester on the field

The scenarios were defined according to the following scheme:

Description: Contains all scenes that belong to the scenario, including all settings and actors. That includes parameters such as velocity and involved actors or even equipment as well as the environment description. The termination of the scenario is also defined. Here, it ends with fulfilling the scenario objective or failing to do so.

Successful end condition: defines a successful termination of a scenario, e.g. drone finished scouting of the field, transmitted the collected data and returned to the base.

Failed end condition: defines failure of the scenario, e.g. drone did not return, collected data was lost, tractor returned before offloading the harvester's grain cart.

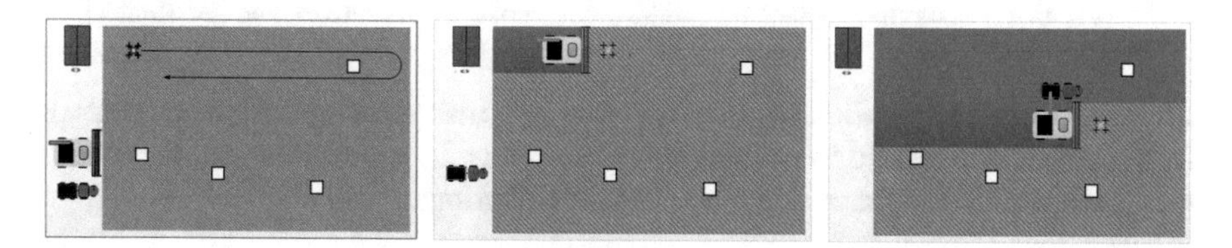

Fig. 11 Harvesting scenarios: drone pre-flight (left), harvesting (middle), grain cart offloading (right)

9 Discussion

All definitions lack a proper definition of what a 'relevant' scenario should be. While 'relevant' can be understood as a critical scenario. In that case, criticality has to be defined on a scale of material damage up to fatal accidents, its meaning changes with the testing purpose in which the scenario is used. Therefore, the relevance needs to be defined individually for test systems.

Although [6] does not define explicitly a scenario, it provides many underlying considerations that are an important basis for the Ulbrich definition. The definition of Ulbrich et al. has a descriptive and behavioral component. At one side, a scenario is described by scenes that are like momentary snapshots at one timestamp (descriptive). This is relevant for open loop sensor testing (e.g. camera frames). On the other hand, the definition talks about actions and goals, focusing more on behavior and causality between events. The behavioral component is more relevant for closed loop system testing.

Another important point in scenario definitions is the point of view and perspective. In automotive scenarios, usually an ego vehicle is defined, which is the vehicle under test in the sense that the ego vehicle defines the perspective of the scenario. For mining scenarios, the ego vehicle is instrumented and collects the data. An equivalent ego vehicle definition may be difficult in other domains, such as health care.

A common discussion is whether a scenario should include a desired outcome. Both the medical and agriculture scenarios include positive or negative outcomes. In automotive, performance indicators are usually separated from the scenario as they may change depending on the function under test.

10 ENABLE-S3 Working Group

Within the scenario working group of the ENABLE-S3 project, the following multi-domain definition was developed based on all definitions shown above.

> A scenario class is a formalized description of the multi-actor process, including its static environment, its dynamic environment and environmental conditions.
>
> In a scenario class, the parameters are described and may have parameter ranges or distributions.
>
> A scenario class may include activities, events, goals of the activity and decisions of actors.

The ENABLE-S3 definition focuses more strongly than other definitions on formalized scenario classes and its determining parameters. It emphasizes that parameters may have variance expressed in ranges and distributions.

Most importantly, the definition is a multi-domain definition that by distinguishing static, dynamic and condition aspects suitable for diverse application domains such as aerospace, maritime or agriculture. Environmental conditions are very prominent in sensor perception research in automation nowadays and therefore made explicit in the definition.

The definition contains many optional components. Goals drive the activities of the actors in the scenario class, which can be observed externally as events or can be known internally as decisions. The use of decisions connects the definition to the mathematical approaches of Pflug [18, 19] and Kaut.

11 Conclusion

Scenarios play a central role in capturing the diversity of situations that automated systems need to deal with. The ENABLE-S3 working group on scenarios proposes to create a scenario class that includes both the typical in the scenario ('overtaking on a highway') but also the variation in its parameters ('with a lateral speed of 3.1, 6.3 or 7.6 m/s'). In this way, the complexity of the real-world environment of the automated system can be broken down in manageable tests.

It is clear that massive virtual testing will complement physical testing. By learning scenarios from system use in the market, the quality of virtual testing can be improved over various system generations. At the same time, the systems can be improved quickly using a large body of application knowledge captured in scenarios. In this way, the costs of physical testing for certification can be kept at reasonable levels.

Various partners describe their application of scenarios, from usability of medical systems, testing of automated driving to annotation of automotive camera data at the pixel level. Interestingly, scenarios may also play an important role in explainable~AI.

This section described and discussed various existing definitions for scenarios and their components. It turns out that the static elements (scenery or physical environment), momentary elements (scene) and dynamic elements (actors, activities or actions, events) and environmental conditions (weather, lighting) can be separated. Pegasus, TNO and ENABLE-S3 all distinguish a level of quantification of scenarios as a separate stage (concrete scenario) or through parametrization. The proposed ENABLE-S3 definition clearly distinguishes static, dynamic elements and environmental conditions. For automated use in simulation, it proposes to use parameters to capture the variation of scenarios while keeping the number of scenario classes manageable.

Acknowledgements Major contributions to the ENABLE-S3 working group were given by Oliver Zendel (AIT), Andrea Leitner (AVL), Gustavo Garcia Padilla (Hella Aglaia), Jakub Marecek (IBM), Pavlo Tkachenko (JKU), Harald Waschl (JKU), Jinwhei Zhou (JKU), Daniel Reischl (LCM), Stefan Bernsteiner (Magna Steyr Engineering), Angelique Brosens (Philips), Martijn Rooker (TTTech), and many others.

References

1. US DOT, Systems Engineering for Intelligent Transportation Systems (PDF). US Department of Transportation, p. 10. https://ops.fhwa.dot.gov/publications/seitsguide/seguide.pdf. Accessed 21 Nov 2018
2. IABG: Das V-Model, referenced 21 November 2018. https://v-modell.iabg.de/index.php/vm97-uebersicht
3. Kalisvaart, S.: Using the StreetWise scenario base for virtual testing; method, workflow and tooling, Autonomous Vehicle Test & Development Symposium, presentation, Stuttgart, 5 June 2018
4. SAE, J3016 Surface Vehicle Recommended Practice, (R) Taxonomy and definitions for terms related to driving automation systems for on-road motor Vehicles, SAE International, Revised 2018-06
5. Ulbrich, S., Menzel, T., Reschka, A., Schuldt, F., Maurer, M.: Defining and substantiating the terms scene, situation, and scenario for automated driving. In: IEEE 18th International Conference on Intelligent Transportation Systems (2015). https://doi.org/10.1109/ITSC.2015.164
6. Geyer, S., Baltzer, M., Franz, B., Hakuli, S., Kauer, M., Kienle, M., Meier, S., Weissgerber, T., Bengler, K., Bruder, R., Flemisch, F., Winner, H.: Concept and development of a unified ontology for generating test and use-case catalogues for assisted and automated vehicle guidance. IET Intell. Transp. Syst. **8**(3), 183–189 (2014)
7. SOTIF draft standard, ISO/PAS 21448 Road vehicles – safety of the intended functionality. https://www.iso.org/standard/70939.html
8. Bagschik, G., Menzel, T., Maurer, M.: Ontology based scene creation for the development of automated vehicles. IEEE. https://arxiv.org/abs/1704.01006 (2018)
9. Elrofai, H., Paardekooper, J.-P., de Gelder, E., Kalisvaart, S., Op den Camp, O.: StreetWise scenario-based safety validation of connected and automated driving. White paper. http://publications.tno.nl/publication/34626550/AyT8Zc/TNO-2018-streetwise.pdf (2018)
10. Ekkel, R., et al.: ENABLE-S3 D4.5.1 v2 specification of the health demonstration. Deliverable (2018)
11. Kaut, M.: Scenario generation for stochastic programming, A practical introduction, Håholmen, June 10–12. http://work.michalkaut.net/papers_etc/scen-gen_intro.pdf (2006)
12. Metzner, S., et al.: ENABLE-S3, D4.1.1 v2 specification of the automotive demonstration. Deliverable (2018)
13. Leitner, A.: Experiences gathered from the application of the ENABLE-S3 V&V architecture. Autonomous Vehicle Test & Development Symposium, presentation, Stuttgart, 5 June 2018
14. TEST PROTOCOL – AEB VRU systems, v2.0.2, Euro NCAP protocol. https://cdn.euroncap.com/media/32279/euro-ncap-aeb-vru-test-protocol-v202.pdf (2017)
15. TEST PROTOCOL – AEB systems v2.0.1, Euro NCAP protocol. https://cdn.euroncap.com/media/32278/euro-ncap-aeb-c2c-test-protocol-v201.pdf (2017)
16. OpenSCENARIO, open standard for exchange of scenarios for simulation. http://www.openscenario.org/index.html
17. Rooker, M., Horstrand, P., Salvador Rodriguez, A., Lopez, S., Slavik, Z., Sarmiento, R., Lopez, J., Lattarulo, R.A., Perez Rastelli, J.M., Matute, J., Pereira, D., Pusenius, M., Leppälampi, T.: Towards improved validation of autonomous systems for smart farming. Workshop on Smart Farming at CPS Week 2018
18. Pflug, G.: The generation of scenario trees for multistage stochastic optimization. http://dinamico2.unibg.it/icsp2013/doc/Pflug.pdf (2013)
19. Pflug, G.: Scenario tree generation for multiperiod financial optimization by optimal discretization. Math. Program. **89**(2), 251–271 (2001)

Traffic Sequence Charts for the ENABLE-S_3 Test Architecture

Werner Damm, Eike Möhlmann, and Astrid Rakow

1 Introduction

Scenarios are considered a key element for acceptance testing of highly autonomous vehicles (HAV) within the German automotive industry [3]. A traffic scenario describes how a traffic situation evolves over time. It captures the behavior of the involved traffic participants and describes the static infrastructure as well as environmental conditions. For example a *"circumvent construction site"* traffic scenario may describe how a car approaches a construction site at its lane, then changes onto the lane to its left and circumvents the construction site. Such a scenario can then be analyzed wrt to certain criteria (e.g., criticality measure for a collision during the scenario, fuel consumption of the HAV, profiles of involved traffic participants).

A collection of traffic scenarios can be used to characterize the application domain of an HAV. It be used as an input to derive requirements at the early stage of an HAV's development process as well as for the specification of test cases. Scenario catalogs systematically compile traffic scenarios and aim to span the space of relevant traffic scenarios. The approach of using scenario catalogs for acceptance testing is currently taken in the German PEGASUS project[1] and also the German

[1] www.pegasusprojekt.de.

W. Damm · E. Möhlmann
OFFIS – Institute for Information Technology, Oldenburg, Germany
e-mail: damm@offis.de; e.moehlmann@offis.de

W. Damm · A. Rakow (✉)
Carl von Ossietzky University of Oldenburg, Oldenburg, Germany
e-mail: a.rakow@uni-oldenburg.de

A. Leitner et al. (eds.), *Validation and Verification of Automated Systems*,
https://doi.org/10.1007/978-3-030-14628-3_6

Fig. 1 A traffic scenario where a car circumvents a construction site. In the first snapshot the car approaches the construction site. While being at a distance greater d to the site, it changes lanes. Then, the car stays on the neighboring lane at least as long as it is less then 0.5 d apart from the construction site. After having passed the site, it changes back to the original lane

Ethics Commission envisions a catalog of scenarios of a state-certified body as part of the overall architecture of cooperative road traffic [9].

We observed in several projects like PEGASUS, CSE[2] or ENABLE-S_3[3] that in order to boost the comprehensibility of such scenarios and to thereby facilitate discussions among experts, the scenarios are often visualized. Such a visualization of "*circumvent construction site*" may then look like in Fig. 1.

The ECSEL project ENABLE-S_3 establishes a scenario-based verification & validation (V&V) methodology. The project develops procedures and tools to combine simulations with real world tests for practical testing of highly and fully automated systems. Scenarios are used to capture the intended application domain of the system under test (SUT) in a structured way. Note that in order to characterize the domain, it is necessary to capture infinitely many (combinations of) environmental situations and evolutions. Hence, the collected scenario(description)s have to focus on the relevant/characteristic aspects.

As part of the V&V methodology ENABLE-S_3 develops a generic test architecture as in Fig. 2 to orchestrate its tools and within which the test procedures are implemented. The test architecture defines three layers (1) V&V management *(concerned with general information on the application domain and requirements that have to be established for the SUT and test blue prints)*, (2) test management *(concerned with test case generation, monitoring and evaluation)* and (3) test platform *(concerned with execution of tests)*. Regarding the usage of TSCs, note that the ENABLE-S_3's V&V methodology characterizes the application domain of the SUT by a scenario data base as part of the V&V management. This scenario data base serves as the reference for test environment (How should the environment of the SUT behave during a test or simulation?).

So for *V&V management* scenarios are used to characterize the application domain. To this end, real world observations are collected in a database. These observations are analyzed and critical traffic scenarios are derived in an abstraction process. The abstract scenarios are entered into the scenario database. The content of the scenario database spans the space of all possible scenarios that the SUT might face during its lifetime. A process of compiling a scenario catalog is summarized in Sect. 3. Along with such a characterization of the application domain, the V&V

[2]www.offis.de/en/offis/project/cse.html.

[3]www.enable-s3.eu.

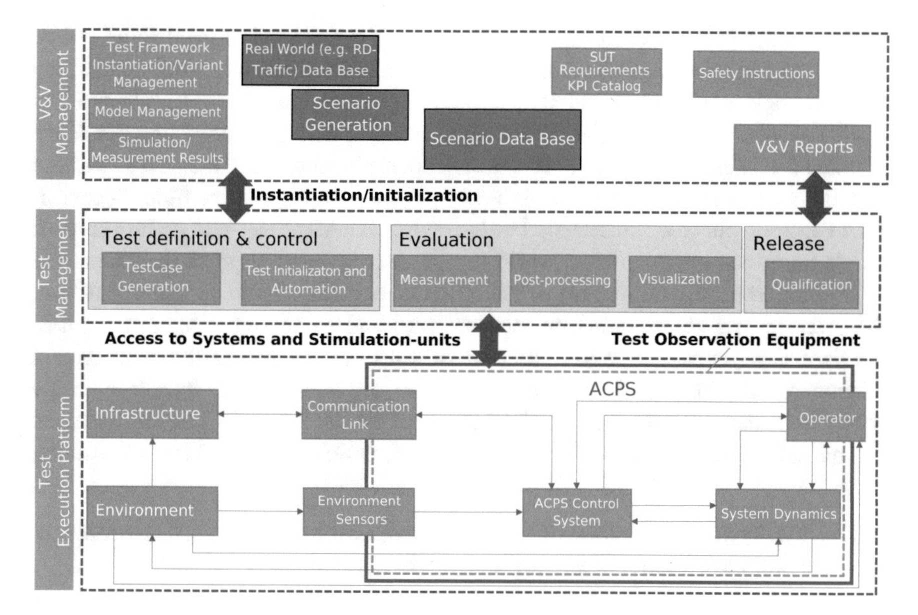

Fig. 2 The generic test architecture

management includes capturing the safety instructions, SUT requirements, different (environment simulation) models, as well as blue prints for test frameworks. For simulation-based test systems these information are used to instantiate the test platform.

The *test management* performs and evaluates tests using the test platform. The management of tests includes activities such as (1) deriving configurations for the individual components of the test platform and the evaluation monitors, (2) executing the generated test cases on the test platform, (3) collecting the evaluation data (e.g., via monitors for satisfaction of the safety instructions or for measuring the key performance indicators (KPIs) for a given test scenario), and (4) deriving or inferring knowledge (e.g., test coverage, remaining risk) from the evaluation data.

To execute a test case, the *test platform* provides models—or even specialized simulators—for different aspects of the test scenario. The test platform provides for instance simulation models of the weather conditions, the infrastructure, the communication between entities, the sensors, the ACPS (automated cyber-physical system) control system, the system dynamics, and the operator. At a test scenario, the SUT has to satisfy its requirements (like KPIs, safety instructions) while its environment is driven to expose the behavior as specified by the scenario. Along a scenario several test cases might be derived for different test criteria. The test case specifies which parts are counted as the SUT (from only the ACPS control system up to everything but the environment).

To summarize, ENABLE-S_3 aims to reduce the test environment setup effort, reduce the test execution effort, allow ensuring a high safety level (malfunctioning behavior below 10^{-9}/h),[4] and reduce the re-qualification effort. To achieve this (fully) automated scenario-based testing combined with the above described test architecture is employed. Using scenarios allows characterizing the possible traffic situations and evolutions that an ACPS might face during its live time. Scenario-based testing plays a fundamental role for the ENABLE-S_3 test architecture. It builds on the collection of real world traffic scenarios and requires the generation of abstract scenarios representing the space of possible traffic evolutions as well the management of scenarios on the one hand and on the other hand it requires the derivation, execution, and evaluation of test cases as well as the management of their results. Finally, the results are used to, e.g., statistically infer knowledge about the remaining risk that the SUT does not satisfy the requirements. To make this approach viable, there are several challenges which must be addressed. Among these are:

(C1) How do we capture the space of all possible traffic situations and environmental factors relevant for safe trajectories for autonomous vehicles?

   (a) How can we determine whether a traffic evolution is covered by a scenario?
   (b) How can we represent the infinitely many traffic evolutions within a finite traffic catalog?

(C2) How can we decide, whether the reaction of the SUT is compliant to the scenario catalog?
(C3) How can we assure, that the interpretation of scenarios and thus interpretation of test results is unambiguous across all test platforms?
(C4) How do we determine which tests need to be run (in a virtual, semi-virtual, or physical environment)?

   (a) How do we determine which tests need to be re-run after changing the SUT?
   (b) How can we (iteratively) generate tests in a way that maximizes coverage, minimizes the remaining risk, or maximizes the chance of finding implementation bugs?

All these are important challenges within the ENABLE-S_3 project and can only be addressed by using a language for capturing scenarios, which is intuitively easy to understand, and, most prominently, which is equipped with a formal (declarative) semantics. In this paper we summarize what TSCs can address the challenges C1 to C4, thereby surveying [5, 1, 10]. We elaborated how TSCs contribute to C1 to C3 in [5, 1] while [10] has a stronger focus on testing (C4). Moreover, we sketch in this paper our vision on using TSCs especially for C4.

[4] 10^{-9}/h corresponds to the safety assurance level A in aviation which classifies catastrophic failure conditions (cf. [13, p. 14]).

Outline In the next section we give a short overview of the TSC formalism. We then describe how we envision that TSCs are used for the construction and representation of a scenario catalog. We briefly sketch in Sect. 4 the use of TSCs along the development process of a system where they can be used to capture system requirements as well as for testing. In Sect. 5 we describe the use of TSCs for quantitative verification as in [10]. Finally in Sect. 6 we sketch our vision on using TSCs for test management.

2 The TSC Formalism

In this section, we briefly introduce the TSC formalism. TSCs are a visual formalism tailored for describing traffic scenarios. TSCs combine two formalisms synergetically: (1) Snapshot Charts (SCs) focusing on the visual specification of the continuous evolution at a scenario, and (2) LSCs [2] supporting the visual specification of communications between entities.

A TSC consists of a *header*, an SC and optionally an LSC part and refers to a *world model*, which reflects the physical aspects of the application domain. The meaning of symbols used in a TSC is declared in terms of the world model by a *symbol dictionary*. Note that we do not require a certain formalism for the world model.[5] A TSC represents a multi-sorted real-time logic formula [6] and has hence a platform independent semantics. A TSC specification consists of a collection of TSCs that conjunctively constrain the behavior of the considered world model. In this paper we will present a special form of TSCs that do not have an LSC part, so only use SCs. SCs are a visual mean to specify in a scenario-based way families of trajectories. They describe an abstract scenario (an evolution over time) via a sequence of *snapshots* (pictures that describe a traffic situation) like in Fig. 1. Moreover it is possible to describe branching scenarios within one SC. A snapshot visualizes invariant constraints characterizing a traffic situation, that have to hold until the next snapshot along a TSC sequence. For visualization we currently focus on a *spatial view*. Example 1 illustrates this visualization concept.

Example 1 In the following example we illustrate the spatial view by a list of simple exemplary snapshots. Please note, TSCs provide a flexible formalism for visual specifications. Often the user has a choice of adapting the representation to his/her own needs.

We can express the existence of an object by placing the object symbol within a snapshot. So Fig. 3a expresses that there is a car. We can also rule out that there is a certain object or an object of a certain kind. In Fig. 3b we specify that within a certain area there is nowhere a car. Attribute values like having active indicator lights can be visually specified like in Fig. 3c. Alternatively, we allow the use of predicate labels.

[5] You may for instance use hybrid automata to model the car dynamics.

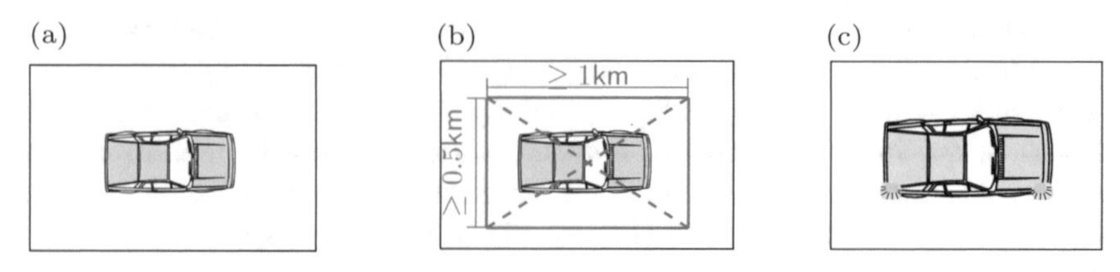

Fig. 3 Visualization of objects and their attributes. (**a**) a car (existence of objects). (**b**) no car (non-existence). (**c**) signal (object attributes)

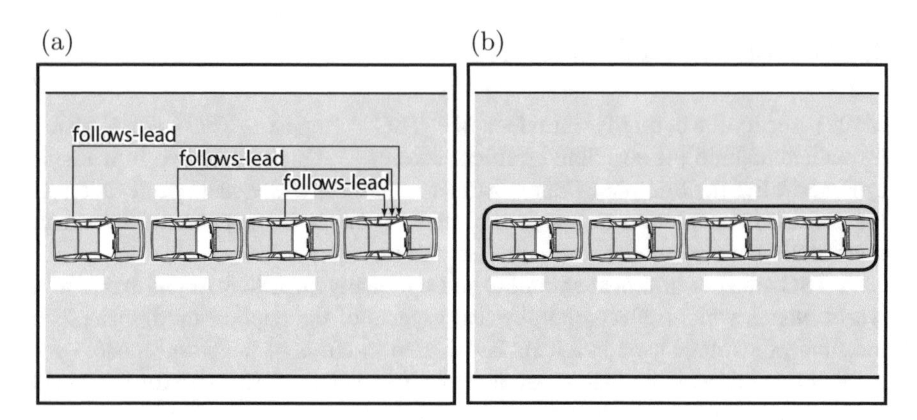

Fig. 4 Visualization of relations between objects. (**a**) platoon (relations via predicate label). (**b**) platoon (relations via symbol)

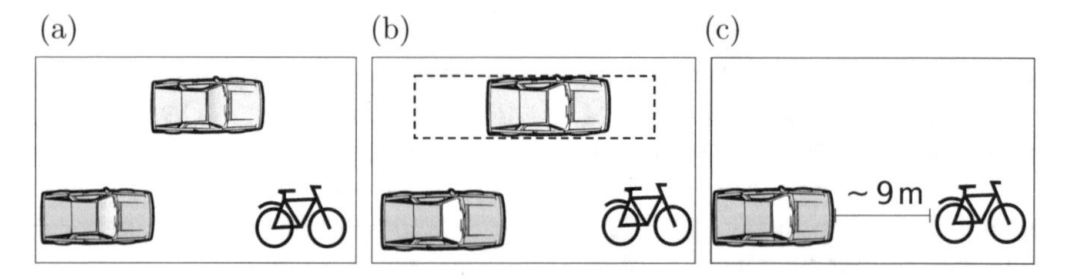

Fig. 5 Placement of objects. (**a**) relative placement. (**b**) somewhere box. (**c**) distance line

Similarly, relations among objects can be specified via predicate labels. For instance we can express that cars are in a follow-lead relation within a platoon (cf. Fig. 4a). TSCs also allow introducing a dedicated platoon symbol (cf. Fig. 4b). Both representations of a platoon can be used within a TSC specification interchangeably.

The core idea of the spatial view is that we place the symbols relative to each other and then derive constraints reflecting this placement. So Fig. 5a means that there is a white car, and there is a grey car right of it and further behind. In front of the grey car there is a bike. The bike is also right of and preceding the white car. So the relative positioning of each symbol to any other is considered. To weaken this translation scheme, we introduce the concept of *somewhere boxes* as in Fig. 5b.

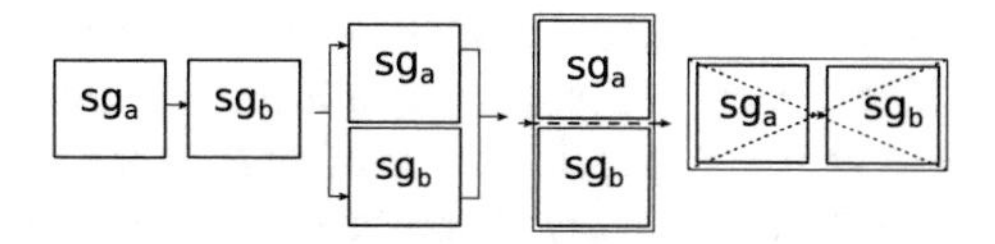

Fig. 6 Building snapshot graphs: (**a**) Concatenation: sg_a holds then sg_b (**b**) Choice: sg_a holds or sg_b holds (**c**) Concurrency: sg_a holds and sg_b holds (**d**) Negation: It does not hold, that first sg_a holds and then sg_b holds

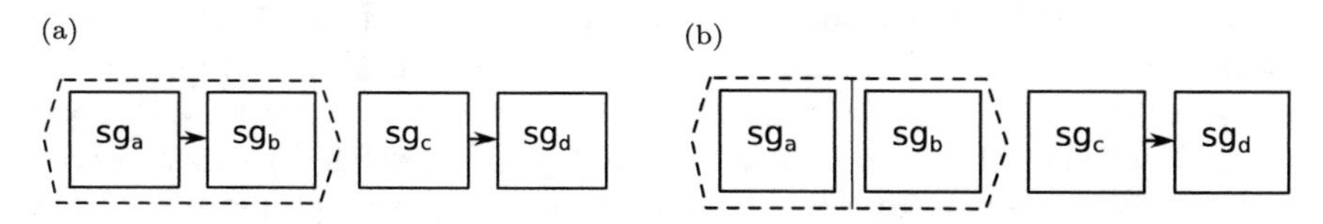

Fig. 7 (**a**) specifies: "If first sg_a and next sg_b happened, then next first sg_c will happen and then sg_d." (**b**): "If first sg_a and in the future sg_b will hold, then next first sg_c will happen and then sg_d, concurrently sg_b will hold". (**a**) Premise consequence SC. (**b**) Future-premise consequence SC

The white car is somewhere right of the grey car and the bike, but not necessarily horizontally in between them. We can even specify absolute distances via distance lines as in Fig. 5c. Note that the spatial view has a formal semantics, so that any snapshot is directly translatable into a formula. For instance, the snapshot of Fig. 5c is translated to $\exists car \in \texttt{Cars} : \exists bike \in \texttt{Bikes} : \mathsf{dist}(car.x\_pos, bike.x\_pos) \sim 9m \wedge car.x\_pos < bike.x\_pos \wedge car.y\_pos = bike.y\_pos.$[6] □

Snapshot graphs arrange *snapshots* within a directed graph via concatenation, concurrency and negation (cf. Fig. 6).

SCs basically consist of snapshot graphs but may have additional annotation and may implement a premise-consequence pattern: An SC's snapshot graph can be annotated with timing and synchronization constraints along a path. Furthermore, SCs provide a distinct syntax to specify premise-consequence rules (cf. Fig. 7).

We can translate an SC to a first-order multi-sorted real-time formula, by composing the snapshot formulas—the formulas encoded by the snapshots—according to the graph's structure and annotations. The header of a TSC holds additional information like the quantification mode and the activation mode. An existential TSC expresses that *there is* one evolution of the world model satisfying the SC while a universal TSC express that *all* evolutions satisfy the TSC. The activation mode is either initial or invariant. An initial TSC has to hold right at the start of an evolution, while an invariant TSC has to hold at all times.

World Model and Ontology The world model is a formal model, that specifies the assumed application domain. It can be seen as the test environment at the analysis and validation phase. A TSC is interpreted on a world model (or on a ontology referring to the world model, respectively), i.e., everything a TSC talks

[6] assuming an appropriate world model and symbol dictionary.

about has a semantics within the world model/ontology. The world model/ontology defines classes of objects (cars, bikes, ...) with a set of attributes (position, size, velocity, ...) and the dynamics of moving objects (for instance via hybrid automata). Each object belongs to one of finitely many classes. The symbols and predicate annotations used in snapshots refer to the objects and their states within the world model/ontology.

Symbol Dictionary The symbol dictionary links the visual symbols—e.g., —used in TSCs to the world model/ontology objects (objects of class Car). So the s-dict lists atomic predicates and their interpretation. It declares symbol properties like equivalences (i.e. "Are symbols different if they are rotated, scaled or of changed color?"). It defines for a symbol modification (e.g.,) which object features are represented (*the car is indicating to its left*). To link symbol positions to object positions, an anchor has to be defined for the symbol and the correspondence to the object's anchor has to be declared.

3 Building a Scenario Catalog

In the following we describe the steps of constructing a scenario catalog from a database of real world observations and how TSCs are used in this process. Note that this step is also required as part of the ENABLE-S_3's V&V methodology within the V&V management. This section mainly summarizes [5, 1] with a focus on using TSCs for the scenario catalog construction.

In the sequel, we briefly list the steps of building a scenario catalog (cf. Fig. 8).

(Scenario Screening) First, data from risk analysis, accident databases (like GIDAS (German In-Depth Accident Study)) and virtual and real long term testing etc. are screened. The German PEGASUS identifies a list of characteristics of critical situations (like time to collision). Concrete scenarios will be produced for each of these, either by simulation or by field tests. The concrete scenarios along with the original recorded data is stored in a real world database.

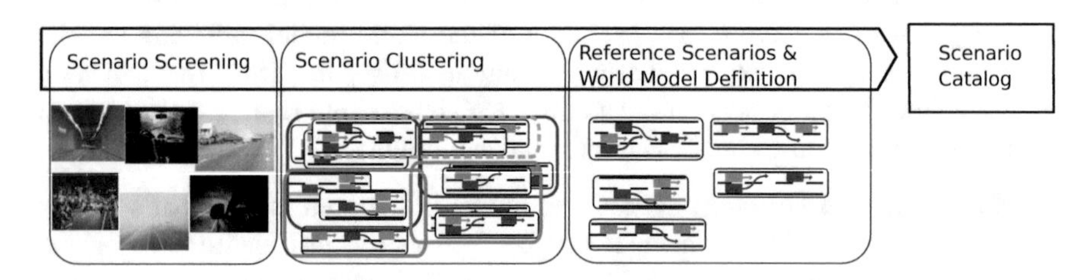

Fig. 8 Steps for compiling a scenario catalog

(Scenario Clustering) Scenario clusters are built of the concrete scenarios that are similar wrt context and evolution. Relevant parameters, influencing factors that characterize critical situations (like time to collision) are identified at this step.

(Reference Scenarios and World Model Definition) After that, the concrete scenarios will be abstracted to obtain entries for the catalog and a world model is defined that consists of the relevant objects and that reflects relevant real world phenomena. Based on this world model, abstract scenarios are derived for the collected concrete scenario(cluster)s.

Each of the steps for generating a scenario catalog is far from trivial. In [1] we presented a series of methods based on learning key structural properties from a data base of real world traffic observations and on statistical model checking that lead to the construction of a scenario catalogue that captures requirements for controlling criticality for highly autonomous vehicles. Moreover, we sketched the mathematical foundations to derive confidence levels that a vehicle tested by such a scenario catalogue will maintain the required control of criticality in real traffic matching the probability distributions of key parameters of data recorded in the respective reference data base.

Our approach [1] analyses the space of all real-world traffic evolutions and environmental factors, identifying reasons that allow collapsing infinitely many evolutions into a parameterized set of finite equivalence classes. To build the equivalence classes, we intent to learn all indicators relevant for judging the criticality of driving situations. The criticality depends on the observability (perception) of a vehicle, the predictability (cognition) of the environmental conditions including its future and the controllability (motion) of the vehicle. The learning methods are used for (1) the dynamics of all relevant classes of traffic participants (e.g. through parameter learning in probabilistic hybrid automata of vehicle dynamics), (2) the controllability of the dynamics of the ego vehicle, and (3) the perception errors along the complete perception chain.

We initially focus our learning process on identifying all causes for accidents or near-accident situations and also learn probabilistic hybrid automata as models of dynamic of all classes of traffic participants. Using these, we learn a catalog of scenarios allowing us to test whether vehicles are able to control criticality in the presence of hazards. In all these scenarios—i.e., all combinations of hazards, traffic evolutions—the vehicle under test is required to reduce the risk to an acceptable level. For testing, we follow the principle of separation of concerns and, hence, reduce the complexity by first testing the vehicles capability of risk mitigation under perfect perception and perfect control, and then testing for robustness in perception and control failures. For this, we learn (a) hazardous environmental conditions that explain perception failures and (b) statistical models of the confidence level of the complete perception chain assuming absence of the learned hazards. A formalism like TSCs is a prerequisite for our approach [1]. As such, we need a representation of scenarios which is expressive enough to cope with the plethora of environmental conditions and traffic evolutions, while yielding concise and unambiguous interpretations based on a formal semantics. To represent infinitely many traffic evolutions within a finite scenario catalog (C1b) demands using a declarative specification language like the TSC formalism, where one single scenario specification abstracts a possibly extremely large set of real world traffic evolutions.

A formal satisfaction relation (as we have it for TSCs) is a prerequisite to formally define, whether an observed evolution is covered (or not) by the scenario catalog (C1a), which is minimally required when checking for completeness. Further, it is a basis for playing out the current scenario catalog generating traffic flows, which an expert can assess, as experienced in the play-out approach for Live Sequence Charts (LSCs) [2, 11].

In general if we use TSCs to specify a scenario catalog, we can (a) list required scenarios and also (b) express that a set of scenarios covers all possible evolutions (coverage). For (a) we use existential TSCs while we express (b) via universal TSCs. Whereas *listing scenarios* is the minimum requirement for a formalism to specify scenario catalogs, *expressing coverage* is a distinguishing feature of TSCs. Figure 9 illustrates how TSCs can be used to express coverage. Expressing coverage via TSCs allows checking that no scenario is missing. The relevant scenarios can be modelled in more detail while the less critical can be treated abstractly. This can help to ensure that the test efforts are invested where they are necessary and allow to relieve the test effort for other scenarios.Via TSCs we can also describe relations among scenarios (like disjointness or inclusion). Further we can use the formal satisfaction relation to check whether the concrete scenarios observed are actually covered by the abstract scenarios (which are the result of an abstraction and modelling process).

For instance at step *(Reference Scenarios and World Model Definition)* the right level of abstraction has to be ensured. If we generalize a concrete scenario s_c to an abstract scenario s_g, we will have to investigate whether all concretizations of s_g still reflect the relevant criteria of s_c. If for instance only some concretization of an abstracted critical scenario lead to a critical situation while other concretization do

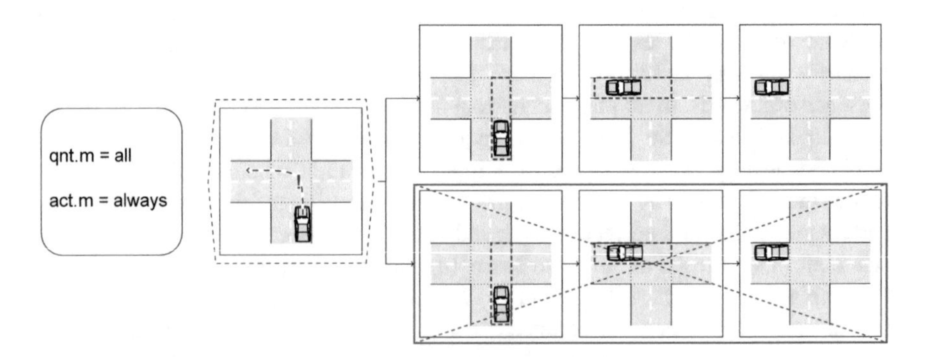

Fig. 9 A trivial example of coverage: "If a car wants to do a turn, it either does a turn or not." The TSC use a premise consequence pattern to describe that if a car wants to turn left (snapshot within the dashed hexagon), then (upper snapshots) the car successfully turns left, or (lower snapshots) it does not

not, the scenario specification has to be refined yielding more determinate abstract scenarios.

4 TSC for Scenarios at the Development Process

In [5] we illustrated the use of TSC scenario catalogs along a development process of a safety concept following the V-Model. Here, we summarize of [5] mainly the use of TSCs for verification and validation phase of the V due to its immediate relation to ENABLE-S_3. Nevertheless, we note that the requirement specification is also strongly influenced by the scenario space, since it reflects the necessities, risks and optimization potentials of the system within the targeted context. Moreover, system requirements can also be captured in a scenario based way.[7]

In the verification and analysis phase (right arm of the V), analysis methods (such as virtual testing or model checking) are triggered on different levels of the abstract system and finally on the realized system with the goal to ensure that the realized system fulfills the system requirements. The realized system is tested component-wise with increasing system complexity, and finally acceptance is tested.

For this activities the scenario catalog provides the system context and targeted use cases, based on which test runs can be generated. So scenario catalogues provide a reference for the test environment that supports virtual tests already at abstract system specifications up to the definition of concrete tests for acceptance testing. In Sect. 5 we describe one such approach.

5 TSCs for Quantitative Verification

In [10] we have presented an approach for quantitative verification of complex stochastic systems such as driver assistant systems with a high dependence on the environmental situation. Firstly, the environmental situations as well as the desired and undesired behavior of the system are characterized via an TSC dialect. Secondly, information about specified environmental situations as well as the SUT is used for a particular simulation of the combined system, *environment+SUT*, to statistically check whether the SUT operates safely in all specified situations.

We extended the model checking procedure for solving satisfiability modulo theory (SSMT) problems to a more scalable version based on simulations and a corresponding statistical analysis, i.e., statistical model checking (SMC) [14], to generate quantitative statements about the performance of the system with respect to a criticality measure. In contrast to classical model checking, the SMC method

[7] To this end we usually use the premise-consequence pattern. Within such TSC we refer to a dedicated ego car, representing the vehicle under design and thereby constrain its behaviour[6].

is based on samples using an underlying simulator and therefore generates results which can be guaranteed with a certain level of confidence only, i.e., a residual probability of being wrong. Note that, in an SSMT formula we allow not only purely random influences, but also pure non-deterministic influences. These are resolved either optimistically or pessimistically as reflected by the maximum and minimum operator. Both the environment and the SUT are allowed to contain abstract characterizations, thereby providing such safety statements for a large set of possible implementations (of the system) and specific situations entailed by the abstract description. We already presented in [8] a simulation based evaluation strategy that can handle under-specified systems. There we incorporated methods from noisy optimization which requires detailed knowledge about the probability distributions associated with the randomized quantifier as well as about the underlying dynamics. Whereas in the approach of [10] we extend the evaluation method, so that it assumes less knowledge about the system and is therefore easily applicable in a black-box setup, in which some parts of the system are only given in terms of available simulator components. These simulations might still contain randomness, for example the behavior of other vehicles within a scenario might be subject to a particularly designed probabilistic dynamic model which is only available via a black-box simulation.

Our statistical model checking of the *SUT+environment* system provides a confidence statement about the satisfaction of the requirements. We thereby focus on problems of the following form: Is there a bound on the likelihood that a system satisfies a certain property, that holds for all environmental scenarios? This accounts to problems of possibly multiple existential or universal quantifiers followed by randomized ones. To use a statistical model checker for answering such problems we need to construct the input SSMT-formula. To this end, we proposed *stochastic TSCs* (sTSCs)—basically TSCs with additional annotations regarding the distribution of variable values. From these we can derive formal requirements for statistical model checking. Note, that such a translation alleviates the burden for engineers of getting the formulas of nested (universal, existential and random) quantifiers right. At the time of the original publication [10] the TSC formalism was developing and early work on a TSC dialect [12] was published. Now (at the time of writing this contribution) the TSC formalism as in [6, 4] has been developed, but the stochastic extension as visioned in our work is still not completed, but it is planned to become part of the TSC formalism. We hence summarize in the following, how we imagined the extension of TSCs. We present an example of an sTSC, that is slightly modified wrt to [10]. The example adopts in particular the current TSC formalism. We proposed to base the TSC extension on stochastic satisfiability modulo theory (SSMT). So we can capture requirements involving dynamics using differential equations, switching between different dynamics, and setting new variable values using either stochastic transitions and valuations or completely non-deterministic versions.

To support a semi-automatic construction of the input formula we annotate the quantification, the domain, and the distribution—if available—to further constrain uncertain behavior within the environment. In Fig. 10 we describe a maneuver,

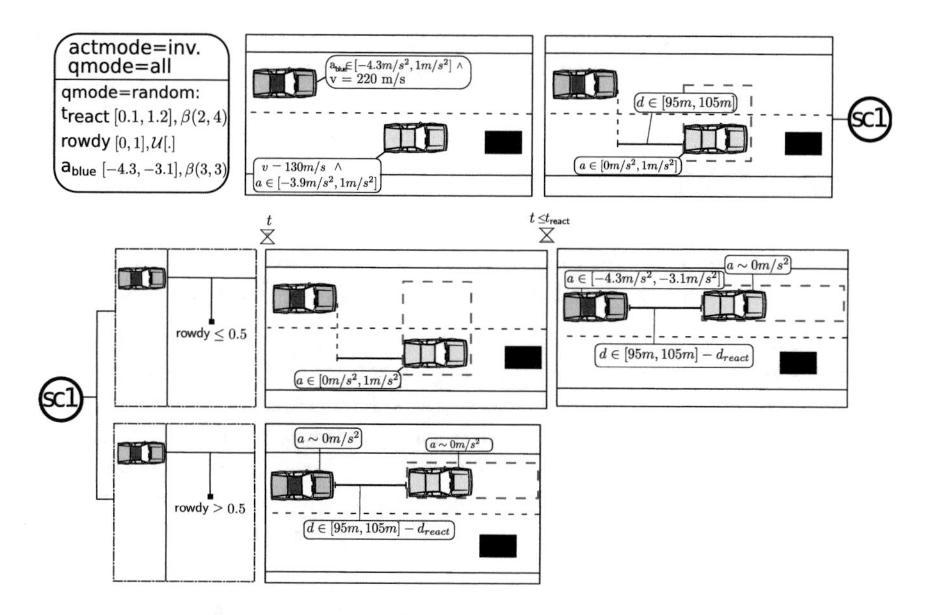

Fig. 10 An sTSC describing a scenario where ego (grey car) is forced to circumvent an obstacle. The sTSC uses additional syntax to specify the quantification, domain and distribution of random variables at the TSC header. The TSC body describes that first ego car is approaching an obstacle (black box), while at the neighboring lane a blue car is approaching from behind at the neighboring lane. The velocity and acceleration of the two cars is constrained via predicate labels. Then the blue car gets into a distance d∈ [95 m, 105 m] and ego stops accelerating. Then, depending on whether the blue car behaves as rowdy, the maneuver evolves differently. The first node after the sc1 circle is an LSC node, in which the value of the rowdy variable is tested. If the blue car is not a rowdy (upper branch), the blue car will decelerate (last snapshot in the upper branch) after its reaction time has passed (annotated via hour glass above the second node). In case the blue car is a rowdy, it will not decelerate even though ego changed lanes

where the grey car, ego, has to circumvent an obstacle. The blue car either behaves as a rowdy or “normal”. For better readability, we introduced a line break in the TSC.[8] We model the decision of being a rowdy using a continuous uniform distribution of a random variable *rowdy* and then test whether the variable value is below or above a certain threshold. In addition to the standard TSC syntax, we here annotate the distribution and the domain for the reaction time of the blue car (t_{react}), the expected acceleration (a for the grey car, a_{blue} for the blue car) at the TSC header.

Figure 10 captures visually constraints on the maneuver and specifies quantification and distribution of the scenario parameters. Together with a world model

[8]A TSC one liner can be retrieved by gluing the two parts the circles inscribed sc1 and remove the circles.

that specifies dynamic models of the vehicles, we have all information at hand to generate a formal representation for our verification approach.

6 Future Work

Moreover, we envision TSCs to be useful for testing, managing test cases and evaluation data. In the following, we sketch ideas regarding the application.

Description of Requirements We distinguish between the ego car, which is the car under design and the environment including other traffic participants. Referring to ego in a TSC, specifies a requirement on ego. Referring to other vehicles in a TSC, specifies an assumption on ego's environment. That way we can define the desired and forbidden behaviour of an HAV within its assumed environment. Usually this is done by universal TSC rules and patterns like "if ego is in SC_A, then ego has to do SC_B". For the following let $\mathbb{TSC}_{\mathsf{ego}}$ be the set of such TSCs specifying constraints on ego.

Testing and Simulation Using existential TSCs, we can specify abstract test cases, like "We want to see one evolution where ego circumvents the obstacle with high speed and low friction and the other car behaves like a rowdy." The underlying TSC world model defines the behaviour of the environment and also provides a model of the physical dynamics of the ego car. We can hence simulate the specified ego by playing out the behaviour that accords to TSCs. LSCs have already been successfully used for such an approach [11]. If we want to run a test along a TSC scenario, we simulate the implementation of ego within its assumed environment and monitor for instance whether the implementation accords to the specification $\mathbb{TSC}_{\mathsf{ego}}$.

Analysis of Test Case Coverage Given a test coverage requirement (like executing all operational modes of some subsystem), we can specify a set of TSC scenarios $\mathbb{TSC}$ that ensure that all modes are visited. Testing several subsystems simultaneously, we can use this information to minimize the number of test runs. One idea is that we can check whether a scenario tsc_1 of $\mathbb{TSC}$ can be run concurrently with a scenario tsc_2 of $\mathbb{TSC}$, the test set of another subsystem. Formally, we can express this combined scenario easily as one TSC.

Reporting and Management of Evaluation Data We imagine that the intuitiveness of TSCs helps documenting test results (including bugs or requirement violations) and reporting back to the engineers. In our companion paper [7] we describe how we generate a TSC, an abstract representation, of a concrete test run. Such a TSC visualizes a concrete evolution in terms of the notions of the referred symbol dictionary (chosen to the needs of the engineers). Similarly we imagine that TSCs are a suitable formalism to identify and represent common features of several concrete test runs. In order to alleviate the traceability of test results to requirements, we advocate associating concrete test runs (specified via, e.g., OpenSCENARIO) to abstract scenarios. This also eases the interpretation of test results by the engineers. A formal representation of a test case also alleviates regression tests, since we might already judge from the formal characteristics of the abstract test case whether a changed subsystem might be affected.

7 Conclusion

TSCs are a visual specification formalism for specifying abstract traffic scenarios. They have a formal semantics ensuring platform independence and offering the possibility to apply formal methods examining the scenarios. Since scenarios are at the heart of ENABLE-S3, we presented a survey of [5, 1, 10] to highlight the features of TSCs that are especially relevant for ENABLE-S3.

TSCs allow expressing system requirements in a scenario based way as well as capturing observed traffic scenarios. Since TSCs also allow specifying relations among the different scenarios (complete coverage, disjointness, inclusion), we promote them as formalism for the construction and specification of scenario catalogs. For the same reason they seem a good candidate for managing test scenarios, since they allow to determine formally whether certain coverage criteria along a scenario are satisfied.

TSCs are a versatile specification formalism especially suited for the early and late phases of the development. In the early phase it can serve as mean to structure the application domain of the HAV by distinguishing relevant use cases and critical scenarios. When finally the developed system is tested, the scenario catalog specifies the test cases, in which the vehicle under test is examined to show the expected behaviour. With additional annotations, we see moreover a great potential for using TSCs to manage test evaluation data.

References

1. Damm, W., Galbas, R.: Exploiting learning and scenario-based specification languages for the verification and validation of highly automated driving. In: Proceedings of the 1st International Workshop on Software Engineering for AI in Autonomous Systems, pp. 39–46. ACM, New York (2018)
2. Damm, W., Harel, D.: LSCs: Breathing life into message sequence charts. Formal Methods Syst. Des. **19**(1), 45–80 (2001)
3. Damm, W., Heidel, P.: Recommendations of the SafeTRANS working group on highly autonomous systems (2017). www.safetrans-de.org/en/Latest-reports/management-summary-for-highly-automated-systems/192
4. Damm, W., Kemper, S., Möhlmann, E., Peikenkamp, T., Rakow, A.: Traffic sequence charts—from visualization to semantics. Reports of SFB/TR 14 AVACS 117, SFB/TR 14 AVACS, 10 2017
5. Damm, W., Kemper, S., Möhlmann, E., Peikenkamp, T., Rakow, A.: Traffic sequence charts—a visual language for capturing traffic scenarios. In: Embedded Real Time Software and Systems - ERTS2018 (2018)

6. Damm, W., Möhlmann, E., Peikenkamp, T., Rakow, A.: A formal semantics for traffic sequence charts. In: Principles of Modelling Essays dedicated to Edmund A. Lee on the Occasion of his 60th Birthday. LNCS. Springer, Berlin (2017)
7. Damm, W., Möhlmann, E., Rakow, A.: A scenario discovery process based traffic sequence charts. Technical report, 2018. submitted to ENABLES book
8. Ellen, C., Gerwinn, S., Fränzle, M.: Statistical model checking for stochastic hybrid systems involving nondeterminism over continuous domains. Int. J. Softw. Tools Technol. Transfer **17**(4), 485–504 (2015)
9. Federal Ministry of Transport Ethics Commision and Germany Digital Infrastructure: Automated and connected driving (2017). www.bmvi.de/SharedDocs/EN/publications/report-ethics-commission.html
10. Gerwinn, S., Möhlmann, E., Sieper, A.: Statistical Model Checking for Scenario-Based Verification of ADAS, pp. 67–87. Springer, Berlin (2019)
11. Harel, D., Marelly, R.: Come, Let's Play: Scenario-Based Programming Using LSC's and the Play-Engine. Springer, Berlin, (2003)
12. Kemper, S., Etzien, C.: A visual logic for the description of highway traffic scenarios. In: Proceedings of the Fourth International Conference on Complex Systems Design & Management CSD&M 2013, pp. 233–245. Springer, Berlin (2013)
13. S-18 Aircraft and Sys Dev and Safety Assessment Committee: Guidelines and Methods for Conducting the Safety Assessment Process on Civil Airborne Systems and Equipment, vol. 12. SAE International, Warrendale (1996)
14. Younes, H.L.S., Kwiatkowska, M., Norman, G., Parker, D.: Numerical vs. statistical probabilistic model checking. Int. J. Softw. Tools Technol. Transfer **8**(3), 216–228 (2006)

A Scenario Discovery Process Based on Traffic Sequence Charts

Werner Damm, Eike Möhlmann, and Astrid Rakow

1 Introduction

Scenarios are considered a key element for acceptance testing of highly autonomous vehicles (HAV) within the German automotive industry [4]. For instance the ECSEL project ENABLE-S_3[1] bases its generic test architecture on scenario catalogs to capture the application domain and, hence, the test environment. A traffic scenario catalog lists possible traffic evolutions. In the early phase such a scenario catalog can serve as mean to structure the application domain of the HAV by distinguishing relevant use cases and critical scenarios. When finally the developed system is tested, a scenario catalog specifies the test cases, in which the vehicle under test is examined to show the expected behavior (like avoiding near crash situations).

To support the systematic development of a scenario catalog for the early system specification phase, we present in this paper a tool-supported process that determines scenarios violating a given safety requirement φ like collision-freedom. The generated list of scenarios describes every possible evolution of the world model that leads to a violation of φ. The process employs model checking, and derives from a concrete counterexample an abstract scenario, that possibly represents (infinitely) many concrete evolutions. This scenario generation procedure

[1]www.enable-s3.eu.

W. Damm · E. Möhlmann
OFFIS – Institute for Information Technology, Oldenburg, Germany
e-mail: damm@offis.de; moehlmann@offis.de; a.rakow@uni-oldenburg.de

W. Damm · A. Rakow (✉)
Carl von Ossietzky University of Oldenburg, Oldenburg, Germany

A. Leitner et al. (eds.), *Validation and Verification of Automated Systems*,
https://doi.org/10.1007/978-3-030-14628-3_7

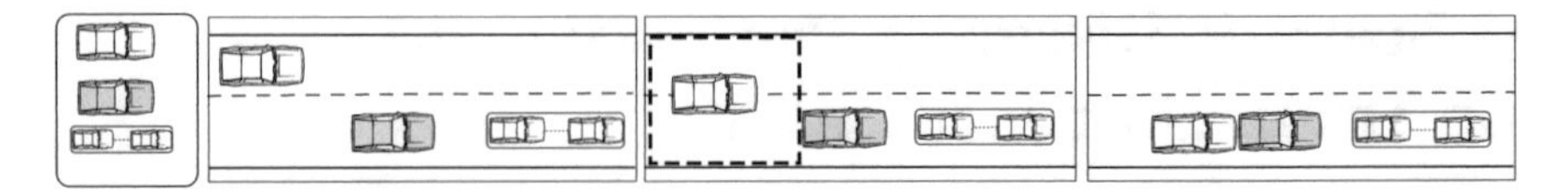

Fig. 1 Snapshots telling a story: The snapshot graph describes a story of a platoon and two cars. First the grey car follows the platoon and the white car is behind and left of the grey car. Then the white car changes lanes (is somewhere within the dashed box). Then white car violates the grey car's safety distance

faces several research challenges such as:

1. How can we define a catalog of finitely many abstract scenarios to represent the infinite space of traffic scenarios?
2. How can we assure that we capture in our catalog the relevant aspects of our scenario? How do we find a suitable abstraction level for the abstract scenarios?
3. How do we identify the interesting scenarios as representatives for covering the scenario space?

Our scenario discovery process is based on the TSC formalism, which is a new formal and visual specification language [5] tailored for the specification of scenario catalogs of traffic maneuvers. TSCs visually define constraints on a world model capturing possible evolutions of the application domain. While a TSC tells a story in snapshots, the world model ought to capture the physical aspects. TSCs can be translated–snapshot by snapshot–to a real time first order logic formula. The objects referred to by snapshots move (or more generally behave) as specified in the world model. The link between the visual symbols used in TSCs and the world model is defined in the symbol dictionary (s-dict). Figure 1 presents a small example of how scenarios are captured via a TSC and we give a short introduction of TSCs in the companion paper [6].

The abstraction step of our process generalizes the counterexample in terms of the symbol dictionary's vocabulary. An expert has to choose the appropriate level of abstraction.

TSCs Listing Scenarios (TSC-Cat) When TSCs are used to specify scenario catalogs, mainly existential TSC scenarios are used to express that the respective scenario occurs. These TSCs may be derived from accident databases like GIDAS,[2] analytical processes, natural driving studies or the like by for instance machine learning as outlined in [3]. Besides listing existence of scenarios, TSCs allow to specify the relation of two scenarios, that is, whether they are exclusive or may happen simultaneously. We can also describe via TSCs that a list of scenarios characterizes all possible evolutions within a window of interest. Having a world model that satisfies the scenario catalog means that the world model exposes the required evolutions.

[2]www.gidas.org.

TSCs for Specification (TSC-Spec) TSCs can also be used for system specification. Let us assume that we specify requirements on an ego vehicle for the nominal case. Note that additionally emergency routines for the exceptional cases will have to be developed. In a "nominal case", all traffic participants but ego will show nominal behavior. We assume that we derived constraints on the nominal behavior of the environment when defining the scenario catalog of (TSC-Cat). The behavior of a car in the nominal case will not only capture physical laws like how the speed changes with acceleration, but will also be restricted by assumptions like "if the car follows another car in close distance, it will not suddenly start to strongly accelerate". So we assume a "nominal controller" for other traffic participants. This is different for the ego car. Since we want to formulate requirements on the ego car, we only make minimal assumptions constraining ego's behavior. In the extremest case, ego has only to obey to the physical laws and apart from that we allow any behavior. We hence will have to deal with trajectories where our ego car suddenly starts to strongly accelerate, even though it is following a car in close distance and that this leads to a collision. If we now specify rules for distance control, restricting ego's behavior, we can formally analyze whether all collision scenarios of the world model have been ruled out by our TSC specification. To therefore structure the scenario space, we present in this paper a scenario discovery process that generates a list of scenarios, such that all evolutions leading to a collision (violating the safety property) are captured by a scenario.

We present our scenario discovery process in Sect. 2 on a small running example listing collision scenarios between two cars and a platoon at a highway (a first such scenario was presented in Fig. 1). We draw conclusions in Sect. 3 and finally we discuss related work in Sect. 4.

2 Scenario Discovery

In this section, we propose a scenario discovery process based on TSCs. Our process generates a list of scenarios that characterizes possible evolutions of the world model that lead to a violation of a given safety property φ (cf. Fig. 2). Such a process can be used to cover the scenario space (of scenarios violating φ) in a structured way for subsequent specification or testing steps.

First, we introduce the basic procedure on a simple example in Sect. 2.1 and, than, discuss the steps in more detail in Sect. 2.2 highlighting intricacies, restrictions and future work.

2.1 The Discovery Process

We start the scenario discovery on a world model WM, a symbol dictionary s-dict that provides the vocabulary for the specification task, and a safety property

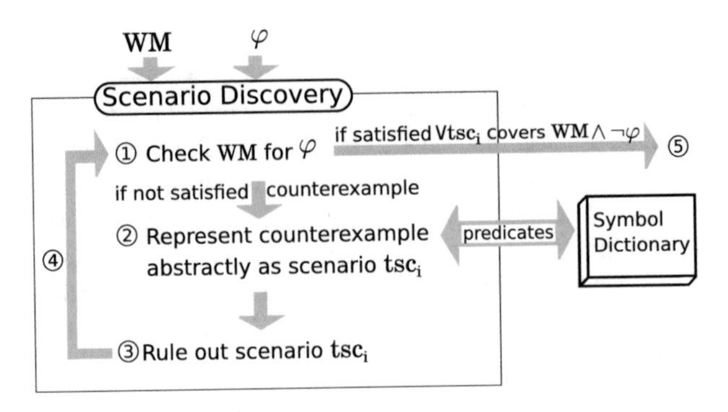

Fig. 2 Scenario discovery process

φ describing an aspired property like "collision freedom" or an initial scenario list like "tsc1: Within the next 7 min, a car approaches another car and then overtakes it. tsc2: Within the next 10 min a car approaches another car and then follows it within close distance.".

Example 1 In order to illustrate our approach, we consider as running example the scenarios arising at a two lane highway when two cars approach a platoon that stays at the right lane. The scenarios arise from our discrete, simple world model which is defined via a (composite) finite automaton modeling the two cars and the platoon. We illustrate how our approach allows to study which scenarios lead to a collision between the two cars when we do not restrict their behavior (e.g., to obey traffic rules). Our world model allows a car to nondeterministically choose to increase its position by zero up to three regions or to change lane or to join the platoon. The platoon has a spatial extend of three regions and does not move. ■

As the first step of our scenario discovery process, cf. ① of Fig. 2, we check whether the world model satisfies φ (collision-freedom for our running example). If WM satisfies φ right away, cf. ⑤, no scenario is generated, since φ covers WM and no evolution of WM violates φ. If otherwise WM does not satisfy φ, there is at least one evolution that does not satisfy φ. We call this evolution a counterexample τ.

Example 2 Let us consider step ① for our running example. In our example, we consider a run as collision free, if after initialization the two cars are never simultaneously at the same lane and road position. We capture this formally by φ_{col}. For our running example we used the Uppaal model checker. We checked whether WM models φ_{col} and got from the model checker a counterexample trace τ_1 where first $car1.pos = 0 = car2.pos$, $car1.lane = 1$ and $car2.lane = 2$ holds. Then *car1* moves forward by one region and *car2* changes onto the next lane, resulting in $car1.pos = 1 = car2.pos$ and $car1.lane = 2 = car2.lane$, the collision state. ■

The next step, ②, of our scenario discovery process is to generalize the concrete counterexample trace τ to an abstract scenario based on the predicates of the symbol dictionary s-dict. To this end, we derive a snapshot chart (SC) for τ. This SC,

sc_τ, will abstractly describe the counterexample via a sequence of snapshots. This sc_τ describes τ visually—like a flip book—and also formally, i.e., τ satisfies the temporal logic predicate encoded by sc_τ. An SC describes system runs based on predicates listed in the s-dict, which in turn are visualized in snapshots.[3] Step ② is not deterministic, since a concrete counterexample can usually be represented by many different abstract scenarios of varying degree of abstractness. So we derive candidate scenarios from τ and an expert chooses the desired level of abstractness. At its heart, ② is a creative step, where the expert decides which scenarios have to be considered distinctly for the system.

Next, we will discuss how to derive sc_τ candidates in case the counterexample τ is derived from a discrete-time WM as in our running example. This can be seen as the simplification of a case with a hybrid WM. Our discrete-time WM reflects discrete actions (like communication) and continuous-time evolutions (like driving). In order to derive a sc_τ we examine which state predicates hold along τ's state sequence, $s(\tau) = s_1 s_2 s_3 \dots s_n$. We then summarize via snapshots those subsequences where the same set of (positive and negative) predicates of s-dict holds. So we formally describe s(τ) by a contiguous conjunction of predicates of the form $\Box_{[j,k)} \bigwedge p_i$,[4] where p_i either equals an s-dict state predicate q or equals its negation $\neg q$ and concrete times are assigned to the time variables j and k. The visualization (the snapshot picture) of $sc_\tau = \Box_{[j,k)} \bigwedge p_i$ can be generated from the s-dict, where visual symbols are declared and mapped to predicates.

Example 3 Let us now illustrate step ② on our example. Therefore we consider τ_1 and assume we only have as predicates of our s-dict the positional predicates $c \bowtie c'$ referring to positions of the two cars, the platoon's start and end position and the left and right border of the two lanes, with $c, c' \in$ {*car1.pos, car2.pos, p.spos, p.epos, lane1.l, lane1.r, lane2.l, lane2.r*} and $\bowtie \in \{\leq, <, \geq, >\}$. Figure 3 shows the derived visualization of sc_1, that is a derived abstract SC representing τ. ■

For the generation of SCs for a given counterexample τ, the expert can select predicates that are not relevant for the scenario according to his opinion. He can iteratively generate SCs and adapt the selection of s-dict predicates. He may also discover that the s-dict does not list the relevant predicates to capture the essence of the scenario. After adding new predicates to s-dict, a new SC can be generated for τ. The expert many also discover that WM does not reflect all relevant phenomena, for instance when he is aware that τ represents a relevant real world evolution whose causalities are not sufficiently reflected by WM. If WM gets updated, the discovery process starts anew and previously discovered scenarios may have to be dismissed as they refer to an old version of WM. The regeneration of sc_τ for a given τ can be iterated until the expert is satisfied.

Such a derived SC still denotes when exactly a snapshot node starts to hold and up to which state/time it holds. Hence the expert also has to decide to what

[3]Note that in order to constitute a TSC at least a header has to be added to an SC.

[4]Read as for all times t, $j \leq t < k$, the conjunction of all p_i holds.

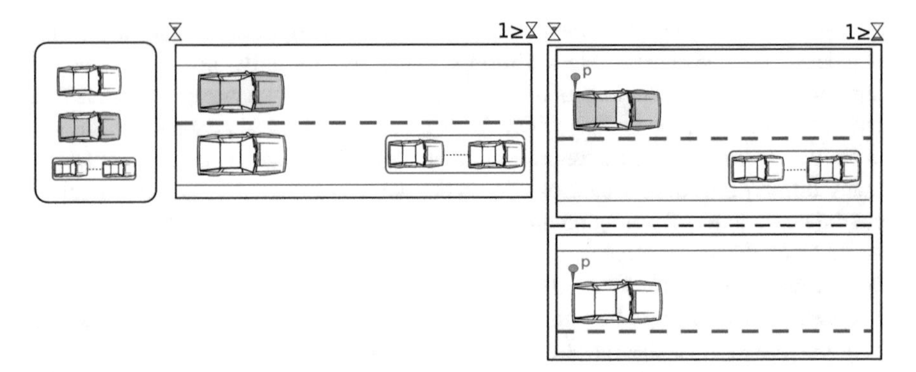

Fig. 3 sc_1 describes τ_1 in terms of the predicates/visual symbols of s-dict. The bulletin board (box with rounded corners) fixes the actors of the scenario, the two cars and the platoon. The first snapshot expresses that initially the two cars are side by side at different lanes and the platoon is in front of the white car. Next the grey car is still behind the platoon and the grey and the white car are at the same position, as expressed via two concurrent snapshots and the pin symbol, that "pins" the same position (both cars are at position p)

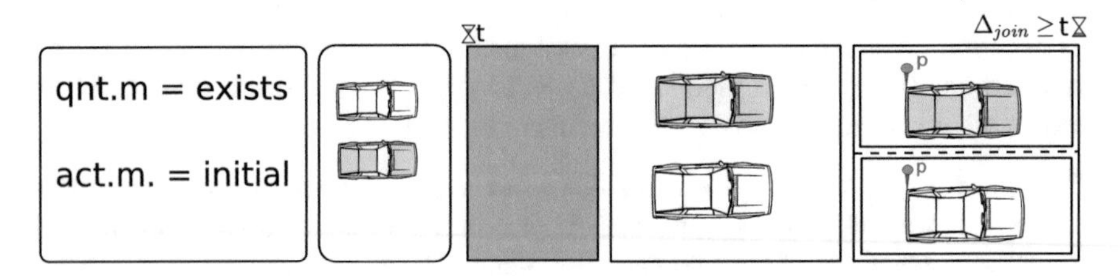

Fig. 4 The TSC expresses that there is a run where the cars eventually are side by side and then at the same position (collision). To the very left is the TSC header specifying that there has to be a run as specified in the following. Next comes the bulletin board—like in sc_1. Then next, the grey snapshot expresses "eventually" (in other words, unconstrained behaviour is allowed to happen for some finite time). The next snapshot means, that then the cars are side by side. At last, the two cars are at the same position. The hour glass annotation on top of the snapshot sequence denotes that from the start of the run to the collision less than the time for a join maneuver elapsed, which we assume here as our time window of interest

extend timing constraints should be lifted. For technical reasons we require that the resulting scenario has a maximal extend in time.

Given the expert has determined a sc_τ that describes abstractly the relevant aspects of τ, he extends the scenario list now by this newly discovered scenario. Therefore we derive a TSC, tsc_τ, from sc_τ: tsc_τ is like sc_τ but prefixed with a header declaring that the TSC is initially activated and existial, so tsc_τ expresses that there is a run satisfying sc_τ.

Example 4 Let us illustrate how the selection of s-dict predicates influences the level of abstractness of the scenario for τ_1. Please note, that the platoon and also the lanes are mentioned by sc_1. Reflecting the opinion that neither is relevant for the collision, we remove the predicates of sc_1 that refer to the platoon or the lanes and thereby get the existential TSC of Fig. 4. Note that we also decided to use weaker timing constraints in Fig. 4 than in Fig. 3. Here we allow some unconstrained behaviour to happen (grey snapshot) before the cars are side by side. ■

Let us now consider step ②, the abstraction of a counterexample to a scenario, for a hybrid WM. We now distinguish between discrete actions that take zero time and continuous flows. A predicate's validity can change infinitely often within finite time during a flow of a formal hybrid WM. But we here aim to describe such flows as finite sequences of s-dict predicates captured by snapshots. To this end, we partition a flow into segments. The partitioning can be refined, if necessary. We then assign $\square_{[j,k)} \bigwedge p_i$ to flow segments similar to the discrete case for states. But at a flow segment a predicate q may be neither invariantly true nor false. So that we can neither assign q nor $\neg q$. Instead, we assign `True`—expressing no constraint (or "$\neg q \vee q$"). After this step, the expert may decide to examine certain flow segments more fine grainedly, by splitting it up into subsegments of even smaller size. This may also be done automatically up to a predetermined window size, that reflects for instance how long an effect has to be persistent to reflect a relevant physical effect. As a result of this transformation step, we get a contiguous conjunction of snapshot nodes. Similarly to the discrete-time case, we can now merge sequences of snapshot nodes with the same snapshot.

Note that the existential TSC, tsc_τ, that extends the scenario catalog as result of step ② usually represents a bundle of trajectories (not only τ), since τ is abstractly captured via predicates of s-dict.

③ We now know that there are evolutions of WM that do not satisfy φ but satisfy tsc_τ. *Step ③ means for our running example, that we extend the scenario catalog by tsc_1, the TSC of Fig. 4.* To discover further scenarios of WM that are not yet covered, we, technically, either modify WM to rule out trajectories satisfying tsc_τ or we relax our property φ to $\varphi \vee sc_1$.

④ We then continue examining whether our world model WM satisfies the (relaxed) property φ. In that way, we iteratively rule out bundles of trajectories that spoil achieving φ and thereby create a list of "spoiling" scenarios, where all the prefixes are represented that cause violation of φ.

Example 5 We know that our WM satisfies tsc_1. In order to detect "new" collision scenarios that are not like τ_1, we chose to relax the property φ_{col}, ③. Technically, we used Uppaal to check invariant properties. So we modified our Uppaal model by adding an observer automaton *obs*(sc_1) that enters a distinct state, called *fin* here, iff the run satisfies the sc_1. Additionally, we relaxed the examined property. Instead of φ_{col} we examined $\varphi_{col} \vee obs(sc_1).fin$. The construction of the observer quite naturally follows from sc_1, which defines a sequence of invariants that have to be observed. ■

For a finite automaton the process will terminate. There, at worst, merely all runs within the considered bound will be listed. But even for a hybrid automaton the process may terminate. Here, termination might depend on choosing the right abstractness of scenarios (including the granularity of flow segmentation). Note, that the expert can decide to complete the list using negation. She can express via TSCs

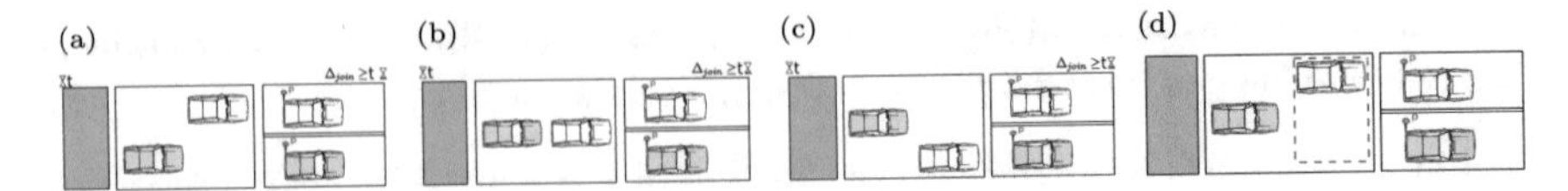

Fig. 5 SCs generalizing counterexamples τ_1 to τ_4. **(a)** SC for τ_2. **(b)** SC for τ_3. **(c)** SC for τ_4. **(d)** SC for τ_2, τ_3, τ_4

that within the considered bound, a set of unexplored scenarios remains by stating roughly *"There is at least a scenario that is not like one of the discovered scenarios"*.

Example 6 At our running example, we next iterate the discovery process, ④, and get in step ① a new counter example, τ_2, where the one car is at the lower lane and approaches the other, then changes lane and thereby causes the collision with the simultaneously moving other car on the upper lane. Figure 5a shows sc_2, a possible generalization of this counterexample. After ruling out runs satisfying sc_2, we get in step ① τ_3, generalized to sc_3 in Fig. 5b. There a car approaches the other on the same lane. After also ruling out sc_3, we get τ_4 generalized to sc_4 in Fig. 5c, which is like sc_2 but the grey car is left of the white car. Then the model checker does not discover any more collision scenarios, and the discovery process terminates ⑤. We hence know that all runs of WM are covered by $\varphi \vee sc_1 \vee sc_2 \vee sc_3 \vee sc_4$. ■

So the scenario discovery process terminates, if all possible evolutions of the world model that violate φ are represented by at least one existential TSC in the scenario list.

2.2 Notes and Discussion

We sketched our scenario discovery process above. Here we provide more details on individual steps and discuss their limitations and extensions.

Property φ We consider as examined property φ (cf. Fig. 2) a safety property and moreover, if we relax φ by discovered scenarios then also the relaxed property $\varphi \vee sc_i$ has to be a safety property. This ensures that trajectories violating (the relaxed) φ can be characterized via their finite prefix. Such a finite prefix can be represented as SC quite straightforwardly, as demonstrated in this paper. For practical applications, the restriction to safety properties is not utterly severe though, one could argue. By adding an upper time bound to the scenario representation sc_i, it already expresses a safety property and at practical applications such a limiting time window can usually be determined (a drive ends at some time).

As we have shown, the discovery process can be used to list all scenarios that spoil a given safety property like collision-freedom. We can express via a TSC this completeness of spoiling prefixes. *This is demonstrated in Fig. 6 for collision-freedom. The TSC uses an SC with premise consequence pattern, where the premise is the SC within the dashed hexagon and the consequence follows right of the*

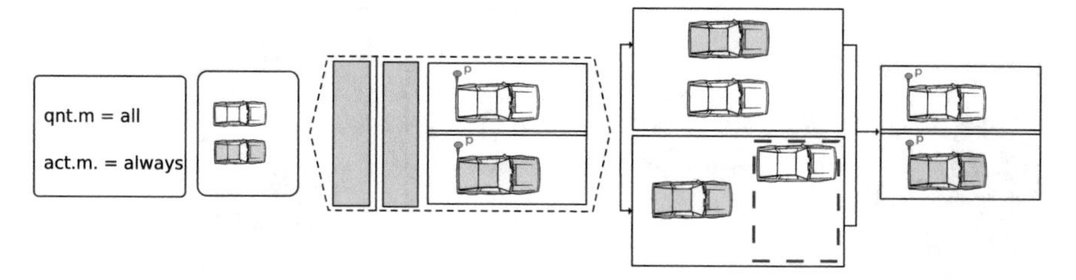

Fig. 6 TSC describing complete list of prefixes of a collision

hexagon. The premise uses true snapshots (grey boxes) to express that anything may happen initially and in the future eventually there will be a collision. The consequence expresses that before a collision, the cars are either side by side (lateral collision) or one follows the other (rear-end collision).

The presented scenario discovery process can also be used to complete a list of (time-bounded) scenarios. For the case that we have a list of existential TSCs, $\mathsf{tsc}_1, \ldots \mathsf{tsc}_n$, we check if each run satisfies at least one scenario. Furthermore note, that if the discovery process is started with $\varphi = \Box_{[0,k]}\mathtt{False}$, the generated scenarios describe possible evolutions of WM up to time k.

Abstracting Counterexamples to Scenarios In this paper, the domain expert is responsible for determining an abstraction of the counterexample to a scenario. He is supported by the visualization of candidate scenarios but has to decide in terms of which predicates the counterexample is represented, whether additional predicates have to be added or whether the world model needs to be extended. We consider it future work to explore how much more support is possible via knowledge mining techniques, analysis of the world model—there also profiting from model based development and model finding as, e.g., in [14]—and also additional data sources including real world databases.

If our discovery process is used to define a scenario catalog for testing an already implemented system, we can derive from sensor limits and the required robustness of the implemented control [8], that bundles of trajectories are equivalent for the system. The systematic derivation of such scenarios is also considered future work.

Example 7 We saw in Fig. 5a, b and c, quite similar scenarios. Each of them represents several runs of our world model. But we can further summarize these scenarios like in Fig. 5d, where the SC generalizes τ_2, τ_3 and τ_4 by a somewhere box (cf. [7]). The white car is allowed to be somewhere within this dashed box. Depending on further usage of the scenario list, this might be a better, more succinct representation or not. Suppose the list is used for testing. "The white car being on the right" may be substantially different from "the white car being on the left" from the point of the grey car's view due to its sensor equipment. ■

The right level of abstraction is strongly influenced by domain knowledge. Nevertheless, we can use formal method tools to analyze the relation among the generated scenarios, as done for LSC in e.g. [12], to support the expert. For instance,

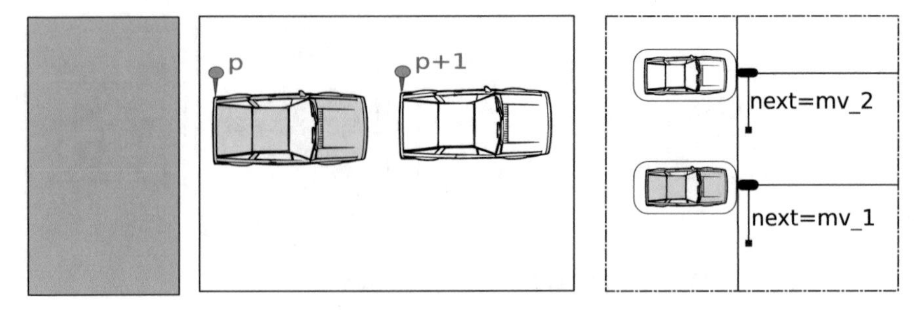

Fig. 7 SC describing a prefix that definitely leads to a collision. The action of deciding is visualized according to the proposed syntax of LSCs within TSCs

we can express via TSCs that two scenarios sc_i and sc_j overlap or even describe the same the runs, although visually different. Depending on the world model, the analysis may not be feasible or even not decidable though.

In particular, we can analyze whether the abstracted scenario sc_τ of a concrete counterexample τ violating φ allows any run τ' that satisfies φ. Since we overapproximate τ by sc_τ, this can be the case. If we are not able to characterize in what respect the scenario prefix corresponding to τ can be distinguished from τ', we might need to extend the set of predicates of the s-dict or we even need to extend the world model (similarly to [2]) to create a mean to express this distinction.

Example 8 In our running example we explicitly included the collision as part of the scenario. If we omit the collision (so we omit the last snapshot of Fig. 5a, b and Fig. 5c), we obviously have not separated the scenarios leading to a collision from those not leading to a collision. In fact, a separation is not possible at our minimalist world model. For instance, we can separate the decision to move from the move itself. Then we extend the symbol dictionary, so that we can also refer to the action of deciding on a move, allowing to specify prefix scenarios like in Fig. 7, where the grey car is one region behind the white car, then both simultaneously decide on their next move, which will then lead to a collision. ■

Steering Scenario Generation by Criticality Lists Our procedure to discover scenarios relies on a model checker to determine the next counter example in step ①. The expert generalizes this counterexample to a scenario. Since the process does not terminate in general and also because the order of generated scenarios influence what counterexamples will be produced in the future, the possibility to influence what counterexamples are generated first, is advantageous.

We can employ the model checker to find counterexamples satisfying additional criteria. Therefore, we imagine that a prioritized list of safety properties $\psi_1, \ldots, \psi_n, \texttt{True}$ specifies these additional criteria—for instance saying we are more interested in "no heavy collisions" than in "no medium collisions" than in "no mild collisions" than in arbitrary collisions (we end the list with $\psi_{n+1} = \texttt{True}$). Given such a prioritized list we can then first determine whether there is a counterexample to $\varphi \wedge \psi_1$, in case there is no (more) counterexample, then we check whether there is one for $\varphi \wedge \psi_2, \ldots$ and finally of $\varphi \wedge \texttt{True}$.

Discrete World Model We took the liberty to illustrate the discovery process on a case study with a discrete time world model, although TSCs have been defined with a continuous time world model in mind [7]. We summarize briefly how we used a discrete model to reflect the continuous aspects like driving and sketch related problems. We modeled a continuous evolution by (an interleaving of several) discrete actions and disregarded the intermediate states of the interleaving as "auxiliary". For step ②, "describing the continuous evolution via snapshot nodes", we summarize sequences of states that satisfy the same set of s-dict predicates. In order to derive meaningful snapshot nodes, there should be a continuous evolution in the real world though.

We believe that discrete time world models can also be beneficially used for traffic scenarios e.g. at the early phase of an agile model development and for scenario planning against tunnel vision. Such coarse models will only reflect a few aspects of the application domain. But there are a lot of works for e.g. software development [11] that demonstrate that using micro-models and inspecting limited scopes only, can provide valuable insights. So working with coarse models already highlights some intricate insights and deepens problem awareness, while developing only a coarse model will save model development and analysis time. To what extend the results gained on the coarse model are valid for the later models and how discrete and continuous models can be linked by formal approaches in our context, is future work. In [15], Olderog et al. demonstrated how a discrete and continuous model of traffic maneuvers can be linked.

3 Conclusion

To support the definition of scenario catalogs for system development or testing we presented an approach of scenario discovery. Our approach generates a list of abstract scenarios that covers a set of evolutions of a world model WM violating a given safety property φ. We use as seed for building the abstract scenarios the concrete counterexamples to φ. The abstraction step from a concrete counterexample to an abstract scenario is driven by the predicates of the symbol dictionary s-dict. Although the abstraction step is strongly based on expert knowledge, we identify several formal analyses tasks that additionally support the expert determining the right level of abstraction.

The TSC formalism is in particular apt for this process, since TSC scenarios are formal, concise and yet visual. Visualization of a scenarios and play-out [9] might—as we strongly suppose—ease the experts' analysis and their discussions. Further TSCs provide means to support the formal analysis of candidates, e.g., it allows to express relations between generated scenarios (like intersection or overlap).

The list of discovered scenarios can serve as mean to structure the application domain of the HAV by distinguishing relevant use cases and critical scenarios, which is an important step in the early phase of system development. Such a list of scenarios can also be used to specify test cases, in which the vehicle under test is examined to show the expected behavior.

As future work, especially the use of domain knowledge (like functional characteristics of sensors or other components) for the generation of candidate scenarios, seems to promising. This knowledge might be used for the generation of predicates in the s-dict or to guide the counterexample generation.

4 Related Work

Closely related to our work are the works of Harel et al. on the play-in/play-out approach for LSCs [9]. The *smart play-out* also uses model checking to smartly execute the LSC scenarios, that is to find an evolution that does not violate other LSCs. In [10] a methodology for the verification of Java programs and a supporting model-checking tool is presented. Besides direct model checking of behavioral Java programs, they show how model-checking can be used during development for early identification of conflicts and under-specification. Therefore the program is specified in a scenario based way. They observe, that counterexamples themselves are event sequences, which they can use directly for refinement and corrections. In contrast, our work aims to define abstract scenarios covering all possible evolutions of a world model. Counterexamples (additionally satisfying criticality criteria) are the seeds for the derived scenarios.

There is a lot of literature on scenarios, and even the term scenario is ambiguous. In this small section, we relate our approach to notions used in the scenario literature. Scenario planning [1] is a planning process where future scenarios are developed and analyzed to support strategic decisions. Scenario planning originated in military but has also been successfully used by companies. We imagine that our approach of discovering scenarios driven by the TSC formalism—even on a very coarse world model—can be used beneficially for scenario planning as a means to discover otherwise overlooked scenarios. The term scenario discovery is used in e.g. [13] for approaches that aim to characterize uncertainties of the simulation model. In contrast our process works also for models where no probabilities are provided.

References

1. Bradfield, R., Wright, G., Burt, G., Cairns, G., Van Der Heijden, K.: The origins and evolution of scenario techniques in long range business planning. Futures **37**(8), 795–812 (2005)
2. Damm, W., Finkbeiner, B.: Does it pay to extend the perimeter of a world model? In: 17th International Symposium on Formal Methods, FM 2011. LNCS, vol. 6664, pp. 12–26. Springer, Berlin (2011)

3. Damm, W., Galbas, R.: Exploiting learning and scenario-based specification languages for the verification and validation of highly automated driving. In: Proceedings of the 1st International Workshop on Software Engineering for AI in Autonomous Systems, SEFAIS. ACM, New York (2018)
4. Damm, W., Heidel, P.: Recommendations of the SafeTRANS working group on highly autonomous systems (2017). www.safetrans-de.org/en/Latest-reports/management-summary-for-highly-automated-systems/192
5. Damm, W., Kemper, S., Möhlmann, E., Peikenkamp, T., Rakow, A.: Traffic sequence charts—from visualization to semantics. Reports of SFB/TR 14 AVACS 117, SFB/TR 14 AVACS, 10 2017
6. Damm, W., Möhlmann, E., Rakow, A.: Traffic sequence charts for the enable-s_3 test architecture. Technical report, 2018. Submitted to ENABLES book
7. Damm, W., Möhlmann, E., Peikenkamp, T., Rakow, A.: A formal semantics for traffic sequence charts. In: Principles of Modelling Essays dedicated to Edmund A. Lee on the Occasion of his 60th Birthday. LNCS. Springer, Berlin (2017)
8. Fränzle, M.: Analysis of hybrid systems: an ounce of realism can save an infinity of states. In: Computer Science Logic, pp. 126–139. Springer, Berlin (1999)
9. Harel, D., Marelly, R.: Come, Let's Play: Scenario-Based Programming Using LSC's and the Play-Engine. Springer, Berlin (2003)
10. Harel, D., Lampert, R., Marron, A., Weiss, G.: Model-checking behavioral programs. In: Proceedings of the Ninth ACM International Conference on Embedded Software, EMSOFT '11, pp. 279–288. ACM, New York (2011)
11. Jackson, D.: Software Abstractions: Logic, Language, and Analysis. MIT Press, Cambridge (2006)
12. Kam, N.A., Harel, D., Kugler, H., Marelly, R., Pnueli, A., Hubbard, J.A., Stern, M. J.: Formal modeling of *c. elegans* development: a scenario-based approach. In: Computational Methods in Systems Biology, pp. 4–20. Springer, Berlin (2003)
13. Kwakkel, J.H., Jaxa-Rozen, M.: Improving scenario discovery for handling heterogeneous uncertainties and multinomial classified outcomes. Environ. Model. Softw. **79**, 311–321 (2016)
14. Macedo, N., Cunha, A., Guimarães, T.: Exploring scenario exploration. In: Fundamental Approaches to Software Engineering. LNCS, vol. 9033, pp. 301–315. Springer, Berlin (2015)
15. Olderog, E.-R., Ravn, A.P., Wisniewski, R.: Linking discrete and continuous models, applied to traffic manoeuvrers. In: Provably Correct Systems, NASA Monographs in Systems and Software Engineering, pp. 95–120. Springer, Berlin (2017)

Digital Map and Environment Generation

Gerald Temme, Michael Scholz, and Mohamed Mahmod

1 Introduction

In order to build a simulation environment for automated driving functions, at least a 3D model of the environment and a detailed road network are required. To obtain these datasets there are many challenges which will be outlined by DLR's experiences in use case 2 "Validation of left turning with automated vehicle".

2 Satisfying the Demands of Driving Simulation and Vehicle Functions

Different road network applications exist today within the domains of driving simulation and vehicle automation functions. Both domains share a common, intersecting set of demands such as, for example, accurate lane models in terms of topography and topology, linked road elements and intelligent infrastructure (e.g. signals and signs). Nevertheless, generating solely one road database for both domains is not trivial. On the one hand, a complete virtual environment for driving simulation purposes bases upon various visual properties, such as certain textures or 3D models, which in turn may not be relevant for vehicle automation functions. On the other hand, vehicle automation functions form strict requirements on absolute and relative coordinate errors, which could be of less importance for driving simulation applications.

G. Temme · M. Scholz · M. Mahmod (✉)
Deutsches Zentrum für Luft-und Raumfahrt e.V. (DLR), Institut für Verkehrssystemtechnik, Braunschweig, Germany
e-mail: Gerald.Temme@dlr.de; Michael.Scholz@dlr.de; Mohamed.Mahmod@dlr.de

A. Leitner et al. (eds.), *Validation and Verification of Automated Systems*,
https://doi.org/10.1007/978-3-030-14628-3_8

To cope with these demands a common data format has to be defined which is capable of representing all desired attributes and can be interpreted by both domains. During the last decade OpenDRIVE evolved as the de facto standard in driving simulation environments. It is now standardized in the automotive domain through the Association of Standardization of Automation and Measuring System (ASAM). Toolchains are available to derive virtual 3D models form OpenDRIVE datasets. Since OpenDRIVE is also used to cover the (mostly smaller) subset of automation function demands, it has been chosen by DLR for the purpose of this project.

3 Geodata-Based Approach for ENABLE-S3

For the use case 2 "Validation of left turning with automated vehicle" in ENABLE-S3, a traffic training area[1] in Brunswick has been chosen as a test ground. This section gives a background on common approaches in derivation of simulation environments from real-world geodata and summarizes the approach which has been chosen for ENABLE-S3.

In contrast to imaginary simulation environments the derivation of environments from real-world geographic data becomes more and more popular in times where not only the driver's behavior has a scientific focus but also the vehicle's interaction with its environment through built-in sensors. Such geodata-based settings are still a challenge to obtain because either the underlying digital source data is not widely available, or it is not available in the desired formats and the explicit acquisition is time-consuming and expensive. Through the projects SimWorld, SimWorldURBAN and Virtuelle Welt[2] (Virtual World), the DLR gained experience in processing of heterogeneous spatial data into 3D virtual environments.

The map basis to support vehicle functions and virtual simulation environments lies in vector data[3] which allow the representation of certain environmental features as discrete objects with an arbitrary number of attached attributes. This attribution allows depiction of complex relationships between such objects which in turn can be used to build interconnected, navigable road networks with linked infrastructure. Vector data can be modelled on database level, easily exchanged, procedurally modified and visualized through standardized computer graphics tools. For road network representation in ENABLE-S3, OpenDRIVE has been chosen which acts as a database capable of representing vector-based objects and their relationships combined with visual properties.

The following question arises: How to obtain an OpenDRIVE dataset for a certain geographic location to be used in a simulation environment? The answer

[1] Verkehrsübungsplatz Braunschweig e.V. http://www.vp-bs.de/

[2] https://www.dlr.de/ts/en/desktopdefault.aspx/tabid-11264/19778_read-46760/

[3] https://2012books.lardbucket.org/books/geographic-information-system-basics/s08-02-vector-data-models.html

is that today it still has either to be ordered from a specialized map provider or to be generated manually because local authorities do not provide such detailed geodata yet. Local authorities such as cadastral administrations or infrastructure maintainers were not faced until recently with the specific requests of driving, traffic or sensor simulations which impose strong requirements on data accuracy and logical relationships between elements. The local authorities' databases often describe the planned state but not necessarily the actual state of an environment. This kind of data serves for basic mapping, planning and visualization and thus does not require its elements to expose complex semantics and relationships. The consequence is that such datasets cannot be used for simulation purposes ad hoc and cannot be transformed into the desired target formats easily due to missing information.

Also, different element categories which are required for the simulation environment can be maintained by different authorities. In the city of Brunswick in Germany, for example, the raw street topography is maintained by the public cadastral administration whereas traffic infrastructure, such as signs and signals, is maintained by a private company. Both authorities work with different standards, different data backends and different processes. To generate a coherent OpenDRIVE network it would be necessary to transform and fuse those heterogeneous data sources accordingly. This approach was followed in the DLR-project Virtuelle Welt (Virtual World) with the result that a general, automatic conversion is possible but mostly will not satisfy the strong requirements due to imprecise raw data.

The designated test ground for ENABLE-S3 is maintained by the Verkehrsübungsplatz Braunschweig e.V. which does not have spatial reference data itself and because of the location being a private ground, public authorities provide sparse data only. This leads to mainly two feasible approaches to still obtain the desired test track in form of a linked and smart road network dataset.

Firstly, explicit data acquisition through a surveying company can be instructed. Nowadays the most common and feasible way is to perform road feature collection through mobile mapping which uses regular cars equipped with special sensor setups (e.g. camera, Lidar and IMUs). Afterwards the collected raw point cloud data is classified and annotated to extract the discrete road network elements and to transform them into the desired output data format. This is done partly automatically but still mostly manually by the provider. The latter is the reason for unforeseeable project duration and high costs, being the main disadvantage of this approach. One advantage is the scalability and the accuracy of the obtained road data though. The used sensor setups and mature post-processing allow relative and absolute coordinate errors to be in sub-decimeter magnitude.

Secondly, digitalization of the desired road network elements can be done manually on base of commonly available spatial reference data, such as aerial imagery or cadastral datasets. This approach has been chosen for the use case 2 of ENABLE-S3 due to the availability of high-resolution aerial images and its practicability. In this case the element annotation and extraction can be done in a geographic information system (GIS) which allows overlaying of different,

Fig. 1 Exemplary annotation of Road2Simulation vector elements (green, yellow, cyan) of the traffic training area in QGIS, using aerial imagery as reference background map

geo-referenced datasets (see Fig. 1). For small areas of interest this approach is fast, but its accuracy depends on the used reference data.

Derivation of the required road elements could also be done automatically from aerial imagery which would speed up data acquisition in large areas. But today this is still cumbersome due to missing of a well-developed and closed capturing and processing chain. Such challenges are focused in the projects AeroMap[4] and LOVe.[5]

4 Road2Simulation as Approach for Digitalization of the Traffic Training Area

Plain identification and annotation of relevant road and infrastructure elements, as described in the second approach before, is not enough. Those features must be transformed into the desired output format for simulation purposes accordingly, which requires having the appropriate processing toolchain at hand. The project

[4]http://www.bmvi.de/SharedDocs/DE/Artikel/DG/mfund-projekte/ableitung-von-strassenraumbeschreibungen-aus-luftbildern-aeromap.html

[5]https://www.bmvi.de/SharedDocs/DE/Artikel/DG/mfund-projekte/objekterkennung-durch-deep-learning-verfahren-love.html

Road2Simulation[6] defines guidelines on how to annotate such road features and proposes a light-weight data model for storage and simplified post-processing into domain-specific simulation formats such as OpenDRIVE. A prototypic processing toolchain into OpenDRIVE was developed in the project's scope as well.

As mentioned before, public authorities do not provide cadastral data for the private training area as test ground. But aerial imagery is available for the whole area of Brunswick with a spatial resolution of around 10 cm per pixel, also covering the test ground. Those orthophotos in conjunction with the Road2Simulation-approach served as a first starting point to derive an "intelligent" road network for the simulation and vehicle function development.

4.1 Choosing Aerial Imagery and Other Base Layers

As described above, in ENABLE-S3 the Road2Simulation guidelines have been chosen for test track generation because of their practicability. Aerial images with a ground resolution of 10 cm and a positioning error of comparable magnitude were acquired from the department of geo-information of Brunswick's city administration and used as a base layer for manual data digitalization. Generally, such aerial imagery is not available for free in Germany or only with worse spatial resolution and accuracy, which is not sufficient for visual recognition of the required road network elements. The DLR conducted additional surveying flights to obtain aerial imagery with even better spatial resolution and more recent recording date.

To create a road network with elevation information additional digital elevation models (DEM) and digital terrain models (DTM) are required and can also be obtained from public authorities with spatial resolutions of around 1 m. Interpolation between such coarse data points should be sufficient in many cases but applicability has to be analyzed in advance. For the case of ENABLE-S3 it was decided to keep the complexity as low as possible and thus to only create two-dimensional road network data.

4.2 Applying the Road2Simulation Guidelines for Enable-S3

The manual and visual element annotation on base of the acquired aerial imagery was carried out in the geographic information system QGIS[7] (see Fig. 1), working directly on the Road2Simulation data model v1.2, which in turn was deployed in a

[6]https://dlr.de/ts/en/road2simulation

[7]https://qgis.org/

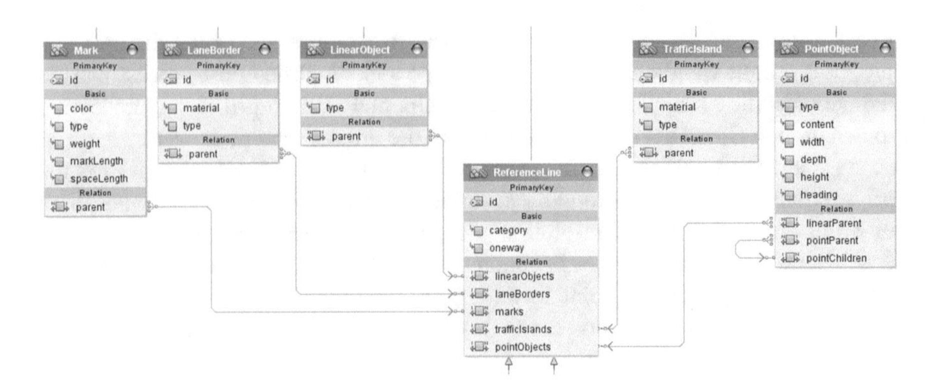

Fig. 2 Extract of element classes used in Road2Simulation

PostGIS[8] database as part of AIM[9] Data Services,[10] situated in the AIM Backend of DLR's Institute of Transportation Systems.[11]

The Road2Simulation guidelines describe a huge, generalized amount of possible road network elements which, at first, have to be condensed into a subset which is relevant for the test track in ENABLE-S3. Due to its limits in space the test track contains only a small variety of different Road2Simulation elements (see Fig. 2) which can be classified into: ReferenceLines, LaneBorders, Marks and PointObjects of the types "traffic light post", "signaling device", "sign post", "road sign" and "pictogram" (see Road2Simulation guidelines v1.2, section 3.6.15). Additional photo and video material of road signs was collected on-site because the actual road sign types cannot be determined from two-dimensional orthophotos alone. As an extension for later 3D model generation trees around the test ground were also included as PointObjects, with values for their height and crown diameter derived from digital elevation and digital surface models.

The actual digitalization of relevant Road2Simulation elements is illustrated in Fig. 1. Simple LineStrings[12] and Points were created in QGIS with added tabular attributes, such as actual element types, element properties and reference links to other elements. In the end this resulted in around 90 ReferenceLines, 140 LaneBorders, 160 Marks and 100 PointObjects as database for further conversion into OpenDRIVE. It is advisable to use different background image layers (having an older/younger time stamp) for digitalization because sometimes road features such as road marks may be obscured through shadows or vegetation in one but not in the other layer. In the worst case, reference data has to be collected locally.

[8]http://postgis.net/

[9]Application Platform for Intelligent Mobility https://www.dlr.de/ts/aim

[10]https://jlsrf.org/index.php/lsf/article/view/124

[11]https://www.dlr.de/ts/en

[12]http://www.opengeospatial.org/standards/sfa

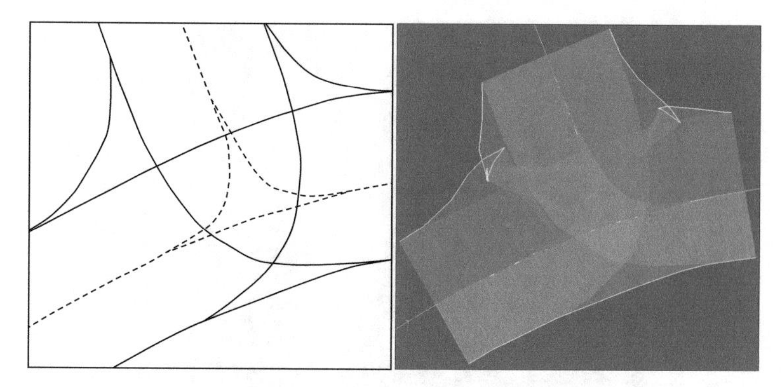

Fig. 3 Discrete, continuous gray lane borders in Road2Simulation (left) and their erroneous mathematical/continuous representation in OpenDRIVE (right)

4.3 Conversion into OpenDRIVE

During the project period of Road2Simulation, the DLR developed a prototypic post-processing toolchain to export generalized Road2Simulation geodata into OpenDRIVE. This toolchain was used for the ENABLE-S3 use case and mainly supported simple urban and extra-urban settings but had to be extended and enhanced to process the traffic training area data. Point object data was not yet included in the OpenDRIVE output and junction connections were partially generated erroneously. Further, the special, very compact road layout at the test ground led to geometrical problems in the resulting OpenDRIVE representation where discrete curves are represented as polynomial functions of third order. The road layout on the test ground mostly does not comply with road construction regulations, showing sharper curves as usual, squeezed intersections and unconventionally wide or narrow driving lanes. In OpenDRIVE, all lane information including the course of lane boundaries is represented in relation to a reference line which can lead to mathematical ambiguities in inner corners of sharp curves having small radiuses (see Fig. 3). These problems were not fixed completely until the end of the project but the generated OpenDRIVE road network served well for the purpose.

In order to create a complete 3D virtual simulation environment, the obtained OpenDRIVE road network was manually extended in the software suite Vires Road Designer[13] by surrounding elements such as grass fields, vegetation, additional road objects and simple buildings. A final 3D database was exported which, together with the OpenDRIVE road network, served as simulation environment for both the driving simulation and development of vehicle automation functions. Figure 4 shows a side by side composition showing the final 3D environment model used within Enable-S3.

[13]https://vires.com/vtd-vires-virtual-test-drive/

Fig. 4 Comparable photo composition of the proving ground intersection with (left) the real intersection and (right) a view into the complete 3D virtual simulation environment

5 Use Case "Validation of Left Turning with Automated Vehicle"

Use Case 2 of the project Enable-S3 addresses scenarios in urban intersections, where turning left with oncoming traffic poses a special challenge for automatic assistance system. Since wrong decisions can directly lead to a catastrophic result (i.e. a fatal accident), a decision-making component of automation should be well validated before allow use of the system. Validating the decision-making component of automation in real-world is a real challenge. This is because not only certain reproducible scenarios are rarely available, but also because dangers to humans and technology could be significant. Moreover, real-world validation is associated with considerable costs. A feasible approach to these problems could be to validate the assistance system in simulation. First results in the project Enable-S3 show that a cost reduction of up to 70% can be achieved by using simulation to validate the assistance system in crossing scenarios.[14]

Unfortunately, it is unclear if validation results from simulation are meaningful or not. This is due to the fact that the results can vary widely depending on many factors such as the chosen scenarios, the environmental model being used, the sensor models and vehicle model used, the simulation platform used and many others. Especially with complex assistance systems that have to cope with complex situations in a complex environment, any simulation component used which deviate from reality can lead to inaccurate results. On the one hand, errors which do not exist in the real world could occur during validation by simulation. These non-relevant errors would then lead to additional development costs while eliminating them. On

[14]Evaluation Result of the Automotive Domain. D4(D1.4)—D1.1.3 V1 ENABLE-S3 Project.

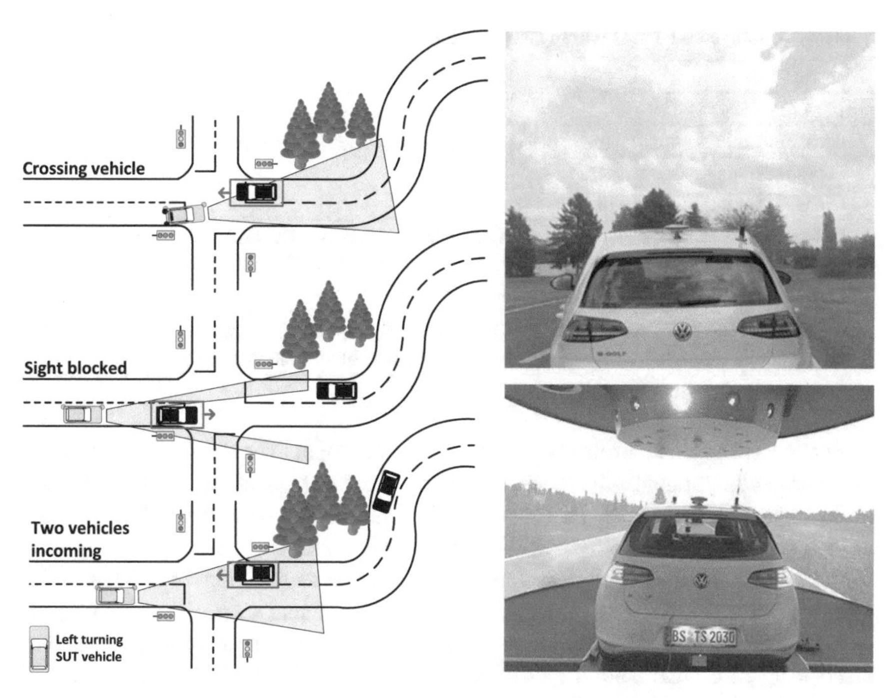

Fig. 5 Addressed scenario groups (left), research vehicle on proving ground (right above) and in simulation as 360° field of view (right below)

the other hand, there might be also errors existing in real world that will not occur during a validation by simulation. If such errors only occur in real world, this can have bad consequences for manufacturers and users.

In order to approach the question on how comparable validation results in simulation with reality are, DLR carries out a comparative test campaign as part of the left-turn use case. Within the scope of this test campaign some left-turn scenarios are tested (Fig. 5 (left) and Table 1). The same automatic left-turn assistance system is used in simulation (simulator Fig. 5 (right below)) and in reality (proving ground Fig. 5 (right above)). The goal is to make the tests in both environments as comparable as possible to allow a direct comparison of the validation results using Key Performance Indicators (KPI) defined within Enable-S3.[15]

When evaluating collected data from the test campaign in simulation and proving ground, special focus will be given to results which show differences between the two validation environments. These differences will give a better understanding of the used models and simulation components as well as their utility of validating the

[15]Demonstration Report + Feedback from the Automotive Use Cases. D71(D4.3)—D4.1.2 V1 ENABLE-S3 Project.

Table 1 Addressed scenarios within test campaign

Scenario	Description
1. Baseline	No traffic.
2. Crossing vehicle—small gap	One incoming vehicle. They meet at the crossroads.
3. Crossing vehicle—medium gap	
4. Crossing vehicle—large gap	
5. Laser blocked—small gap	One incoming vehicle. They meet at the crossroads. One vehicle blocks the laser.
6. Laser blocked—medium gap	
7. Laser blocked—larger gap	
8. Two vehicle incoming—small gap	Two incoming vehicles. They meet at the crossroads. One blocks the view to the other.
9. Two vehicles incoming—large gap	

assistance system. Accordingly, additional requirements for the used models and simulation components will be derived and new research questions can be collected.

In the test campaign, the DLR uses an assistant (prototype version of an assistant to the DLR for research purposes), which has the function of automatically turning left at an intersection. This was integrated into a research vehicle of DLR. This research vehicle is designed for a test of assistance in real world (Proving Ground & Public Street). At the same time, the research vehicle can also be integrated as a mock-up in a DLR 360° Field of View (FoV) simulator laboratory (as part of the simulation). The integration into a 360° FoV simulator is especially necessary when transitions between humans and assistants are to be investigated. In the setup selected here, the driver is involved as a supervisor before the left turning procedure. This is done by transitioning from SAE4 to a temporary SAE3 level[16] just before the turning procedure. For that, the driver receives information from the System about the planned turning maneuver and has to agree on the planned maneuver before the system is allowed to turn left automatically (again in SAE4). However, in the event of non-clearances, the system stops in a safe position (before entering the oncoming traffic lane) in the intersection and waits for an approval before continuing the left turning maneuver.

5.1 Sensors Setup

On proving ground, the assistant uses a digital map and GPS for localization on the route and to stay in the lane. The laser scanners (see Fig. 6) are used to detect the incoming traffic as base for the decision unit to choose a good left turning behavior (Fig. 7).

[16]https://www.sae.org/standards/content/j3016_201806/

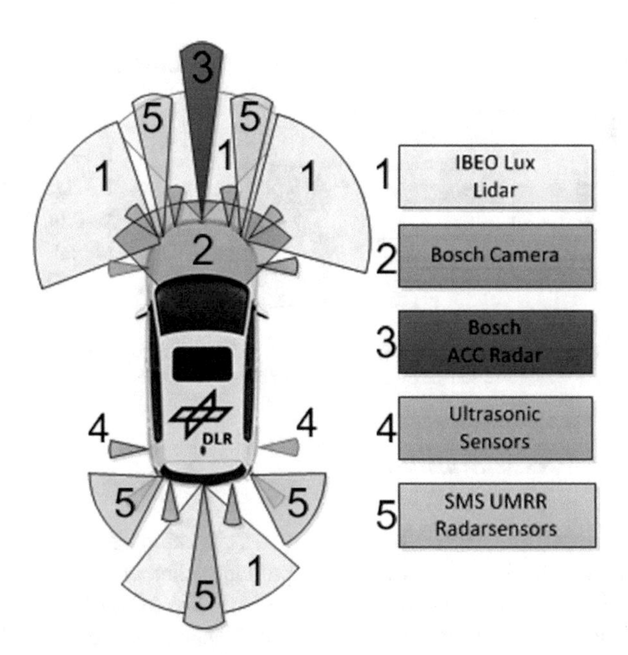

Fig. 6 Overall available research vehicle sensor configuration

Fig. 7 Bird view of proving ground with test vehicle route overlays

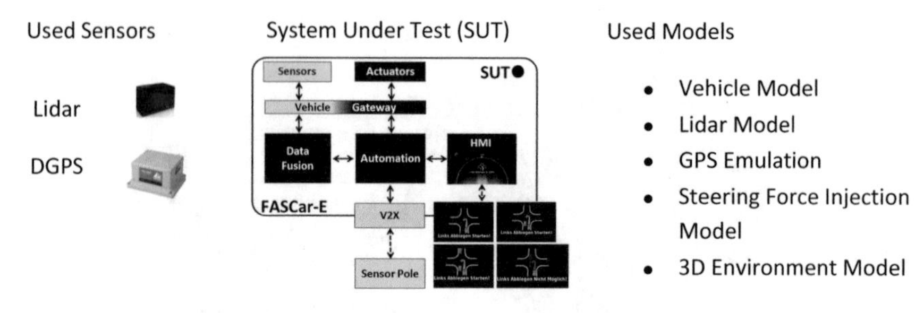

Fig. 8 Architecture overview of the System Under Test with the used components in both environments (in black), used sensor on proving ground (left) and used models in simulator (right)

5.2 Reality Versus Simulation

The overall simulation approach used in this test campaign is a human-in-the-loop combined with vehicle-in-loop test approach. When changing from reality to simulation, all assistance functions (in Fig. 8 (middle) the SUT is marked by the blue boxes) are applied unchanged. The vehicle was also integrated directly into the simulator. The vehicle movement dynamics as well as the sensors of the vehicle sensors was replaced by models in the simulation. Moreover, the street layout and the surrounding environment must also be available as models (Fig. 8 (right)). The models used for the simulation part of the test campaign are developed or adapted for the chosen scenarios. Especially the 3D environment visualization (as described before) of the test area is necessary for tests in simulation where the driver is involved in the approval of left turn maneuver of the system.

The test campaign was carried out by the DLR in October–December 2018 as part of the use case left turning with automated vehicle in the project Enable-S3 in cooperation with other Enable-S3 partners.

To validate the SUTs and to compare the performance on the two validation environments, the defined KPIs were derived from the collected data. The analysis of the collected data is still under progress. A first preview of one KPI namely the driven trajectories is shown in Fig. 9 below. As can be seen, the simulator and proving ground trajectories are fairly similar. However the routes of the vehicle on the proving ground are a bit wider. This could be due to differences in the used models in the simulation.

6 Conclusion

In this chapter a first evaluation of shifting scenario-based ACPS validation from reality to simulation was presented. The evaluation was used by DLR within use case 2 “Validation of left turning with automated vehicle”. Two campaigns were

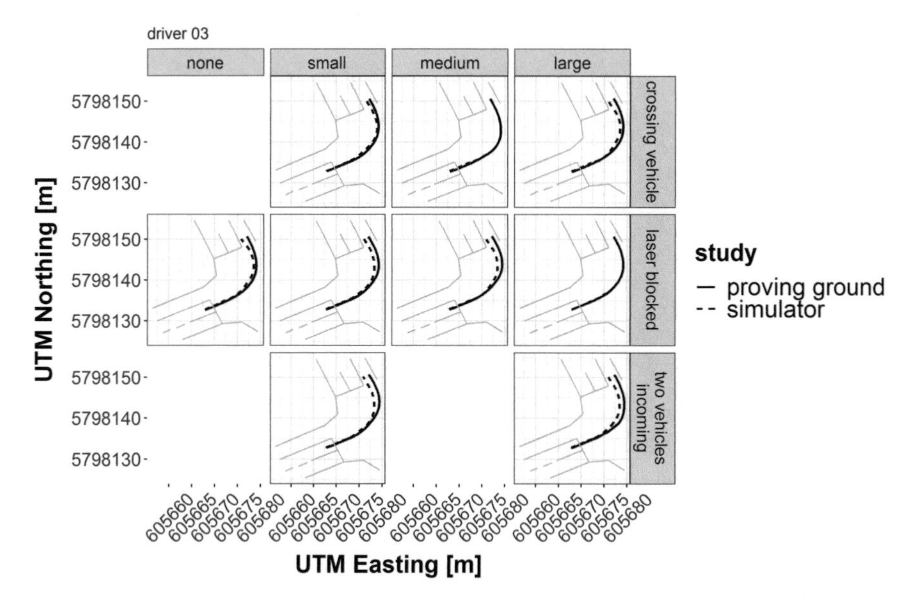

Fig. 9 Comparison of driven trajectories for driver 03 in simulator and proving ground. Gap size is shown on top and driven scenarios on the right side

conducted in simulation and on proving ground. In order to make the two campaigns as comparable as possible, models for different components in simulation were implemented. First, a geo-referenced 3D environment model for the proving ground was implemented. This turned out to be difficult if no publicly available cadastral data is obtainable and manual modelling is necessary. Still, OpenDRIVE shows limitations when used for real-world scenarios because curve modelling can be more complex than it usually is in synthetic simulation setups. After generation of the road network another challenge was the visual enhancement of the 3D model to suit visualization requirements. Vegetation, traffic signals and other peripheral object models had to be optimized accordingly. Other models were implemented which include vehicle dynamic model, Lidar sensor model and a force injection model to simulate the correct driving torque at the steering wheel in simulation. Looking at the first results, it was concluded that our simulation was able to repeat the trajectories of automated vehicle and the number of stops in a fairly manner. When implementing scenarios in simulation, a lesson learnt was the importance of implementing the speed-time profiles of incoming vehicles as close as they were driven on the proving ground. This can be done by recording the speed-time profiles of incoming vehicles and feed them into the simulation to allow a better comparison of KPIs from both validation environments.

Systematic Verification and Testing

Dana Dghaym, Tomas Fischer, Thai Son Hoang, Klaus Reichl, Colin Snook, Rupert Schlick, and Peter Tummeltshammer

1 Introduction

Highly dependable systems require a coordinated approach to ensure safety and security properties hold and that the behaviour is as desired. Formal, mathematical modelling enables a rigorous system-level analysis to ensure that the specification is consistent with important properties such as safety and security. Using formal verification techniques, e.g., theorem proving or model checking, such properties can be proven to hold. However, the human-centric processes of understanding a natural language or semi-formal requirements document and representing it in mathematical abstraction is subjective and intellectual, leading to possible misinterpretation. Formal verification focuses on safety properties rather than 'desired' behaviour which is more difficult to verify as a proof obligation. Domain experts need to validate the final models to show that they capture the informally specified customer requirements and ensure they fit the needs of stakeholders.

A widely-used and reliable validation method is acceptance testing, which with adequate coverage, provides assurance that a system exhibits the informal customer requirements. Acceptance tests describe a sequence of simulation steps involving concrete data examples to exhibit the functional responses of the system. However,

D. Dghaym · T. S. Hoang · C. Snook
ECS, University of Southampton, Southampton, UK
e-mail: D.Dghaym@soton.ac.uk; t.s.hoang@soton.ac.uk; cfs@soton.ac.uk

T. Fischer · K. Reichl · P. Tummeltshammer
Thales Austria GmbH, Vienna, Austria
e-mail: Tomas.Fischer@thalesgroup.com; Klaus.Reichl@thalesgroup.com; Peter.Tummeltshammer@thalesgroup.com

R. Schlick (✉)
AIT Austrian Institute of Technology GmbH, Vienna, Austria
e-mail: rupert.schlick@ait.ac.at

A. Leitner et al. (eds.), *Validation and Verification of Automated Systems*,
https://doi.org/10.1007/978-3-030-14628-3_9

acceptance tests can also be viewed as a collection of scenarios providing a useful and definitive specification of the behavioural requirements of the system. Behaviour-Driven Development (BDD) [12, 14] is a software development process based on writing precise semi-formal scenarios as a behavioural specification and using them as acceptance tests.

The approach presented in this chapter combines the principles of BDD with formal systems modelling and validation. We propose a process where manually authored scenarios are used initially to support the requirements and help the modeller. The same scenarios are used to verify behavioural properties of the model. However, the manually written tests may have limited coverage. To address this, the model is used to automatically generate further scenarios that have a more complete coverage than the manual ones. The additional scenarios should be accepted or rejected by domain experts to ensure they, and hence the model, represent the desired behaviour. These automatically generated scenarios are used to animate the model in a model acceptance stage. For this acceptance stage, it is important that a domain expert decides whether or not the behaviour is desirable. The accepted model then drives the implementation which is verified in a final conformance test.

The abstract, conceptual approach is described in the next section as a so called V&V pattern. An instantiation of this pattern with concrete tools is then detailed in Sect. 3. We discuss the pattern usage and conclude in Sect. 4.

2 Systematic Verification and Testing Workflow

2.1 Pattern Concept

V&V patterns [7] were introduced as a way to describe and communicate ways to combine different V&V methods or tools in context of the MBAT project.[1] They combine graphical and textual descriptions to specify guidelines for V&V activities in a systematic manner. The concept of V&V patterns is inspired by *design patterns* [4] widely used in the software engineering community. In essence, a V&V pattern provides a generic solution in the form of a workflow for a specific V&V activity.

Figure 1 depicts symbols used to define graphical descriptions of V&V workflows in the form of a pattern. The graphical representation of a V&V pattern is accompanied by a textual description, which includes information about the pattern name, its purpose, situations in which the patterns can be used and aspects to consider when applying the pattern. In addition, the textual description defines the participants in the V&V activity and important artefacts, as well as the description of actions and collaborations. The pattern description discusses its benefits, trade-offs and drawbacks, refers to its known uses and possibly other related patterns. Finally,

[1] http://www.mbat-artemis.eu/home/index.html.

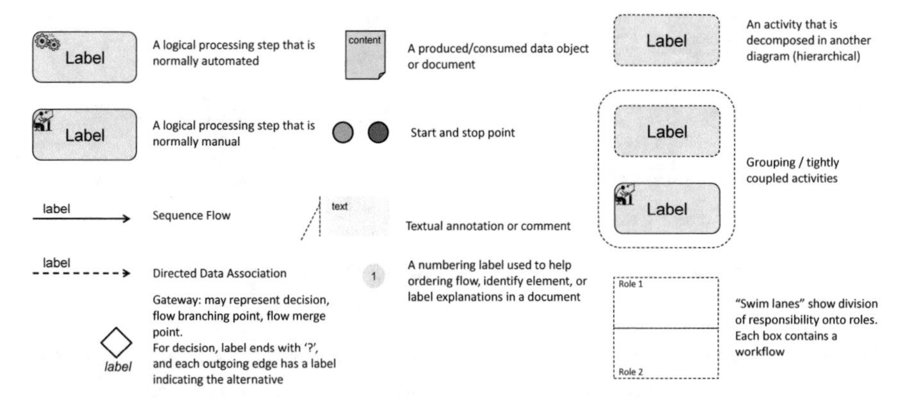

Fig. 1 Symbols used for the graphical description of V&V patterns

a V&V patterns can also contain a discussion and further comments about its use. The pattern catalogue developed in MBAT has been extended in ENABLE-S3.[2]

2.2 *Verification and Testing Method Pattern*

The pattern for 'Systematic Verification and Testing' is a workflow for developing systems into implementations using a behaviour driven approach to formal modelling. The pattern is shown in Fig. 2 and its steps are described below. A discussion of the pattern is part of the conclusion in Sect. 4.

Scenario Modelling Scenarios are derived from the requirements and from experience, user interviews or other sources by domain experts. To support further processing, they are expressed as scenario models. It is important to choose a format that is well readable also for domain experts, while being formal enough to allow automated processing.

Formal Modelling of System In the modelling step, the model is produced from the requirements and the manually written scenarios by a formal modelling expert. The output of the modelling step is a safe model, in the sense that it is rigorously verified to be consistent with respect to its invariants (conditions derived from the requirements that have to hold at any time e.g. safety conditions). Concrete formal methods used in the pattern might use different techniques for verification, e.g. theorem proving and/or model checking. Regarding the outside interface of the modelled system or component, the system model needs to be consistent with the scenario models.

[2]The pattern catalogue of ENABLE-S3 is available at https://vvpatterns.ait.ac.at.

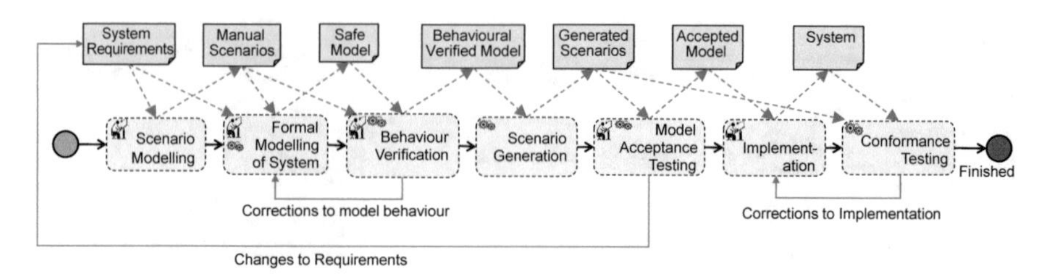

Fig. 2 The pattern for Systematic Verification and Testing Methodology

Behaviour Verification The safe model is behaviourally verified against the manually written scenarios as an automated step. The purpose is to verify that the safe model exhibits the behaviour specified in the requirements which cannot be expressed via invariants. The output of this step is a (safe and) behaviourally verified model.

Scenario Generation The behaviourally-verified model is used as the input for a scenario generator, which automatically produces a collection of scenarios. Manually defined scenarios might focus on the main functionalities or miss at least some special cases. By generating scenarios with a good coverage of the model, it is ensured that all corner cases and critical parameter or data combinations are addressed.

In order to be useful in the next step, it is important that the number of generated scenarios is small enough to allow review with reasonable efforts, while still being thorough enough to substantially increase trust in behavioural correctness of the model.

Model Acceptance Testing The generated scenarios are used for acceptance testing of the behaviourally verified model. This means they are reviewed by domain experts to ensure they represent desired behaviour. If the model still contains undesirable behaviour that was not detected in the behaviour verification step, this will be reflected in the generated scenarios.

Acceptance testing allows stakeholders to assess the usefulness of the model by watching its behaviour. The scenarios are in a form that is understandable for the domain experts and it is easy to see the correspondence between the scenarios and the requirements.

Implementation The model is implemented. This is either done manually by program developers who hand-write code that reflects the model or by automatic code generation tools. Starting from the safe and behaviourally verified model benefits from the clarifications of the requirements that are done to allow formal verification of the invariants.

Conformance Testing The generated scenarios are used for automated testing of the implementation. This ensures that the implementation shows all modelled behaviours. Code coverage analysis of the implementation is done to spot potentially unspecified implemented behaviour. If the code is automatically generated from the model, one could argue that there is no need to test any more—but from generated code to the deployed systems, several elements are introduced that are not considered in the formal verification. These elements could still affect the correctness of the behaviour in the final system.

3 Process Step Details and Tools

In this section, we describe how we instantiated the pattern using a couple of existing, extended and newly developed tools within the project ENABLE-S3. We illustrate the workflow using a motorised lift door as a running example.

3.1 Scenario Modelling with Gherkin

To express the scenarios, we chose Gherkin, a language widely used in BDD. Gherkin [17, Chapter 3] defines lightweight structures for describing the expected behaviour in a plain text, readable by both stakeholder and developer, which is still automatically transformable into an executable test. It uses the following concepts:

Feature A feature is a description of one single piece of business value. There are no technical restrictions about the feature granularity, the structuring of the specification into individual features shall respect the domain model nature.

A feature description starts with the keyword Feature: followed by the feature name and feature description. Neither the feature name nor the feature description are relevant for the test execution, however they are vital for the correct feature understanding. It is recommended to structure the feature description as a story "*As a* «role»*I want* «feature»*so that* «business value»", which gives an answer to three fundamental questions—*who* requires *what* and *why*.

Scenario Each feature contains a list of scenarios, every scenario representing one use case. There are no technical restrictions about the number of scenarios in a feature; yet they all shall be related to the feature being described.

In the simplest case the scenario also contains the test data and thus represents an individual test case. It is however advantageous to separate the general requirement description from the concrete test cases and to describe a group of similar use cases at once. For this purposes a scenario outline with placeholder for the particular test data specified separately as a list of examples can be used.

A scenario starts with the keyword Scenario: or Scenario Outline: followed by the scenario name and scenario description. The list of examples starts with the keyword Examples: followed by the particular test data. Neither the scenario name nor the scenario description are relevant for the test execution, however they are vital for the correct understanding of scenarios.

Steps Every scenario consists of steps starting with one of the keywords Given for test preconditions, When for the tested interaction and Then for the test postconditions. Keywords And and But can be used for additional test constructs.

- Test preconditions describe how to put the system under test in a known state. This shall happen without any user interaction. It is a good practice to check, if the system indeed reached the assumed state.
- Tested interaction describes the provided input. This is the stimulus triggering the execution.
- Test postconditions describe the expected output. Only the observable outcome shall be compared, not the internal system state. The test fails if the real observation differs from the expected results.

Example (Scenario: Closing a Lift Door) The scenario given in Listing 1 describes the closing an open door and stopping to close once an obstruction has been detected. The example scenario uses steps close to natural language.

```
Scenario: Close an open door then detect an obstruction
Given A "door" is open
And can start closing the "door"
When start closing the "door"
Then the "door" is closing
And can detect an obstruction
And can complete close the "door"
And cannot start closing the "door"
When detect an obstruction
Then the "door" is not closing
And the "door" is opening
And can complete open the "door"
And cannot complete close the "door"
And cannot detect an obstruction
```

Listing 1 Test scenario for a motorised lift door

3.2 Formal Modelling in Event-B with Rodin

Event-B [1] is a formal method for system development comprising a formal language and an approach how to build the models making use of *refinement* to introduce system details gradually into the formal model and formal proof to ensure the refinements are consistent. An Event-B model contains two parts: *contexts* and *machines*. Contexts contain *carrier sets*, *constants*, and *axioms* constraining the carrier sets and constants.

Example (Context Statemachine_d_state) The context in Listing 2 declares the set of door states. The axiom states that there are exactly 4 distinct states for a door, i.e., CLOSED, OPENING, OPEN, and CLOSING.

```
context Statemachine_d_state
sets door_STATES
constants CLOSED OPENING OPEN CLOSING
axioms
 partition(door_STATES, {CLOSED}, {OPENING}, {OPEN}, {CLOSING})
end
```

Listing 2 Context Statemachine_d_state

Machines contain *variables* v, *invariants* I(v) constraining the variables, and *events*. An event comprises a guard denoting its enabled-condition and an action describing how the variables are modified when the event is executed. In general, an event e has the following form, where t are the event parameters, G(t, v) is the guard of the event, and S(t, v) is the action of the event.

any t **where** G(t, v) **then** S(t, v) **end**

A special event INITIALISATION without parameters and guards is used for initialising the variables.

Example (Machine m0_door) Consider the machine m0_door modelling a door (Listing 3). A variable d_state is used to model the current states of the door and is initialised to CLOSED (the INITIALISATION event). Variable d_obst indicates whether there is some obstruction or not. Invariant @inv1 specifies the safety condition that, if the door is CLOSED then there must be no obstruction. As an

```
machine m0_door sees Statemachine_d_state
variables d_state d_obst
invariants
 @inv1: (d_state = CLOSED) ⇒ (d_obst = FALSE)
events
INITIALISATION                  door_closed_complete
begin                            when
 d_state := CLOSED                   d_state = CLOSING
 d_obst := FALSE                     d_obst = FALSE
end                              then
                                  d_state := CLOSED
...                              end
end
```

Listing 3 Machine m0_door

example, event door_closed_complete describes the change of the door's state when it changes from the CLOSING position to (fully) CLOSED.

A machine in Event-B corresponds to a transition system where *variables* represent the states and *events* specify the transitions. Note that invariants I(v) are inductive, i.e., they must be *maintained* by all events. This is more strict than general safety properties which hold for all reachable states of the Event-B machine. This is also the difference between verifying the consistency of Event-B machines using theorem proving and model checking techniques: model checkers explore all reachable states of the system while interpreting the invariants as safety properties.

Refinement Machines can be refined by adding more details. Refinement can be done by extending the machine to include additional variables (*superposition refinement*) representing new features of the system, or to replace some (abstract) variables by new (concrete) variables (*data refinement*). More information about Event-B can be found in [8].

Example (Refinement m1_motor) Consider the refinement m1_motor of machine m0_door introducing the door motor (Listing 4). Variable m_state is introduced to capture the status of the motor (either stopped, closing or opening). Additional details are added to the model. For example, event door_closed_complete has an additional guard stating that the motor state is currently closing. New events controlling the motor are also introduced. For instance, stop_motor_closing changes the motor state from closing to stopped. The guards ensure this only happens when the door is closed or there is an obstruction.

```
machine m1_motor refines m0_door sees Statemachine_d_state
variables d_state d_obst m_state
invariants
 @inv1: m_state ∈ m_STATES
events
door_closed_complete
when
 ...
 m_state = CLOSING
then
 ...
end

stop_motor_closing
when
 m_state = CLOSING
 d_state = CLOSED ∨ d_obst = TRUE
then
 m_state := STOPPED
end

...
end
```

Listing 4 Machine m1_motor

Machine Inclusion In the Enable-S3 project we have developed a *machine inclusion* mechanism [9] for reusing machines. This allows models to be developed from verified parts, complementing the top-down development approach of refinement. Machine inclusion allows a machine to include one or more copies of other machines.

All variables and invariants of the included machine B become also part of the including machine A.

In order to modify B's variables, an event of A synchronises with an event of B. Different instances of the same machine can be included using prefix renaming where variables and events of each included machine instance are prefixed accordingly. Machine inclusion enables efficient reuse of formal models and increases the support for teamwork. To build a model for a lift, we can include machine m1_motor to model the lift's door.

Tool Support Event-B is supported by the Rodin [2], an extensible toolkit which includes facilities for modelling, verifying the consistency of models using theorem proving and model checking techniques, and validating models with simulation-based approaches.

3.3 Graphical Formal Modelling with UML-B

UML-B [13, 15, 16], an extension of the Rodin Platform, provides a 'UML like' diagrammatic modelling notation for Event-B in the form of class-diagrams and state-machines. The UML-B makes the formal models more visual and thus easier to comprehend.

State-machines can be structured hierarchically (nested) and this is a natural way to add detail in refinements. State-machines can be *lifted* to the instances of a class so that the behaviour of each instance of the class is modelled by an independent instance of the state-machine. A state-machine automatically generates Event-B data elements (sets, constants, axioms, variables, and invariants) to implement the states, and guards and actions representing state changes are added to events that represent transitions. If the state-machine is lifted, an additional event parameter represents the state machine instance and is used in guards and actions to access the state of the instance.

Example (Door State-Machine) Figure 3 shows the state-machine of a door. State-invariant of the CLOSED state specifies that the door must be free of obstruction when it is fully closed. Note that context Statemachine_d_state and machine m0_door described in Sect. 3.2 was generated from this state-machine.

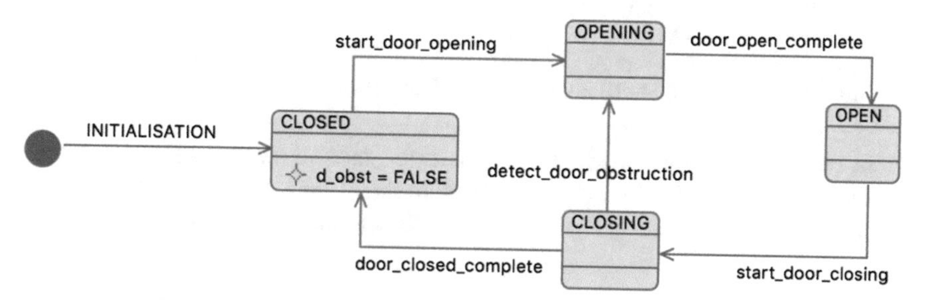

Fig. 3 Door statemachine

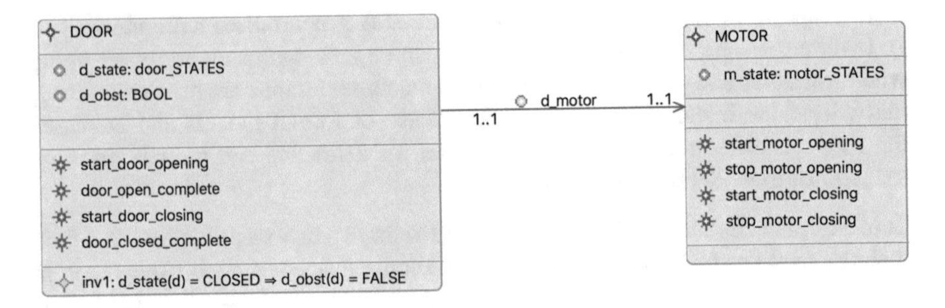

Fig. 4 Door class diagram

UML-B Class diagrams provide a way to visually model data relationships. Classes, attributes and associations represent Event-B data elements (carrier sets, constants, or variables) and generate constraints on those elements. Class methods elaborate Event-B events and contribute an additional parameter representing the class instance. Class diagrams can be developed step-wise by adding new classes, and/or new details to existing classes, using refinement.

Example (Door Class Diagram) Consider a generalisation of the Door example, where we want to model many doors and their corresponding motors. Figure 4 represents a simple class diagram with two classes DOOR and MOTOR, both of them are given sets representing the set of doors and motors, respectively. Each door has a variable attribute d_state to indicate the current state of the door (open/opening/closed/closing). Similarly, each motor has a variable attribute, m_state, to indicate the current status of the motor (opening/stopped/closing). A constant association d_motor represents the relationship between doors to their corresponding motors. The translation of the attributes and the association into Event-B as total functions as follows.

---

constants d_motor
axioms
$d\_motor \in DOOR \rightarrowtail\!\!\!\!\rightarrow MOTOR$
variables d_state d_obst m_state obst
invariants
$d\_state \in DOOR \rightarrow door\_STATES$
$d\_obst \in DOOR \rightarrow BOOL$
$m\_state \in MOTOR \rightarrow motor\_STATES$
$\forall d \cdot d \in DOOR \Rightarrow (d\_state(d) = CLOSED \Rightarrow d\_obst(d) = FALSE)$

Both classes have some methods such as door_closed_complete, start_door_closing, allowing behaviour to be localised to the classes. Note that the constants, variables and events are lifted accordingly to the instances of the classes.

3.4 Model Behaviour Verification with Cucumber

Cucumber is a framework for executing acceptance tests written in Gherkin language and provides Gherkin language parser, test automation as well as report generation.

The BDD principle aims at pure domain oriented feature descriptions without any technical knowledge. In order to make such scenarios automatically executable as tests, the user must supply the actual step definitions providing the gluing code, which implements the interaction with the System Under Test (SUT). The steps shall be written in a generic way, i.e. serving multiple features. This keeps the number of step definitions much smaller than the number of tests. It is an antipattern to supply feature-coupled step definitions which can't be re-used across features or scenarios.

Compound steps may encapsulate complex interaction with a system caused by a single domain activity, thus decoupling the features from the technical interfaces of the SUT. This defines a new domain-related testing language, which may simplify the feature description. The description of the business functionality shall, however, still be contained in the features.

3.4.1 Cucumber for Event-B and UML-B

We have developed 'Cucumber for Event-B' and 'Cucumber for UML-B'. The tools allow to execute scenarios in Gherkin. They include a number of step definitions that define a traversal of the Event-B/UML-B state space.

In order to be generic and work for arbitrary models, the steps used in the scenarios need to follow the terminology of the model. The step definitions forEvent-B are as following:

Given *machine with* "«formula»"
    Setup constants with the given constraints and initialize the machine.
When *fire event* "«name»"*with* "«formula»"
    Fire the given event with the given parameters constraints.
Then *event* "«name»"*with* "«formula»"*is enabled*
    Check if the given event with the given parameters constraints is enabled.
Then *event* "«name»"*with* "«formula»"*is disabled*
    Check if the given event with the given parameters constraints is disabled.
Then *formula* "«formula»"*is TRUE*
    Check if the given formula evaluates to TRUE.
Then *formula* "«formula»"*is FALSE*
    Check if the given formula evaluates to FALSE.

The following Gherkin syntax can be defined for validating state-machines.

Given *state machine* "«name»:«inst»"
    Preset the given instance of the given state machine.
When *trigger transition* "«trans»"
    Trigger the given state machine transition.
Then *transition* "«trans»"*is enabled*

Check if the given state machine transition is enabled.

Then *transition* "«trans»"*is disabled*

Check if the given state machine transition is disabled.

Then *is in state* "«state»"

Check if the state machine is in the given state.

Then *is not in state* "«state»"

Check if the state machine is NOT in the given state.

Example (Cucumber for UML-B State-Machines) An example scenario for the Door Statemachine given in the Fig. 3.

Scenario Outline: *Close an open door then detect an obstruction*
Given *state machine* "d_state:OPEN"
And *transition* "start_door_closing" *is enabled*
When *trigger transition* "start_door_closing"
Then *is in state* "CLOSING"
And *transition* "detect_door_obstruction" *is enabled*
And *transition* "door_close_complete" *is enabled*
And *transition* "start_door_closing" *is disabled*
When *trigger transition* "detect_door_obstruction"
Then *is not in state* "CLOSING"
And *is in state* "OPENING"
And *transition* "detect_door_obstruction" *is disabled*
And *transition* "door_close_complete" *is disabled*
And *transition* "door_open_complete" *is enabled*

Listing 5 Test scenario using UML-B state-machine

Such a scenario is pretty straight forward. The drawback is, that the stakeholders are confronted with the syntax and semantics of the model, which at least in case of Event-B, is far from the language they use. In addition, the user must use the precise event/transition and variable names, which may differ from the terms used in a problem domain (e.g. because of some additional technical restrictions).

It is possible to build a collection of domain specific step definitions thus raising the feature description to the proper abstraction level, like we used in the example already presented in Listing 1. This can be automated, but in general this means an additional effort to keep it in sync with the model as it evolves.

For UML-B, since the model is more easily kept close to the terminology of the domain, it might in many cases be acceptable to work with the generic step definitions.

'Cucumber for UML-B' provides additional step definitions for treating class diagrams, addressing classes, methods, attributes and associations that follow the same concepts as the ones for state machines. Scenarios usually will use the class and the state diagram related steps in combination.

3.4.2 Data

While the event parameters are often simple values, the attribute values may have complex types. This raises the issue of how to describe such data for the setup of constants on one side and for the attribute value checks on the other side. For now, the implementation does not yet support complex data types.

3.5 Model Based Scenario Generation with MoMuT for Event-B

MoMuT[3] is a test case generation tool able to derive tests from behaviour models. The behaviour model represents a system specification, the generated tests can be used as black box tests on an implementation. They help to ensure that every behaviour that is specified, is also implemented correctly.

As input models, MoMuT accepts Object Oriented Action Systems (OOAS) [10], an object oriented extension of Back's Action systems [3]. The underlying concepts of Action systems and Event-B are both closely related to Dijkstra's guarded command language [5], which makes them mappable to each other with only few restrictions. Within the project ENABLE-S3, a transformation from a subset of Event-B into OOAS has been developed. This enables MoMuT to generate tests for Event-B and UML-B models.

In contrast to other model based testing tools, the generated test cases do not target structural coverage of the model, they target exposing artificial faults systematically injected into the model. These faults are representatives of potential faults in the implementation; a test finding them in the model can be assumed to find its direct counterpart as well as similar, not only identical problems in the implementation [6]. In context of our pattern, we use the sequences of inputs and outputs that MoMuT generates as scenarios.

In UML-B/Event-B models, a fault can be as coarse as disabling a complete transition/event. This would obviously be exposed by a scenario that takes this transition/event and it is very likely that this is already part of the manual scenarios. But a fault can also be as sublime as slightly increasing the increment step for a counter, which, when reaching some boundary value, changes the effect of another event/transition. In order to catch this, a scenario would need to take the incrementing event/transition often enough to reach the boundary value and then use the affected transition/event to check if it's result is correct. So the generated scenarios will cover corner cases and complex co-dependencies that are easily missed when manually defining scenarios.

MoMuT strives to produce efficient tests i.e. keeping the test suite's size close to the necessary minimum. Therefore, it is usually possible to review the generated sce-

[3]http://www.momut.org.

narios with reasonably effort in the following acceptance testing step. The artificial faults introduced into the model can be seen as potential small misunderstandings when interpreting the requirements. A set of scenarios that covers as many small misunderstandings of the requirements as possible is expected to be appropriate to evaluate the correctness of the model's behaviour.

3.6 Acceptance Testing with Model Animation

In order to evaluate if the modelled behaviour is in line with the expectations of the stakeholders, it is important that the model and the generated scenarios are in a form that is understandable to them.

Since the generated scenarios are using the concepts from the model and those concepts were kept close to the requirements terminology during modelling, the generated scenarios should be readable and understandable by the stakeholders. Nonetheless is it helpful to have further support.

Rodin with the ProB Plugin [11] allows to animate Event-B models. The UML-B extension includes support to animate state machines with proB. With help of the tool BMotion Studio,[4] domain specific visualizations for animation can be built.

In all three animation variants, the generated scenarios can be walked through and checked for plausibility. BMotion Studio has been used in the project in the Railway use case (see chapter "Validation of Railway Control Systems"). An initial version of a tool to run generated scenarios with proB has been built in the project, but at the time of writing is not yet connected to the visualisation within Rodin.

3.7 Conformance Testing an Implementation

The manual and generated scenarios can be run on an implementation of the SUT in order to test if the implementation conforms with the modelled behaviour. To do so, the scenarios need to be transformed into a format that can be used by the respective test environment of the implementation. This needs to take into consideration that the model and the test environment have to agree on the mode for observing output, i.e. if they look at status or at events for comparison. Otherwise an additional adaption is needed.

In many cases, the model and the implementation might have different inner structures, leading to bad coverage of the scenarios on the code, while the coverage of expected behaviours is good. This can be addressed by generating more varying tests with MoMuT, e.g. by adding random walks on the model, as long as the cost of running them on the implementation becomes not too high. While these behaviours

[4]https://www3.hhu.de/stups/prob/index.php/BMotion_Studio.

are now not reviewed by stakeholders, they still implicitly check the invariants built into the model, since they were proven on the model.

Even for code generated from the accepted model, used I/O-libraries or resource limitations of the target platform could still introduce unexpected and unwanted behaviour. The manual and generated scenarios should therefore still be run on the deployed systems. MoMuT can also generate robustness tests from the accepted model using smart fuzzying, to test if the system sanely handles unexpected inputs.

An initial framework to run generated scenarios on an arbitrary test environment has been built in the project and has been tested by connecting it back to proB for animating the scenarios.

4 Conclusions

We have presented a model-based approach to systems development that combines the strengths of BDD, formal modelling and model-based testing. The combination of these techniques provides rigorous verification that the dependability properties are upheld in the system while also ensuring that the system requirements are valid (useful) and that they are reflected in the formal models and subsequent implementation. The individual strengths of the approach are summarised as follows:

Scenarios Allow domain experts to express the 'desired' behaviour of the system in a precise and concise way that is linked to the formal models and therefore amenable to tool manipulation.

Formal modelling Provides a precise mathematical expression of the system that deals with complexity via abstraction and facilitates verification of critical properties without instantiation of the system

Model-based testing Automatically generates test cases (scenarios) to ensure full coverage of the behaviour of the model, providing completeness of acceptance tests and hence rigorous validation and conformance testing of the implementation.

Within ENABLE-S3, a concrete instantiation of this pattern has been put together from existing, extended and newly developed, cooperating tools. The approach has been applied to a use case from the railway domain (see chapter "Validation of Railway Control Systems").

The approach helps addressing the issue that formal modelling expertise and domain knowledge are rarely held by the same set of people. It has a high potential to reduce development efforts for critical applications due to the possible degree of automation. Nonetheless, the increased dependability is not for free—all the cross-checking activities come at a cost and as long as code is not automatically generated from the model, this is more work than directly developing from the requirements.

Therefore, future work includes, in addition to completing the method with support for more Event-B language elements and further improved and speeded up test case generation that allows interactive use during modelling, also code generation from the proven model.

References

1. Abrial, J.R.: Modeling in Event-B: System and Software Engineering. Cambridge University Press, Cambridge (2010)
2. Abrial, J.R., Butler, M., Hallerstede, S., Hoang, T.S., Mehta, F., Voisin, L.: Rodin: An open toolset for modelling and reasoning in Event-B. Softw. Tools Technol. Transfer **12**(6), 447–466 (2010)
3. Back, R., Sere, K.: Stepwise refinement of action systems. In: International Conference on Mathematics of Program Construction, pp. 115–138. Springer, Berlin (1989)
4. Buschmann, F., Henney, K., Schmidt, D.C.: Pattern-Oriented Software Architecture, 4th edn. Wiley Series in Software Design Patterns, Wiley (2007). http://www.worldcat.org/oclc/314792015
5. Dijkstra, E.W.: Guarded commands, nondeterminacy and formal derivation of programs. Commun. ACM **18**(8), 453–457 (1975)
6. Fellner, A., Krenn, W., Schlick, R., Tarrach, T., Weissenbacher, G.: Model-based, mutation-driven test case generation via heuristic-guided branching search. In: Proceedings of the 15th ACM-IEEE International Conference on Formal Methods and Models for System Design, pp. 56–66. ACM, New York (2017)
7. Herzner, W., Sieverding, S., Kacimi, O., Böde, E., Bauer, T., Nielsen, B.: Expressing best practices in (risk) analysis and testing of safety-critical systems using patterns. In: 25th IEEE International Symposium on Software Reliability Engineering Workshops, ISSRE Workshops, Naples, Italy, November 3–6, 2014, pp. 299–304. IEEE, Piscataway (2014)
8. Hoang, T.S.: An introduction to the Event-B modelling method. In: Industrial Deployment of System Engineering Methods, pp. 211–236. Springer, Berlin (2013)
9. Hoang, T.S., Dghaym, D., Snook, C.F., Butler, M.J.: A composition mechanism for refinement-based methods. In: 22nd International Conference on Engineering of Complex Computer Systems, ICECCS 2017, pp. 100–109. IEEE, Piscataway (2017)
10. Krenn, W., Schlick, R., Aichernig, B.K.: Mapping UML to labeled transition systems for test-case generation. In: Formal Methods for Components and Objects, pp. 186–207. Springer, Berlin (2010)
11. Leuschel, M., Butler, M.: ProB: An automated analysis toolset for the B method. Softw. Tools Technol. Transfer **10**(2), 185–203 (2008)
12. North, D.: Introducing BDD. Better Software Magazine (2006)
13. Said, M.Y., Butler, M., Snook, C.: A method of refinement in UML-B. Softw. Syst. Model. **14**(4), 1557–1580 (2015)
14. Smart, J.F.: BDD in Action: Behavior-Driven Development for the Whole Software Life cycle. Manning Publications Company, Shelter Island (2014)
15. Snook, C.: iUML-B statemachines. In: Proceedings of the Rodin Workshop 2014, pp. 29–30. Toulouse, France (2014). http://eprints.soton.ac.uk/365301/
16. Snook, C., Butler, M.: UML-B: Formal modeling and design aided by UML. ACM Trans. Softw. Eng. Methodol. **15**(1), 92–122 (2006)
17. Wynne, M., Hellesøy, A.: The Cucumber Book: Behaviour-Driven Development for Testers and Developers. Pragmatic Programmers, LLC, Raleigh (2012)

Reliable Decision-Making in Autonomous Vehicles

Gleifer Vaz Alves, Louise Dennis, Lucas Fernandes, and Michael Fisher

1 Introduction

The deployment of Autonomous Vehicles (AV) on our streets is no longer fictional, but is becoming reality. AV technology will represent a significant change in our economy, society, and daily life [21]. There are already several AV tests that are being undertaken and some of them will lead to practical vehicles on our streets quite soon: in Phoenix (USA) the use of a fully driverless taxi service is expected to commence soon [20]; in Singapore's university district there is the world's first self-driving taxi service, which has been operated by NuTonomy since August 2016 [21]; while there are a large number of cars equipped with autonomous driving technology that are expected on the streets of South Korea by 2020 [21].

Consequently, a lot of work is being carried out concerning different stages of an AV design, for example, detection devices, cameras, sensors, intelligent software, decision-making, liabilities, laws and regulations, and so on. Abstractly, the design of an AV can be divided into two main parts: the high-level and the low-levels [23]. The latter are responsible for the sensors, actuators, detection devices and other

Work partially supported by EPSRC projects EP/L024845 and EP/M027309.

G. V. Alves (✉)
UTFPR – Federal University of Technology, Ponta Grossa, Parana, Brazil
e-mail: gleifer@utfpr.edu.br

L. Dennis · M. Fisher
Department of Computer Science, University of Liverpool, Liverpool, UK
e-mail: L.A.Dennis@liverpool.ac.uk; MFisher@liverpool.ac.uk

L. Fernandes
Samsung R&D (SRBR), Sao Paulo, Brazil
e-mail: lucas.f@samsung.com

A. Leitner et al. (eds.), *Validation and Verification of Automated Systems*,
https://doi.org/10.1007/978-3-030-14628-3_10

similar control elements. The former, however, captures the key decision-making capability that the AV must exhibit now that there is no responsible human 'driver'. This high-level decision-making is software responsible for clearly determining the actions that will be invoked at the low-level.

As highlighted by Herrmann et al. [21] achieving the basic 90% of autonomous driving capabilities is not difficult, but the last 10% is the most challenging task. These 10% includes the most difficult traffic scenarios, especially in urban areas, and the possibility of unexpected or emergency situations. It is here that the AV must make the 'correct' decisions, quickly and reliably. Consequently, we focus here on the high-level decision-making process, where an AV decision-maker is modelled as an intelligent agent [18]. Using this agent-based approach we may write high-level plans for describing the AV decisions and actions and, since these plans are transparent and explicit, we can formally verifying some properties related to this agent's behaviour [11], such as *"it is always true the* AV-agent *will stop in the event of an unexpected emergency"*.

In previous work [16] we have created the first version of a *Simulated Automotive Environment* (SAE), where our AV-agent represented a taxi responsible for driving passengers from a start point to a destination point. Through each taxi route, the agent might face different sorts of obstacles that we divided into two classes: *avoidable* and *unavoidable*. When there is an avoidable obstacle it means the agent can potentially find a detour along its route without crashing in to any obstacle. However, if there is no chance to avoid the obstacle, we say there is an unavoidable obstacle. In such cases, the agent should select the obstacle causing the least (physical) damage to the vehicle, whenever it is possible. The taxi agent in SAE is implemented using GWENDOLEN [8] agent programming language, while the model checker AJPF [9] has been applied towards the formal verification of some related properties concerning the behaviour of the taxi agent. Indeed these same tools (GWENDOLEN and AJPF) are also used in the work here presented.

In this paper, we create an extension of the SAE, named SAE-decision-making, which has a similar approach based on a taxi agent, but where we capture the high-level decisions taking by the AV-agent when there are different levels of emergency (classified here by three different colours): (i) yellow (i.e *avoidable obstacle*); (ii) orange (i.e *harsh environment*); or (iii) red (i.e *unavoidable obstacle*).

As we shall see in this article, our agent uses an approach based on the UK suggestions for driverless vehicles (see [14] and [13]), which state that a human controller has liability concerning the actions taken by the AV. As highlighted by Vellinga [25], in the UK, regulations for testing AVs mention the role of a *test operator*. In [13], it is stated that:

> A test operator will be responsible for ensuring the safe operation of the vehicle at all times (...). The test operator (...), and be able to anticipate the need to intervene and resume manual control if necessary.

So, in our case, the agent implementation will choose, as a safe and reliable procedure, the following: *the AV-agent will release autonomous control and give it back to the human (controller), in case of orange and red levels of emergency (and also when the human is indeed ready to take over).*

The remainder of this paper is organised as follows. In Sect. 2 we provide the necessary background on agents, their programming, and their formal verification. Next, Sect. 3 presents the formal definitions of both the environment and agent used in our work, in addition to the agent implementation. Section 4 highlights the properties which have been formally verified by using temporal logic, while in Sect. 5 we provide final remarks.

2 Agents and Formal Verification

In this section, we briefly present some background concepts used in this work. Here we use the notion of a *rational agent* (or intelligent agent) introduced by Bratman [5] and described in detail by Rao and Wooldridge [28]. Here, rational agents are software entities that perceive their environment through sensing, build a model of its *belief* about this environment, and then reason about it. Based on its own mental state, such as its *intentions*, a rational agent can then take actions that may change the environment [28]. A rational agent can be implemented in a number of ways, but we choose to utilise the popular BDI (*Belief, Desire and Intention*) architecture [24].

In our work we use a particular BDI programming language, named GWENDOLEN [8], which captures the BDI concepts in a goal-directed Prolog-like language. Agents programmed in GWENDOLEN have beliefs, reasoning rules, goals, and plans. A plan has three main components [8]:

1. trigger: is the event responsible to unleash a given plan.
2. guard: is a condition that should be checked in order to have an applicable plan.
3. body: has a set of actions and/or plans that are supposed to be executed.

The syntax to represent plans is the following,

```
trigger : { guard } < − body
```

And a plan example written in GWENDOLEN can be as follows,

```
+!emergency_plan { B unexpected_emergency } < − stop;
```

Agent programs constructed using GWENDOLEN language can then be verified using the AJPF tool [9]. AJPF is an extension of the *Java Path Finder* (JPF) program model-checker, used for the formal verification of `Java` programs [26]. The AJPF system [4] was specifically conceived to work with BDI agent programming languages, so that the agent code can be formally verified against given properties. These properties are written in a *Property Specification Language* (PSL) which is itself based on LTL (Linear Temporal Logic). The syntax for property formulae, ϕ, in PSL is given below, where:

- ag refers to a specific agent in the system;
- f is a classical first-order atomic formula;
- and $\Diamond$ ("eventually in the future") and $\Box$ ("always in the future") are standard LTL operators [17].

$$\phi ::=$$

$$\mathsf{B}_{ag}\ f \mid \mathsf{G}_{ag}\ f \mid \mathsf{A}_{ag}\ f \mid \mathsf{I}_{ag}\ f \mid \mathsf{ID}_{ag}\ f \mid \mathsf{P}(f) \mid \phi \vee \phi \mid \phi \wedge \phi \mid \neg\phi \mid \phi\, \mathsf{U}\, \phi \mid \phi\, \mathsf{R}\, \phi \mid \Diamond\phi \mid \Box\phi$$

Note that,

- $\mathsf{B}_{ag}\ f$ is true if agent ag believes formulae f to be true,
- $\mathsf{G}_{ag}\ f$ is true if agent ag has a goal to make formula f true,
- *and so on* with A representing actions, I representing intentions, ID representing the intention to take an action, and P representing percepts, i.e., properties that are true in the environment.

An example of a PSL formulae is the following (which is based on the example mentioned in Sect. 1):

$$\Box\, \mathsf{B}_{ag}\ \texttt{unexpected\_emergency} \rightarrow \mathsf{A}_{ag}\ \texttt{stop}$$

3 The AV-Agent: Scenario, Plans, and Environment

As previously mentioned in Sect. 1, the SAE-decision-making is mainly concerned with the high-level decisions taken by our AV-agent as illustrated in Fig. 1.

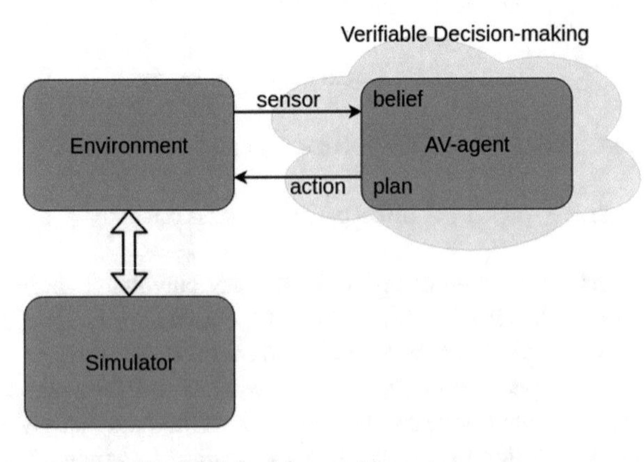

Fig. 1 The general diagram for SAE-decision-making

The SAE-decision-making comprises an environment, the AV-agent, and a (graphical) simulator. The agent captures an autonomous vehicle with basic driving decisions in the environment, where the agent works like a taxi, taking passengers from a starting point towards a destination point and checking its environment constantly for different levels of emergency. This will lead to a range of different actions being considered. Both the environment and the AV-agent can be formally defined as follows.

Definition 3.1 (Environment and Agent) An environment Σ_n is given by the following tuple, such that $n \in \mathbb{N}$:

$$\Sigma_n = (G_r, C_e, P_a, R_d, A_g)$$

where,

- G_r is a square grid[1] of $n \times n$
  - each element from the grid determines a coordinate $(X, Y) \mid (X, Y) \subset G_r$
  - X and Y set the indexes for rows and columns in G_r.
- C_e is a set of i cells, $C_e = \{c_1, c_2, \ldots, c_i\}$, such that $0 \leq i \leq n \times n$.
  - a given cell, c_i, may have one of the following status:
  - `unknown`: it means is an unknown cell, not yet discovered by the agent.
  - `clear`: it means is clear to go through it.
  - `avoidable_obstacle`: the cell has an avoidable obstacle (i.e. it represents a *yellow* level of emergency).
  - `harsh_environment`: the cell has a harsh environment (i.e. it represents an *orange* level of emergency).
  - `unavoidable_obstacle`: the cell has an unavoidable obstacle (i.e. it represents a *red* level of emergency).
- P_a is a set of j passengers, $P_a = \{p_1, p_2, \ldots, p_j\}$, such that $0 \leq j \leq n$.
- R_d is a set of k rides, $R_d = \{r_1, r_2, \ldots, r_k\}$, such that $0 \leq k \leq n$.
  - each ride is given by $r_k = (sp, dp)$
  - sp stands for *starting point*, while dp is the *destination point*
- A_g (or AV-agent) is given by the following tuple:

$$A_g = (B_e, P_l, A_c, D_i)$$

  - B_e is a set of Beliefs of A_g,

[1] To showcase the approach, we take a very simple grid model of the environment.

$$B_e = \{\, \texttt{passenger}, \texttt{ride}, \texttt{position}(X, Y), \texttt{starting\_point}(X, Y), \texttt{destination\_point}(X, Y), \texttt{moving}, \texttt{unknown}, \texttt{clear}, \texttt{avoidable\_obstacle}, \texttt{harsh\_environment}, \texttt{unavoidable\_obstacle}, \texttt{human\_controller\_ready} \,\}$$

* `human_controller_ready`: means the agent believes the human is capable of controlling the AV and so is ready to take over.

- P_l is a set of plans for A_g,
  $$P_l = \{\, \texttt{liability\_controller}, \texttt{autonomous\_control}, \texttt{resume\_manual\_control}, \texttt{finish\_all\_rides}, \texttt{complete\_journey}, \texttt{drive\_to}, \ldots \}$$
- A_c is a set of actions,
  $$A_c = \{\, \texttt{drive}, \texttt{navigate}, \texttt{localise}, \texttt{search\_ride}, \texttt{refuse\_ride}, \texttt{parking}, \texttt{sound\_alarm}, \texttt{brakes}, \texttt{slow\_speed}, \ldots \,\}$$
- D_i is a set of possible directions that A_g may take in the G_r,
  $$D_i = \{\, \texttt{North}, \texttt{South}, \texttt{East}, \texttt{West} \,\}$$

□

As it follows, we describe the intended scenario of our approach, as well as, the emergency plans written for the AV-agent. Moreover, we present our graphical simulator, which has been built to illustrate the actions taken by the AV-agent and how these actions reflect on the environment, as generally pictured on Fig. 1 (previously seen at the beginning of this section).

3.1 Scenario

In order to showcase the approach, we will consider here a basic scenario in which an AV must decide what to do in a situation where it detects an obstacle and/or a dangerous environment (e.g., an icy patch of road). In the case of an obstacle, the AV may decide it can plot a safe path around it whereby it considers the obstacle to be avoidable and so retains control.

However it may be that the vehicle is not able to calculate a safe path around the obstacle—for instance it may detect oncoming traffic in the other direction. In such a case the AV may be faced with having to solve a version of the so-called *trolley problem* [19, 3] in which a choice must be made between a number of dangerous options. Since appropriate solutions to the trolley problem remain under debate and, furthermore, in some jurisdictions (e.g., Germany) it may even be illegal to decide in advance by some algorithm what the solution to the trolley problem should be (see [27], Chapter 6), we here take the view that the AV needs to hand control back to the driver at this point. (Note that work on the formal verification of such *ethical* dilemmas is part of current research [10, 12].)

Unfortunately, while current legislation generally requires the driver of an AV to be ready to assume control at all times, it is widely recognised that it is unrealistic to expect this to always be possible [6]. So, ethically speaking, it is important for the vehicle to have a contingency plan in case the driver does not, or can not, assume control. In this case we follow the recommendation from [7] that instead of attempting to solve the trolley problem itself the AV should instead just engage its emergency stop procedures.

3.2 Emergency Plans

The AV-agent has plans that enable the capability to drive, to undertake rides with passengers, and to successfully complete the journeys. However, we here just describe the plans specifically created to deal with different levels of emergency, which can be seen in Code 1 (which is a fragment from our AV-agent's actual programming).

Notice that if there is a *yellow* emergency level, i.e. the AV-agent believes that there is an avoidable obstacle in the environment, then the action `autonomous_control` should be taken representing the decision that the AV-agent maintains control of the vehicle.

However, if there is an *orange* emergency level, i.e. the AV-agent believes there is a particularly harsh environment and also believes the human who has the liability to control the vehicle is ready, then the autonomous control mode is released and the human regains control. But, if the agent believes the human is not ready to again take over control, actions for sounding an alarm and slowing the vehicle's speed must be taken. In this case, the AV-agent retains autonomous control until it decides the human driver is ready to (safely) regain control.

In the case of a *red* emergency level, i.e. the AV-agent believes there is an unavoidable obstacle and also believes the human who has the liability to control the vehicle is ready to take over control, then autonomous control is released. Nonetheless, if the agent again believes the human is not yet ready to take over, the actions for sounding an alarm and applying the brakes must be taken.

```
1  GWENDOLEN

3  :name: AV

5  :Initial Goals:

7  liability_controller [achieve]

9  :Plans:
   // levels of emergency
11
   +! liability_controller [achieve] {B unavoidable_obstacles, ~B
13     human_controller_ready} <- sound_alarm, brakes;

15 +! liability_controller [achieve] {B unavoidable_obstacles, B
       human_controller_ready} <- +!resume_manual_control[perform];
17
   +! liability_controller [achieve] {B harsh_environment, B human-
19     controller-ready} <- +!resume_manual_control[perform];

21 +! liability_controller [achieve] {B harsh_environment, ~B human-
       controller-ready} <- sound_alarm, slow_speed, +!autonomous_control[
23     perform];

25 +! liability_controller [achieve] {B avoidable_obstacle} <- +!
       autonomous_control[perform];
27


29 +!autonomous_control[perform] <- ...
   // set of actions to keep the AV in autonomous mode
31
   +!resume_manual_control[perform] <- ...
33 // set of actions to the human controller resume the control
```

Listing 1 GWENDOLEN code—fragment of AV-agent's plans

3.3 Agent-Environment Simulator

In order to better represent the environment in which the AV-agent proposed here is inserted, it was developed a graphic simulator to illustrate its actions. Such tool allows to observe the behaviour of an agent in a high-level perspective, by displaying every decision taken and beliefs acquired by the AV-agent in its environment.

Said agent sends messages using a pre established protocol to the simulator, which in turns interprets it translate into new modifications of the simulation. And, to build the simulation, it is used*Java 2D Graphics* library. Meanwhile, all communication is made through a client-server architecture using UDP network protocol.

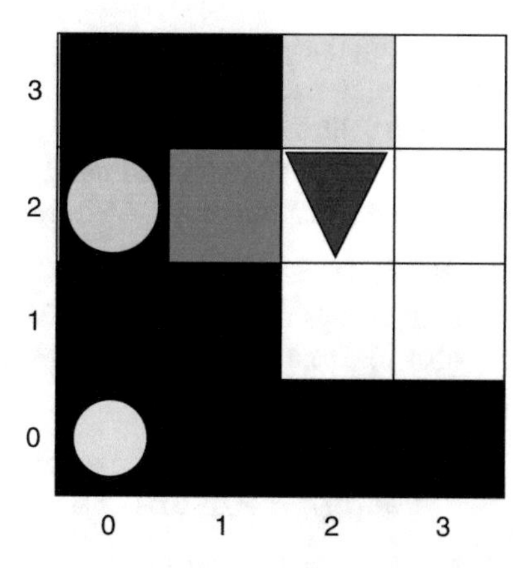

Fig. 2 A possible scenario for the SAE-decision-making simulator

Here, we create a grid to represent the environment and numbered rows and columns to identify coordinates in which the agent can move to. Each coordinate represents a cell where the AV-agent is able to move towards to. As stated before, these cells have states as it follows, `unknown`, `clear`, `avoidable_obstacle`, `harsh_environment`, `unavoidable_obstacle`, represented by, respectively: black, white, yellow, orange and red cells. In turn, the AV-agent is represented by a blue triangle. It is placed in the simulator to show in which direction the agent is moving towards. If a coordinate is the starting point (*sp*) or the destination point (*dp*) for a ride, it will contain a cyan circle or a green circle, respectively.

In Fig. 2 it is displayed a scenario in which the AV-agent located in `(2,2)` is moving south (down), trying to find a way to the starting point `(0,2)` of its next ride. Note that, the agent should be aware of the obstacles in the environment which can be found in its path. As pictured in Fig. 2 there is an avoidable obstacle at `(2,3)` and a harsh environment at `(1,2)`. If everything occur accordingly, after picking up the passenger (at the starting point), the AV-agent will head towards the destination point at `(0,0)`.

Understand that while this simulator provides a high-level representation of the proposed agent and its corresponding actions in the environment, it cannot be used to state its correctness. The AJPF model checker should be used separately from the simulator when the formal verification is conducted.

4 Formal Verifying Agent Decision-Making

In the previous, Sect. 3, we presented our basic AV-agent in GWENDOLEN. Now, using the MCAPL framework we will describe the corresponding formal verification that can be undertaken. As an example, we will concentrate on verification concerning the plan `liability-controller`, which is indeed responsible for actions related to the reliable decision-making process from the AV-agent.

In the following, we state four properties. Each property is both given in natural language and also described using temporal logic operators, which are used in the formal specification of MCAPL framework.

1. **Red level property**:
   It always the case that the AV-agent, when it believes there is an unavoidable obstacle, but believes the human controller is *not* ready, should apply the brakes.
   - $\Box(B_{AV}$ `unavoidable_obstacle` $\wedge \neg$ B_{AV} `human_controller_ready`$) \Rightarrow A_{AV}$ `brakes`
2. **Red/orange level property**:
   It always the case that the AV-agent, when it believes the human controller is ready and either believes there is an unavoidable obstacle or a harsh environment, then should eventually release autonomous mode in order for the human controller to resume manual control.
   - $\Box(B_{AV}$ `human_controller_ready`$\wedge(B_{AV}$ `unavoidable_obstacle` $\vee$ B_{AV} `harsh_environment`$)) \Rightarrow \Diamond G_{AV}$ `resume_manual_control`
3. **Orange level property**:
   It is always the case that the AV-agent, when it believes the human controller is *not* ready to take control and believes there is a harsh environment, will slow down its speed.
   - $\Box(\neg B_{AV}$ `human_controller_ready` $\wedge$ B_{AV} `harsh_environment`$)$ $\Rightarrow A_{AV}$ `slow_speed`
4. **Yellow level property**:
   It always the case the AV-agent, when believes there is an avoidable obstacle, then at some time should keep the autonomous control.
   - $\Box B_{AV}$ `avoidable_obstacle` $\Rightarrow \Diamond G_{AV}$ `autonomous_control`

Notice these properties are described using the colour code. In the red/orange level property we have a situation which is really a mixture of red and orange levels of emergency since there are two possibilities, one when there is an unavoidable obstacle, and a second one when there is a harsh environment. However, for both situations is indeed important to check whether the human driver is ready to regain control. If all the specified requirements hold, then the goal to resume manual control should be achieved. In addition, we could still have a third possibility namely that we have both an unavoidable obstacle and a harsh environment at the same time. But, when this happens the plan for the unavoidable obstacle is firstly selected since the order established in the GWENDOLEN agent programming (previously shown in Code 1).

5 Conclusion

In this work, we have presented an overview of how we can construct a high-level decision-making mechanism based on an intelligent agent approach for an autonomous system, here specifically an Autonomous Vehicle. Moreover, the decision-making process captured by such an intelligent agent can be formally verified by means of model checking techniques. Here we have used the AJPF model checker, where some properties (written in temporal logic) have been defined in order to identify whether the decision-making process of our AV-agent is indeed a reliable mechanism.

To exemplify the approach, we choose a very simple scenario whereby our AV-agent is implemented in a basic urban traffic environment with three different levels of emergency. According to the emergency recognised by the agent, different actions will be triggered.

One possible extension for the AV-agent implementation can be foreseen by the following scenario: an autonomous vehicle is used on autonomous mode if and only if the vehicle is on the motorway when the vehicle is not on the motorway, then a human driver is required. As a matter of fact, this example is mentioned in a recent Consultation paper from the Law Commission and CCAV (*Centre for Connected and Autonomous Vehicles*) from the UK [22]. This document describes that the aforementioned scenario is most likely to happen, at least in the early stages of AV deployment on the streets.

Notice we could easily adapt the AV-agent implemented in our work by means of adding new beliefs and plans in order to capture the requirements from the scenario mentioned above. That is, when the AV-agent has a belief that *it is on the motorway*, then it should engage the autonomous mode control. But, if the AV-agent believes *it is not on the motorway*, as a result the plan responsible to resume the manual control would be selected.

Furthermore, we have used colours (*yellow*, *orange*, *red*) to abstractly represent the different levels of unexpected situation, or emergency, in our urban traffic environment (deployed in the SAE-decision-making). A similar colour code could be adapted and extended for different environments which might also benefit from the use of Autonomous Systems. An example is the application of autonomous systems in nuclear energy management control [1], where an intelligent agent could be in charge of some autonomous function, but also should be aware of how, and when, to ask for human operator intervention.

In previous work [2], we have presented the first steps towards the formalisation of certain road traffic laws (aka the Rules of the Road) from the Department for Transport in the UK [15]. Generally, such Rules of the Road should be embedded into an intelligent agent in such a way that we can verify whether the agent behaves safely according to the urban traffic laws and regulations. In the future, we will consider the merging of both streams of work into a new extension of SAE system, combining the formalisation from the Rules of the Road and the decision-making approach presented here. By establishing such a system, we should be able to formally verify whether the AV-agent will always make safe and reliable decisions.

References

1. Aitken, J., Shaukat, A., Cucco, E., Dennis, L., Veres, S., Gao, Y., Fisher, M., Kuo, J., Robinson, T., Mort, P.: Autonomous nuclear waste management. IEEE Intell. Syst. **PP**(99), 1 (2018)
2. Alves, G.V., Dennis, L., Fisher, M.: Formalisation of the Rules of the Road for embedding into an Autonomous Vehicle Agent. In: International Workshop on Verification and Validation of Autonomous Systems, pp. 1–2, Oxford, UK (2018)
3. Bonnefon, J.-F., Shariff, A., Rahwan, I.: The social dilemma of autonomous vehicles. Science **352**(6293), 1573–1576 (2016)
4. Bordini, R.H., Dennis, L.A., Farwer, B., Fisher, M.: Automated verification of multi-agent programs. In: Proceedings of the 2008 23rd IEEE/ACM International Conference on Automated Software Engineering, ASE '08, pp. 69–78. Washington, DC, USA. IEEE Computer Society, Piscataway (2008)
5. Bratman, M.E.: Intentions, Plans, and Practical Reason. Harvard University Press, Cambridge (1987)
6. Cunningham, M.L., Regan, M.: Driver inattention, distraction and autonomous vehicles. In: 4th International Driver Distraction and Inattention Conference (2015)
7. Davnall, R.: Solving the single-vehicle self-driving car trolley problem using risk theory and vehicle dynamics. Sci. Eng. Ethics 1–19 (2019). ISSN 1471-5546, https://doi.org/10.1007/s11948-019-00102-6
8. Dennis, L.A.: Gwendolen semantics: 2017. Technical Report ULCS-17-001, University of Liverpool, Department of Computer Science (2017)
9. Dennis, L.A., Fisher, M., Webster, M.P., Bordini, R.H.: Model checking agent programming languages. Autom. Softw. Eng. **19**(1), 5–63 (2012)
10. Dennis, L.A., Fisher, M., Winfield, A.F.T.: Towards verifiably ethical robot behaviour. In: Proc. AAAI Workshop on AI and Ethics (2015)
11. Dennis, L.A., Fisher, M., Lincoln, N.K., Lisitsa, A., Veres, S.M.: Practical verification of decision-making in agent-based autonomous systems. Autom. Softw. Eng. **23**(3), 305–359 (2016)
12. Dennis, L.A., Fisher, M., Slavkovik, M., Webster, M.P.: Formal verification of ethical choices in autonomous systems. Robot. Auton. Syst. **77**, 1–14 (2016)
13. Department for Transport—UK. Testing automated vehicle technologies in public. Available at: https://www.gov.uk/government/publications/automated-vehicle-technologies-testing-code-of-practice (2015)
14. Department for Transport—UK. Regulations for driverless cars. Available at: https://www.gov.uk/government/publications/driverless-cars-in-the-uk-a-regulatory-review (2015)
15. Department for Transport—UK. Using the road (159 to 203)—The Highway Code—Guidance. Available at: https://www.gov.uk/guidance/the-highway-code/using-the-road-159-to-203 (2017)
16. Fernandes, L.E.R., Custodio, V., Alves, G.V., Fisher, M.: A rational agent controlling an autonomous vehicle: implementation and formal verification. In: Bulwahn, L., Kamali, M., Linker, S. (eds.) Proceedings First Workshop on Formal Verification of Autonomous Vehicles. Electronic Proceedings in Theoretical Computer Science, vol. 257, pp. 35–42 (2017)
17. Fisher, M.: An Introduction to Practical Formal Methods Using Temporal Logic. Wiley, Hoboken (2011)
18. Fisher, M., Dennis, L.A., Webster, M.: Verifying autonomous systems. Commun. ACM **56**(9), 84–93 (2013)

19. Foot, P.: The problem of abortion and the doctrine of double effect. Oxf. Rev. **5**, 5–15 (1967)
20. Hawkins, A.J.: A day in the life of a Waymo self-driving taxi—The Verge. Available at: https://www.theverge.com/2018/8/21/17762326/waymo-self-driving-ride-hail-fleet-management (2018)
21. Herrmann, A., Brenner, W., Stadler, R.: Autonomous Driving: How the Driverless Revolution Will Change the World, 1st edn. Emerald Publishing, Bingley (2018). OCLC: 1031123857
22. Law Commission–Centre for Connected and Autonomous Vehicles: Automated Vehicles: a joint preliminary consultation paper. https://www.lawcom.gov.uk/project/automated-vehicles/. Accessed 8 Nov 2018
23. Lincoln, N., Veres, S.M., Dennis, L.A., Fisher, M., Lisitsa, A.: An agent based framework for adaptive control and decision making of autonomous vehicles. In: Proc. IFAC Workshop on Adaptation and Learning in Control and Signal Processing (ALCOSP) (2010)
24. Rao, A.S., Georgeff, M.P.: An abstract architecture for rational agents. In: Proc. International Conference on Knowledge Representation and Reasoning (KR&R), pp. 439–449. Morgan Kaufmann, Burlington (1992)
25. Vellinga, N.E.: From the testing to the deployment of self-driving cars: Legal challenges to policymakers on the road ahead. Comput. Law Secur. Rev. **33**(6), 847–863 (2017)
26. Visser, W., Havelund, K., Brat, G.P., Park, S., Lerda, F.: Model checking programs. Autom. Softw. Eng. **10**(2), 203–232 (2003)
27. Wicks, E.: The right to life and conflicting interests, pp. 1–288. Oxford University Press, Oxford (2010)
28. Wooldridge, M., Rao, A.: Foundations of Rational Agency, Applied Logic Series, vol. 14. Springer, Netherlands (1999)

Radar Signal Processing Chain for Sensor Model Development

Martin Holder, Zora Slavik, and Thomas D'hondt

Originally intended as surveillance sensors in maritime and aerospace domain, radar sensors are now key sensors for many autonomous systems [1] across domains with an increasing number of applications and functionalities. Compared to lidar and other optical sensors, radar is more robust against adverse weather conditions. Because of the low specific rain attenuation at 77 GHz within automotive ranges of interest [3, 4], radar shows less vulnerability towards adverse environmental conditions in automotive applications for automated cyber-physical systems (ACPS). Its ability to measure relative velocity to targets adds valuable information to ACPS applications, which include adaptive cruise control (ACC) for the use case "Highway Pilot", blind spot detection in the use case intersection crossing, or fleet coordination for automated farming.

Radar signal processing consists of a series of consecutive steps, from analog waveform processing to machine interpretable representation of information that is ultimately perceived by the sensor. A simulation model of a radar sensor is required for the virtual testing of ACPS. From a modeling point of view, a modular structure of a sensor model is advantageous in order to achieve different simulation goals.

M. Holder (✉)
Institute of Automotive Engineering, Technische Universität Darmstadt, Darmstadt, Germany
e-mail: holder@fzd.tu-darmstadt.de

Z. Slavik
Stiftung FZI Forschungszentrum Informatik, Karlsruhe, Germany
e-mail: slavik@fzi.de

T. D'hondt
Siemens Industry Software NV, Leuven, Belgium
e-mail: thomas.dhondt@siemens.com

A. Leitner et al. (eds.), *Validation and Verification of Automated Systems*,
https://doi.org/10.1007/978-3-030-14628-3_11

This is motivated by the following:

(1) Different applications of ACPS require different data to be reported by the radar sensor. When understanding a sensor (model) as an encoder of information from a (virtual) scene, a modular approach of several different processing steps offers more degrees of freedom in the choice of system under test (SUT).
(2) Model check, verification, and validation: One requirement that arises from safety validation in virtual environments is that a sensor model must capture all potential sensing failures that the security architecture has considered harmful to overall safety of the ACPS. A modular sensor model makes it possible to investigate whether the model correctly captures sensor errors. For this purpose, sanitary checks of components can be performed throughout the data processing pipeline, which is possible with a modular modeling approach.

For a modular description of a sensor model, which is also called a *virtual sensor*, we propose the following definitions of an interface:

Definition 1 *A **functional block** comprises several analog or digital processing steps that transform input signals or data to output signals, hence a functional block is located between two interfaces (see Definition 2). The extent of the processing steps covered by a functional block may vary.*

Definition 2 *An **Interface** is a point of information exchange between two or more components where information can be exchanged on hardware level (e.g. analog signals) but also on software level (e.g. data streams). It is not required to be in a human readable form, but it must be possible to derive meaningful information by evaluating at least one information sample available at the interface. An interface can either represent an input or an output to an algorithm or functional component, but it can also be derived from an intermediate step of an algorithm to gain additional insight into the data streams within an algorithm.*

This chapter is organized as follows: The first part introduces a straightforward approach to subdivide the full radar sensor processing pipeline into separate interfaces, following Definitions 1 and 2. As with camera and lidar, these interfaces are not independent of each other, since they operate in a sequential manner, but serve to develop both simulation and stimulation models that operate at a specified interface level.

In the second part of this chapter, the proposed interfaces are compared against radar models, as found in literature. It will be demonstrated that the mechanism of our proposed interfaces integrates seamlessly into existing models. Exemplary embodiments of radar sensor models are, however, not part of ENABLE-S3.

To begin with, the terms *sensor simulation* and *sensor stimulation* will be defined as follows:

Definition 3 *A perception sensor simulation is a replacement of a real environment perception sensor by a software model, in which sensor data from a virtual scene are faithfully reproduced.*

Definition 4 *A sensor stimulation replaces at least one input of the real sensor with synthetic data, while the remainder of the real sensor remains unchanged. At the same time, stimulation models can also be applied to virtual sensors for stimulating certain inputs.*

Unless otherwise stated, this chapter focuses on frequency modulated continuous wave (FMCW) radars with linear chirp sequence waveforms that dominate ACPS in many areas, such as automotive and robotic systems. Despite differences to radars in other domains like aerospace and maritime, where requirements for the measurement range necessitate other measurement (e.g. Pulse-Doppler) and antenna principles (e.g. mechanical scanning), the proposed interfaces are comparable after signal acquisition and passing low-level radio frequency (denoted as RF) stages. Accordingly, the following interface definitions are described on an abstract level that makes them applicable to a variety of different radar systems.

1 Radar Sensor Principle

Radar sensors that are used in ACPS comprise of both hardware and software components that may be understood as a system, as defined in Definition 5:

Definition 5 *A* ***radar system*** *comprises at least one radar sensor and a signal and data processing pipeline, which both work as a compound of multiple software and hardware components. While the output signals of the radar sensor (referred to as "radar measurements") are considered as "raw data", the signal and data processing transforms the signal in such a way that it becomes interpretable and can be used for a measurement purpose (e.g. identification and tracking of objects in the sensor vicinity).*

In the following a breakdown of a radar system into three different parts according to Definition 5 is derived. A radar system is displayed in Fig. 1. It can be seen that it consists of three parts that are composed as follows:

(1) Radio Frequency (RF) frontend: It comprises at least one transmit and receive antenna, which may be operated in a MIMO (Multiple Input Multiple Output) scheme or as a phased array. The included analog processing contains all required analog processing steps such as low-pass filters, local oscillators for conversion between baseband and carrier frequency, and the mixing of transmit and receive signals to obtain the beat signal representing the spectral difference between transmitter and receiver side. The beat signal is the first signal that can be sampled with reasonable effort due to its low-frequency components. The last processing step is analog to digital conversion.

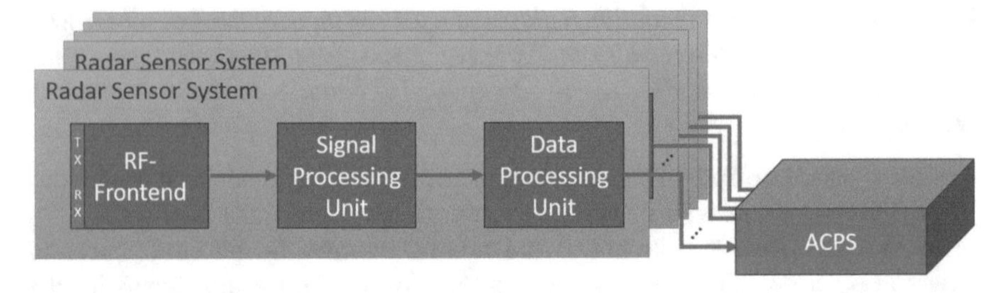

Fig. 1 Several radar systems applied to a single ACPS

(2) Signal Processing Unit (SPU): The discretized beat signal is transformed to the frequency domain, where distance velocity and azimuth range are determined based on the frequency analysis. It is seen as the block that converts radar signals into radar data, with a first loss of information.
(3) Data Processing Unit (DPU): The derivation of information about reflective targets near the radar requires additional processing of the SPU output. Unfortunately, radar data are prone to artifacts caused by aliasing effects, for example. By using different filtering and estimation techniques, the DPU calculates relevant information for the sensor task based on the output signal of the SPU, but also on additional information about the specific application and the use of the radar. This is particularly important for context-dependent classification and interpretation of objects.

An important difference between radar simulation and stimulation is apparent on the system level perspective: Radar stimulation is applied to any sensor interface that is relevant to specific testing objectives, while the real or virtual radar sensor itself remains self-contained. Stimulation can take the form of physical stimulation such as corner reflectors or interfering waveforms, or virtual stimulation with virtual targets provided by environment simulation or filter systems representing virtual environment conditions or scattering. Simulation, in contrast, is a representation of a radar sensor or radar system that can fulfill different levels of fidelity and complexity. The individual components within a radar sensor simulation are represented either with physical-mathematical models or with statistical model descriptions.

In accordance with Definition 1 and 2, the radar system comprises the following interfaces:

Transmitted Signal Being an active sensor, a radar actively transmits electromagnetic waves via at least one transmitting antenna, which illuminates its surroundings. The transmitted signal is denoted as IF1 and an analog, time-domain signal.

Received Signal While the signal propagates through the environment, it is attenuated and reflection occurs on scattering surfaces. The full propagation path between sender and receiver is denoted as radio channel and background material for modeling approaches is abundantly available in the channel model literature.

A mathematical description reads (* denotes convolution): $IF_2(t) = h(t) * IF_1(t)$ and it can be seen that IF2 is also an analog signal in time domain. This signal is of particular importance in sensor stimulation, where it is synthesized with respect to the transmitted signal (IF1) and the channel information $h(t)$.

Beat Signal After being received by at least one antenna, RF processing is performed which includes baseband conversion, filtering and analog-to-digital conversion (ADC). While pulsed radar systems perform correlations within the sensor unit, FMCW-based radar systems compare the received signal with the transmitted signal to provide the so-called beat signal and thus providing the sum of range- and velocity-related frequency shifts. The low-frequent beat signal is sampled by ADCs and is the input to the SPU on IF3, which transforms the signals into data.

A fast Fourier tranformation (FFT) is performed after applying an appropriate window function to the beat signal, where prospective targets, whether real targets or measurement artifacts (e.g. ghost targets or clutter) settle as peaks in the Fourier spectra. In order to distinguish between false and real targets, a dynamic threshold is required. Because the regular constant false alarm rate (CFAR) method cannot handle extended targets that especially occur in the near field, OS-CFAR techniques are widely used [5].

From the point of view of sensor modelling, the threshold formation of the periodogram is a remarkable processing step, since it is the first processing step that has lossy properties, since this step cannot be undone. On the one hand, peaks that are filtered out during threshold detection can no longer be recovered, which can be considered a loss of information. On the other hand, ghost targets or artifacts can emerge, enriching the information from the backscattered radar wave, although there is no assessment of the correctness of the information yet.

Table 1 lists all defined interfaces. They are described in more detail below. Unless otherwise stated, the information available at each interface is updated at each time step, depending on the measurement and simulation cycle. In addition, it

Table 1 Definition of radar interfaces

Interface	Name	Signal domain and brief description
IF1	Transmitting antenna	Analog, time-domain signal, RF-Signal Physical @ GHz
IF2	Receiving antenna	Analog, time-domain signal, RF-Signal Physical @ GHz
IF3	Beat signal	ADC output, Digital, time-domain signal @ MHz
IF4	Spectral periodogram	Digital spectral signal
IF5	Peak list	List with a finite number of entries with the (interpolated) bin number of peaks. List with a finite number of entries, Signal level
IF6	Target list	List with a finite number of entries, Peaks associated to targets, ghosts targets and artifacts are removed by data processing algorithms
IF7	Tracked object list	List with a finite number of entries, Tracked object list

is assumed that the time behavior of the simulation/stimulation model in terms of update rates and measurement cycle duration corresponds to the real sensor.

2 Modeling Considerations for Simulating on Different Radar Sensor Interfaces

Radar perception involves a number of processing steps, as shown above. Depending on the objective of virtual testing of ACPS, there is an increasing interest in radar sensor models that provide either a tracked object list, e.g., for perception testing, but also less processed data such as beat signals or peak lists for testing algorithms that consume such data.

Between environment illumination and a final radar output, several functional stages are passed that affect the characteristics and representation format of information that is contained in an illuminated scene. From electromagnetic waves on the one side, the information is first transformed into a digital signal until it finally becomes high-level data describing the surroundings.

The transformation is represented in several processing steps assigned to interfaces (IF) 1–7. Figure 2 depicts the interfaces that will be described in the following. Only the antenna interfaces IF1 and IF2 are left out as their location in the signal processing chain is already illustrated in Fig. 1.

The sequential radar processing steps are explained in the following using an exemplary radar measurement that illuminates the environment scene depicted in Fig. 3. There are two main targets present in the scene, namely an extended target, which is an iron fence to the left and a car to the right.

2.1 Interface 1/2 (TX/RX Antenna)

Today, antennas for automotive radar systems are usually realized with microstrip patch elements [6], which offer a convenient mechanical integration level. On the transmitter side, IF1 broadcasts transmit waveforms and illuminate the surrounding environment scene. The co-located receiving antennas at IF2 collect the reflected electromagnetic backscatter. Depending on the geometry and orientation of the patch elements, different areas are illuminated, but spatial diversity can also be achieved by using phased arrays or MIMO antennas. Both transmit and receive

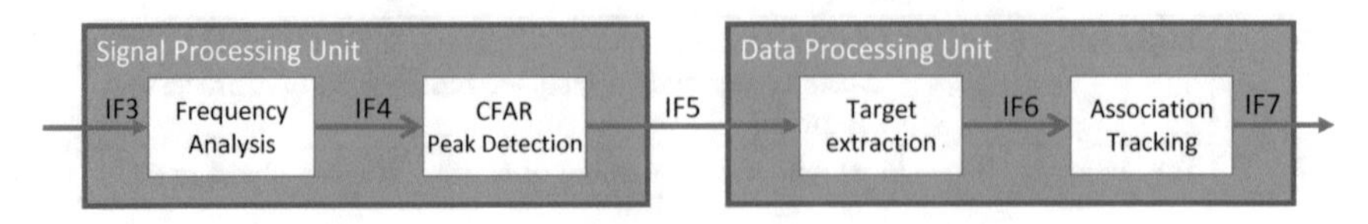

Fig. 2 Block diagram of radar interfaces according to the definitions in Table 1.

Fig. 3 Exemplary environment for interface illustration

antennas are operated at the carrier frequency of about 77 GHz and add an antenna gain to the outgoing and incoming signals. The beam width results from the patch element geometry of an array alignment. This interface is usually hardly accessible and sensor models are more generically defined, including azimuth and treble width, gain, bandwidth and system noise and losses.

The radar backscattering at IF1 contains not only the reflection of targets but also information about the environment, such as clutter inducing surfaces or signal propagation channel characteristics that depend on weather conditions. Accordingly, IF1 and IF2 are of particular interest for stimulation purposes. Before the transition to the next function block, there are several analog processing steps between IF2 and IF3. After filtering, amplification, and baseband down-conversion, the received signal is compared with the transmitted signal by mixing, resulting in the beat signal. After sampling, the beat signal is provided on IF3 to the SPU. In Fig. 4 the beat signal is depicted for a radar system with 4 transmit and 8 receive antennas, resulting in 32 virtual antennas. The sampling rate for all eight receive channels is 80 MHz, hence each receive channel is sampled with 10 MHz. 1024 samples are stored per channel and sweep.

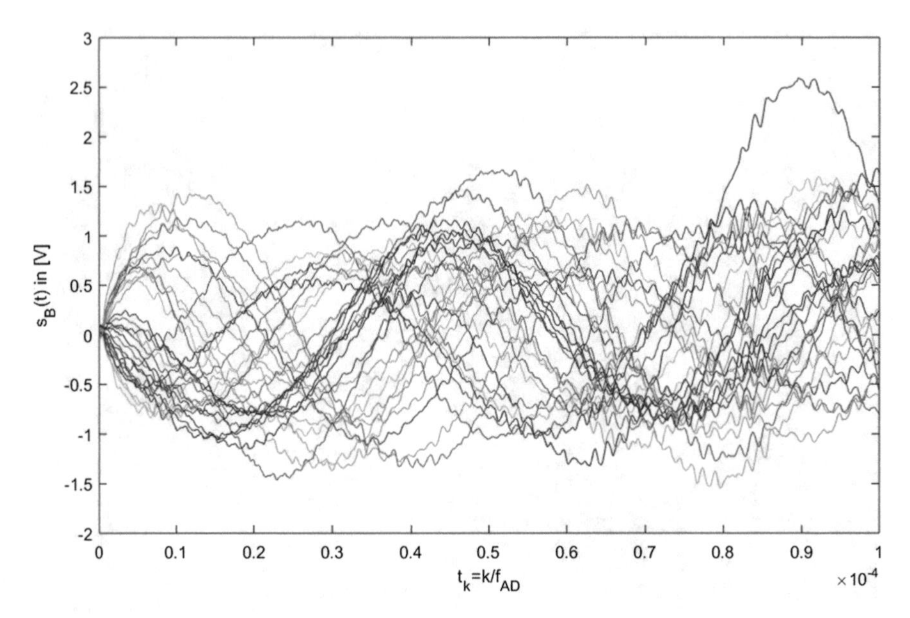

Fig. 4 Beat signal of 32 virtual channels at the ADC output of a 77 GHz radar

2.2 Interface 3 (Beat Signal)

The discrete beat signal is the lowest level digital signal in radar systems. It consists of frequency shifts resulting from the propagation time and a Doppler for moving reflecting targets. As depicted in Fig. 5, the beat signal carries a beat frequency f_B that results from the frequency difference of IF1 and IF2. Its components need to be

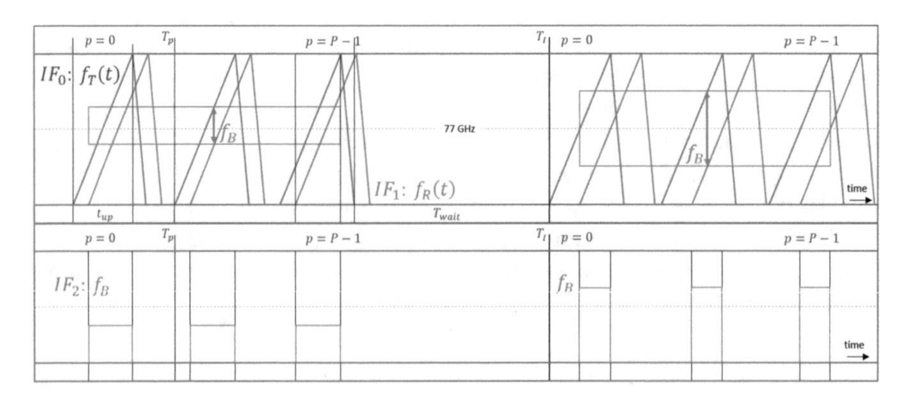

Fig. 5 Computation of beat signal (IF3) using IF1 and IF2: the relation between target distance (i.e. the delay between IF0 and IF1) and the resulting beat frequency (IF2) can be seen

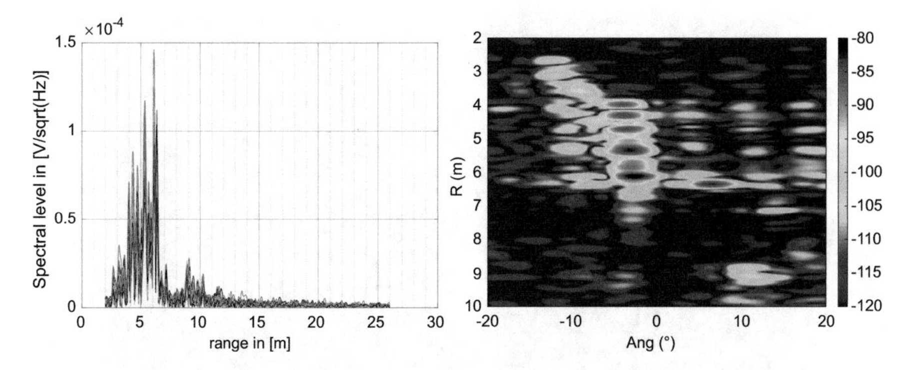

Fig. 6 Results from frequency analysis. Left: Range profile. Right: Range-Azimuth map

resolved regarding range, velocity and azimuth during the frequency analysis within the SPU.

2.3 Interface 4 (Range-Doppler-Azimuth Spectra)

After processing IF2 using frequency analysis, IF3 allows for inference of occupied range, azimuth and Doppler bins. Figure 6 shows a range profile as a result of the first FFT along the modulation ramps. This FFT is also called the "distance FFT". The application of a second FFT above the azimuth antenna array shows the angular position of the spectral peaks, see Fig. 6. The same procedure is used to obtain the Doppler information contained in the beat signal. FFT processing includes window functions that can lead to artifacts and add blur to the periodogram that can affect subsequent peak separation and clustering. In addition, the FFT resolution should be adapted to the bin resolution inherent in the system for range, azimuth and Doppler direction.

2.4 Interface 5 (Peak List)

After frequency analysis, the resulting spectra from IF4 need to be interpreted. From the previous interface, which provides spectral analysis in different directions (range, azimuth, Doppler), intensities at different locations are obtained. As a basis for the interpretation of interpreting the received radar signals, a peak list is generated from the frequency analysis, which extracts relevant spectral signal levels from IF4. This is achieved by threshold algorithms and spectral interpolation, which could, however, introduce aliasing, leading to ambiguous information.

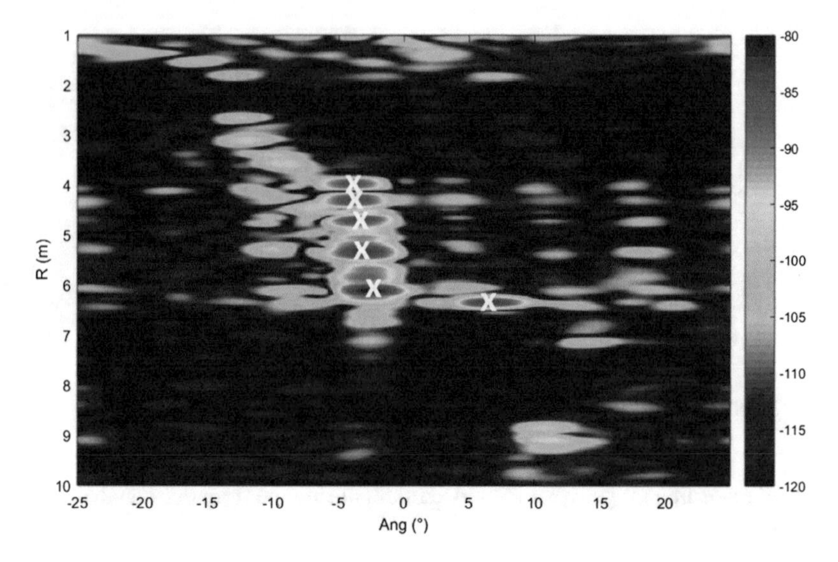

Fig. 7 Extracting targets from the range-azimuth map

2.5 *Interface 6 (Target List)*

Identified peaks in the spectra can be further processed to physical dimensions under consideration of the measurement resolution (i.e. bin width). The peak list contains every target that passed the detection threshold. This naturally results in both real reflections and false alarms or unwanted peaks caused by surface noise, measurement artifacts or noise. After filtering steps, which include de-aliasing, including resolution of ambiguities and compensating of further sensor-specific effects, identified peaks may be arranged in a target list, which has a lidar point cloud as a most similar counterpart. In Fig. 7, targets are marked exemplarily in a range-azimuth map. At this point, the targets are neither clustered nor interpreted as objects. This only happens in the following step, which realizes the transition from a target list to a tracked object list.

2.6 *Interface 7 (Tracked Object List)*

In object tracking, the target list that is available on IF6 is clustered so that target belonging to the same object are assigned correctly. In Fig. 8 on the left, targets for two different objects are clustered, while the structure of the targets in the range-azimuth map reveals that one object is an extended target. Compared to the scenario depicted in Fig. 8 on the right, the extended target can be matched to the iron fence, while the rather point-shaped object turns out to be a car. It should be noted that a

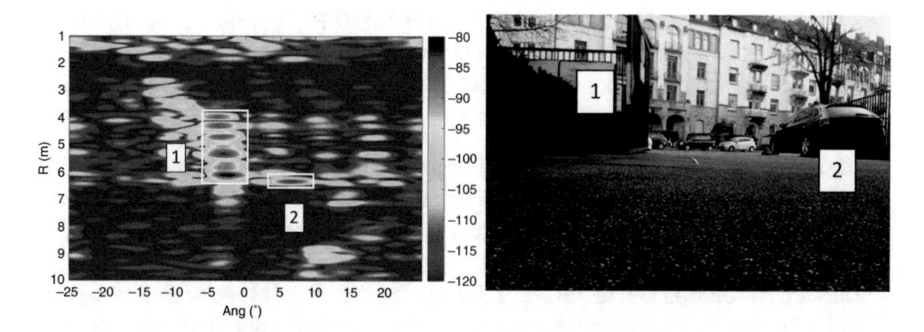

Fig. 8 Ideal output of tracked object list with separated and classified tracked objects 1 and 2

tracking filter algorithm for debugging purposes can output additional information prior to data allocation, which can serve as an "untracked" object list.

3 A Survey on Different Radar Sensor Models and Their Respective Interface

In the last part of this chapter, we review radar sensor models that have been reported outside the ENABLE-S3 context. We evaluate the models with respect to the proposed interfaces that were introduced earlier in the chapter and have been published previously in [7].

Many approaches for radar sensor modeling have been reported in literature. Indeed, in modeling and simulation tasks, the right trade-off between simulation realism, parametrization complexity, and computational speed must be found, see Fig. 9. Therefore, a broad range of simulation models have been studied, each focusing on different applications and fidelity requirements.

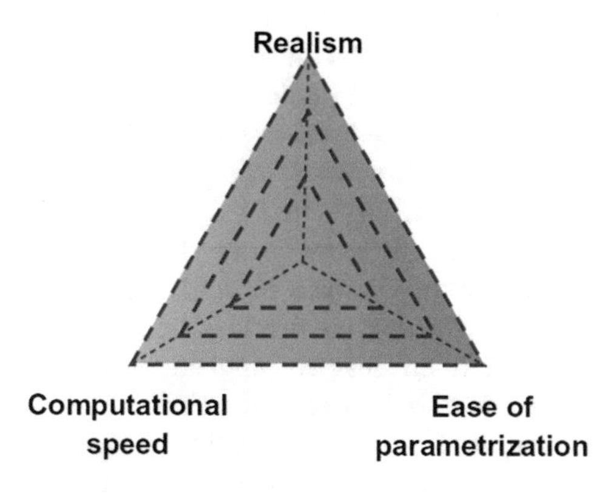

Fig. 9 Design targets for sensor models

Within a project-wide agreement in the ENABLE-S3 project, the following characteristics of a radar sensor are considered as substantial in order to describe a model as "realistic". Without discussing its validity, a realistic radar sensor model should prove that the following effects are captured:

- Multipath Propagation: An occluded object can be detected (either as an object or as a target) due to multipath propagation, which is closely linked to wave propagation and the effects are present on all interfaces after signal perception (IF2 and above).
- Limited Resolution and target separability: The achievable resolution and separability between targets in either measurement domain may not exceed the sensor's specification. This is of particular importance for point targets modeled on IF4 but it is not required on object list level (IF7): Due do filtering algorithms and estimation methods, the reported resolution of an estimated object's pose may be higher than the sensor measurements as estimation basis.
- Aliasing Effects: When exceeding unambiguous measurement intervals, aliasing effects occur, which need to be resolved for correct interpretation of measurements. Aliasing is often respected in IF2 onwards as it allows to lower the data rate and is ultimately resolved on the object list representation (IF7).
- Variation of received power: The received power from a target is subject to a non-stochastic variation composed of multiple effects, such as constructive or destructive wave interference. These effects are already present on the signals on IF2 onwards but the influences are gradually removed during signal and data processing due to smoothing properties of tracking filters.
- Antenna Directivity: The area of detection from the antenna beam width in azimuth and elevation; according aperture and realized gain on transmitter and receiver side. This has an impact on all interfaces, as it basically dictates the probability of target detection with respect to the position of the target relative to the sensor.

Computational speed is understood on an ordinal scale, i.e. "faster than" or "slower than", normalized to real time. Ease of parametrization reflects the ability of a sensor modeling approach to be applied to different radar sensors without the need for exhaustive modifications to the model.

Historically, the first sensor models studied are often simplified abstract models, also called ground-truth or **object list sensor** models. Those models filter ground-truth information on object level, which is directly available in the environment simulation, and emulate idealized behavior of the radar system. Such models do not exhibit errors in detection and measurement of object states, and output ideal simulation values in the format of a radar object list (range, range rate, azimuth, etc.), which corresponds to radar data obtained after undergoing all of the interfaces listed above. Examples of filters that are applied to the ground-truth data [8] include the parametrization with respect to the sensor's measurement range and the geometrical obstruction of different actors in the environment. Even though those models typically lack most of the sensor characteristics observed in real measurements, they are easy to parameterize and fast to execute and do not impose additional, sensor-specific requirements on the virtual environment. This makes them useful in early design phases in order to validate the functional behavior of an ACPS either in ideal conditions or under the assumption that no perception sensor errors must be considered. As indicated by its name, this kind of sensor model produces data on object-list level (IF7).

Phenomenological models offer an intermediate level of simulation fidelity. They try to emulate the physical behavior of the radar through the usage of statistical models, which are inferred from previous observations, expert knowledge, simplified physical equations, or maps and lookup-tables, which are applied to ground-truth simulation data. Therefore, they can be seen as an extension of the ground-truth radar sensor models and include typical sensor characteristics such as measurement noise, clutter detections, the variation of the detection amplitude as a function of the target position and orientation, and limited resolution [9]. They offer some physical insights into the behavior of the system while being reasonable to parameterize and sufficiently efficient to execute in real time or even faster. In contrast to object-list models, sensor-specific properties, such as noise behavior, are captured in the sensor model itself. Additional parameters that are consumed from the virtual environment are typically a (predefined) RCS profile, information about the level of occlusion, overhead clearance of a vehicle (for estimating the likelihood of detecting occluded objects) and information, which allows for a rule-based behavior (e.g. probability of detection varies across different object classes). Since phenomenological models take information about the virtual environment into account, it is possible to simulate effects such as mirror targets occurring at guardrails. Another notable difference to ideal models is that phenomenological models for synthesizing both object list (IF7) but also radar targets (IF6) exist. Phenomenological models relax the assumption of a perfect, error-free environment perception and account for errors in measurement accuracy. A recommended use case is virtual testing if uncertainties in the environment perception algorithms should be considered.

Models that aim for simulating additional physical phenomena in a precise manner under consideration of mathematical models of sensor physics are referred to as **physical models**. They allow for a more realistic modeling of effects that can lead to errors in the environment perception, such as multi-path propagation, ghost targets, and interference. The ray tracing algorithm, as known from image synthesis, is often deployed in that kind of model in order to simulate the propagation of electromagnetic waves in the environment using asymptotic approximations, such as geometrical or physical optics [10, 11]. Their ability to scale well on different kinds of scenarios makes them likely to synthesize a more realistic sensor reading from a virtual scene [12, 13]. However, such methods have several limitations: First, they are computationally expensive, limiting their use in real-time applications. Speed of execution is a limitation when considering the sheer number of scenarios that must be simulated in order to validate an ACPS. It should be noted that recent developments in GPU-accelerated ray tracing are gradually reducing this limitation. Second, the environment model requires a high level of detail, particularly in the geometry and material properties of all static and dynamic objects. This detail is difficult and expensive to measure. Third, to obtain an object list, ray tracing models must execute all procedures in the radar pipeline, which for commercial sensors are typically proprietary and thus not accessible. Physical models usually serve IF2–IF4, while simulation models originating in the domain of computational electromagnetics are able to simulate IF1 and IF2.

Data-driven modeling approaches try to bypass these requirements on model and environment parametrization by using black-box sensor models, which approximate radar outputs from real-world sensor readings. For instance, previous work has studied the computation of a joint probability distribution linking the ground-truth position of a target to the corresponding measured sensor [12]. Sensor-specific characteristics can hereby be learned directly from real-world data, as long as the data set is representative of those phenomena. Furthermore, black-box sensor models do not require computationally expensive ray-tracing techniques. An approach using variational auto-encoders for estimating the radar power field was also investigated and delivers radar data on IF3 [14]. In contrast to phenomenological models, which may be considered as grey-box, data-driven models aim to regress the true distribution by sensor readings given certain inputs and they are often parametrized from large-scale recordings with exhaustive ground truth.

A sensor model category used to a lesser extent are **scattering center models**. They originate from work executed for simulation environments. Real-world measurements show that echoes of a typical passenger car can be summarized by a set of characteristic scattering points, which result in an equivalent but simplified object representation [15]. Therefore, scattering center approaches try to offload the complex computations required to compute this reduced set of reflectors offline and use the simplified object representation in order to enable fast and accurate online computations [16, 17]. While being an attractive choice when serving as a proper measurement model in extended target tracking applications, this type of model is difficult to scale, as it forces the user to compute equivalent models for every type of object in the environment. Additionally, equivalent models of static targets such as trees or sign gantries have yet to be developed. Ultimately, common road objects, such as guardrails, bridges, and tunnels, which are known to have a large impact on wave propagation behavior, are not covered by these models.

References

1. Meinel, H.: Evolving automotive radar – From the very beginnings into the future. In: The 8th European Conference on Antennas and Propagation (EuCAP 2014), The Hague, 2014
2. ITU-R: Propagation data required for the design of terrestrial free-space optical links. ITU, Geneva, 2012
3. ITU-R: Specific attenuation model for rain for use in prediction methods. Recommendation ITU-R P.838-3, Geneva, 2005
4. Hassen, A.A.: Indicators for the Signal Degradation and Optimization of Automotive Radar Sensors Under Adverse Weather Conditions. Technische Universität Darmstadt, Darmstadt (2007)

5. Meng, X.W., Qu, D.C., He, Y.: CFAR detection for range-extended target in Gaussian background. J. Syst. Eng. Electron. **27**, 1012–1015 (2005)
6. Alexandrov, N., Rabinovich, V.: Antenna Arrays and Automotive Applications. Springer, New York (2013)
7. Holder, M., Makkapati, V., Rosenberger, P., D'hondt, T., Slavik, Z., Maier, M., Schreiber, H., Magosi, Z., Winner, H., Bringman, O., Rosenstiel, W.: Measurements revealing challenges in radar sensor modeling for virtual validation of autonomous driving. IEEE International Conference on Intelligent Transportation Systems, 2018
8. Roth, E., Dirndorfer, T. J., Knoll, A., et al.: Analysis and validation of perception sensor models in an integrated vehicle and environment simulation. In: Proceedings of the 22nd Enhanced Safety of Vehicles Conference (ESV), 2011
9. Bernsteiner, S., Magosi, Z., Lindvai-Soos, D., et al.: Radar sensor model for the virtual development process. ATZelektronik Worldwide. **10**(2), 46–53 (2015)
10. Ling, H., Chou, R.C., Lee, S.W.: Shooting and bouncing rays: calculating the RCS of an arbitrarily shaped cavity. IEEE Trans. Antennas Propag. **34**(2), 194–2015 (1989)
11. Herz, G., Schick, B., Hettel, R., Meinel, H.: Sophisticated sensor model framework providing real sophisticated sensor model framework providing realistic radar sensor behavior in virtual environments. Graz Symposium Virtual Vehicle, 2017
12. Hirsenkorn, N., Hanke, T., Rauch, A., Dehlink, B., Rasshofer, R., Biehl, E.: Virtual sensor models for real-time applications. Adv. Radio Sci. **14**, 31–37 (2016)
13. Maier, M., Makkapati, V.P., Horn, M.: Adapting phong into a simulation for stimulation of automotive radar sensors. IEEE MTT-S International Conference on Microwaves for Intelligent Mobility, 2018
14. Wheeler, T.A., Holder, M., Winner, H., Kochenderfer, M.J.: Deep stochastic radar models. IEEE Intelligent Vehicles Symposium, 2017
15. Schuler, K., Becker, D., Wiesbeck, W.: Extraction of virtual scattering centers of vehicles by ray-tracing simulations. IEEE Trans. Antennas Propag. **56**(11), 3543–3551 (2008)
16. Bühren, M., Yang, B.: Simulation of automotive radar target lists using a novel approach of object representation. IEEE Intelligent Vehicles Symposium, 2006
17. Buddendick, H., Elbert, T., Hasch, J.: Bistatic scattering center models for the simulation of wave propagation in automotive radar systems. German Microwave Conference Digest Papers, 2010, pp. 288–291

Functional Decomposition of Lidar Sensor Systems for Model Development

Philipp Rosenberger, Martin Holder, Marc René Zofka, Tobias Fleck, Thomas D'hondt, Benjamin Wassermann, and Juraj Prstek

In this chapter, results of the lidar sensor modeling workgroup within ENABLE-S3 are presented. The main objective is to describe the commonly agreed general functional blocks and interfaces of lidar sensor systems for object detection. Having the in the following described interfaces at hand, requirements for the generation of synthetic lidar data at specific interfaces can be formulated and modeling as well as verification and validation of the simulation can be performed.

P. Rosenberger (✉) · M. Holder
Institute of Automotive Engineering, Technische Universität Darmstadt, Darmstadt, Germany
e-mail: rosenberger@fzd.tu-darmstadt.de; holder@fzd.tu-darmstadt.de

M. R. Zofka · T. Fleck
Stiftung FZI Forschungszentrum Informatik, Karlsruhe, Germany
e-mail: zofka@fzi.de; tfleck@fzi.de

T. D'hondt
Siemens Industry Software NV, Leuven, Belgium
e-mail: thomas.dhondt@siemens.com

B. Wassermann
TWT GmbH Science & Innovation, Stuttgart, Germany
e-mail: benjamin.wassermann@twt-gmbh.de

J. Prstek
Valeo Autoklimatizace k.s., Prague, Czech Republic
e-mail: juraj.prstek@valeo.com

A. Leitner et al. (eds.), *Validation and Verification of Automated Systems*,
https://doi.org/10.1007/978-3-030-14628-3_12

1 Motivation for Interface Definitions by Functional Decomposition of Lidar Sensor Systems

Lidar sensors are seen as one of the key components for automated driving (AD) as well as for highly automated driving functions (HAF), such as a highway pilot. They are active perception sensors, which enable to perceive a vehicle's environment by emitting light and receiving its reflections. The resulting information is currently subject to research for the application in obstacle detection and classification, simultaneous localization and mapping (SLAM), but also ego-motion estimation.

In this chapter, results accomplished by the lidar sensor modeling workgroup within ENABLE-S3 are presented. The partners forming the workgroup all individually work on lidar sensor models of different extent and complexity for different types of lidar sensors and output data. Therefore, the first action of the workgroup was to find a common understanding for describing lidar sensors.

The following chapter presents the definitions of common interfaces and functional blocks in between that were elaborated and agreed upon. Furthermore, the interfaces, which were jointly identified by functional decomposition of lidar sensor systems for object recognition, are listed and described. As the interface specification is derived from inspected, available hardware components and agreed on within the ENABLE-S3 project, the authors propose an interface specification of lidar sensor systems for the application in virtual and real validation and verification. In fact, the standardization of sensor model interfaces is already ongoing, as e.g. the Open Simulation Interface (OSI) [1] originating from the German project PEGASUS [2]. An exchange has been established to this related work to ensure consistency.

Additionally, the definitions serve as a basis for the formulation of requirements, for the implementation, as well as for comparison, benchmarking and validation of sensor models that output data through one of the interfaces. Therefore, the authors propose the sequence of interfaces, as functional decomposition is performed exemplarily on the processing chain for object perception with lidar sensors. The overall goal of the partners forming the workgroup for lidar sensor modeling within ENABLE-S3 is, besides implementing different models, to develop measures for rating model quality and for benchmarking different modeling approaches. Based on the derived commonly agreed interfaces, now, metric definition and model validation against real data as well as benchmarking of different models can be performed.

The following section starts with a general introduction to lidar sensors and their physical measurement and signal processing principle. Then, an overview of the agreed interfaces is given. After this overview, the interfaces are described in more detail. Finally, after a conclusion, possible next steps are proposed.

2 Lidar Sensor Principle

Light detection and ranging (lidar) is an optical measuring technique for the localization and range measurement of surrounding objects. The active sensor emits light—usually from lasers—into the environment at a specific wavelength and measures the resulting echo. The time from transmission of the light (laser) impulse to the time of reception of the reflections is proportional to the radial range between the sensor and the detected object [3]. This principle is called "time of flight" measurement and the radial range r can be calculated by $r = \frac{1}{2}c_0\tau_{of}$, where c_0 is the velocity of light and τ_{of} is the time of flight of the laser beam.

In principle, the operation of a lidar sensor is similar to other active environment sensors like pulse radar, but it uses infrared light instead of e.g. microwaves. This results in an increased angular resolution and a higher atmospherical attenuation due to its shorter wavelength. The attenuation, the properties of the reflective surface, and the power and divergence of the emitted beam determine the ratio of the powers of the received echo over the transmitted power. Depending on the sensitivity of the lidar, this results in a maximum range until which an echo can be detected.

Furthermore, the view angle occupied by a target of a given size decreases as its distance from the lidar increases. Therefore, objects are less likely to be detected at longer distances. Besides, lidar sensors are sensible to particles present in the range between the sensor and the hit obstacle, such as rain, spray or snowfall. Still, automotive lidar sensor systems should be able to recognize that and suppress their effect. Additionally, the emitted laser beam does not stay infinitesimally thin in reality, as illustrated in Fig. 1. E.g., in the case of Ibeo 2010 sensors, the divergence is 0.8° in vertical direction and 0.08° in horizontal direction [5].

Beam divergence causes its energy to be spread over a wider area for higher distances. This spread, together with the already described generally smaller angular coverage by objects and the attenuation by the atmosphere leads to a lower detection probability for objects at higher distances. Furthermore, from a modeling perspective, the number of different materials and the complexity of shapes that get hit by a single beam rise at longer distances, which leads to higher requirements

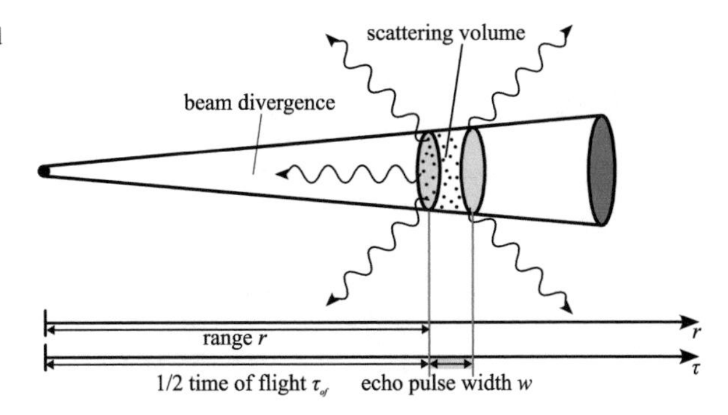

Fig. 1 Beam divergence and echo pulse width, based on [4]

for the sensor model and the modeled environment if beam divergence should be included.

Lidar uses pulse modulation, where the source of the light signal is a semiconductor diode that emits photons when subjected to an electrical current. The sensors can be grouped into two different families depending on the way they spread this light over their field of view (FOV). At the moment, most existing lidars scan the environment by moving complete parts of the sensor system (e.g. Velodyne HDL-32E) or a mirror inside of them (e.g. Valeo SCALA, Ibeo LUX, etc.) using a small electric motor [6]. Moving parts typically increase a sensor's size and cost, while reducing its reliability and robustness for mobile application.

The other group is formed by 3D-flash-lidar or solid-state-lidar sensors, which use non-mechanical scanning elements, such as Optical-Phased Arrays (OPAs). They allow for precise beamforming and steering of the light in order to scan the FOV of the sensor, but also for pointing at or tracking specific targets. OPAs can use liquid crystals [7], MEMS technology [8] and silicon photonics waveguides [9]. This results in a smaller, cheaper and more rugged package. From a modeling perspective, the technical implementation chosen does have an influence on effects like motion blur and rolling shutter, as they are described e.g. in [10].

An example for a signal received by a single photodiode, as it would be available after analog-to-digital conversion, is illustrated in Fig. 2. The first parameter to be considered here is the sampling rate of the signal converter as it can lead to different signal shapes, depending on the ratio of the dominant frequencies in the received signal to the sampling rate. Besides, another solution to collect the echoes can simply be a comparator and a counter. With these, sampling is not needed, whereas intensity measurement is not possible. The counter or the sampling rate, depending on the method to measure the echo, influence the radial resolution of the sensor as resolution in time means resolution in range with the used time-of-flight principle. In the obtained signal in the time domain, peaks are identified by a threshold (here U_{th}) that is applied to the signal. Mostly, the threshold is variable to react to occurring noise and depends on the Signal-to-Noise Ratio (SNR).

Additionally, Fig. 2 shows another signal parameter, which is the echo pulse width (here w_A, w_B, w_C). It directly relates to the radial extent of the hit target, as reflections caused by the target reach the sensor throughout this time, as shown

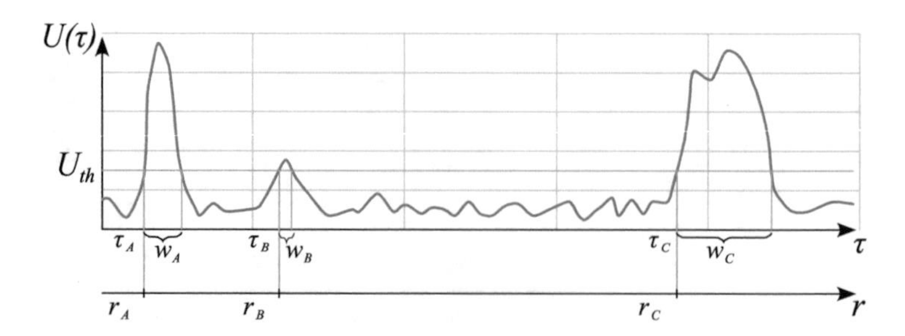

Fig. 2 Shape of the received signal in the time domain, based on [5]

in Fig. 1. By summing up the sampled signal (here $U(\tau)$) over the duration of the echo pulse width, the signal intensity for the particular detection can be obtained. In Fig. 2, the time axis starts at the transmission of the beam, so τ_A, τ_B, τ_C represent the different times of flight for the three targets A, B, C. In consequence, on the second abscissa, r_A, r_B, r_C are the corresponding ranges to the particular targets.

Finally, the number of echoes the sensor keeps has an influence on the provided raw scan, while a scan in this context refers to a completed collection in angular range. If the received signal for a single pulse contains three echoes as in Fig. 2 and the sensor keeps all of them in the provided scan data, no decision has to be made. However, if the scan contains fewer echoes per pulse than are received, there needs to be a rule on which echoes should be kept (e.g. the most intense or the nearest).

3 Overview of Lidar Sensor Systems

The following schematic describes the different processing steps in a generic lidar sensor system (Fig. 3), resulting from a functional decomposition as introduced in the context of safety validation for automated driving in [11]. Among the lidar sensor simulation workgroup within ENABLE-S3, different interfaces have been defined, which are referred to by the abbreviation IF_n. A typical object detection and classification process is chosen to serve as the basis for functional decomposition. The interfaces and the functional blocks in between are further described in the following section.

A generic lidar sensor system is divided into the front-end and a data processing unit. The former handles emission and reception of the laser beams, as described in the previous section about the lidar sensor principle.

Depending on the type of sensor system, the output of the front-end, the so-called raw laser scans, may directly be provided as an output to the user. Therefore, the front-end outputs a raw scan of the environment as 3D-points in spherical

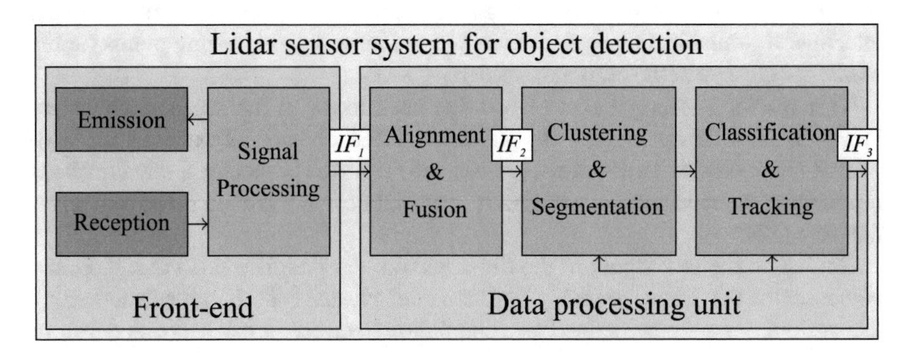

Fig. 3 Overview of the functional blocks of a lidar sensor system

coordinates at interface IF_1. This allows fusion of low-level sensor data with data of other environment sensors like radar detections and is currently subject to research.

It would be possible to simulate the raw signal in the time domain, as shown in Fig. 2. But, as this is not very common right now and mostly not available in current simulation tools, the lidar workgroup within the ENABLE-S3 project decided not to include this as a specific interface of the lidar sensor system to be simulated.

While computing the raw scans, intrinsic calibration of the calculated points is performed, taking into account the alignment of the diodes and their slightly different signal intensities, as well as the timing of the actual sending and receiving process. In addition, these data are used in order to distinguish between different kinds of objects. Nevertheless, they can also be used for further functionalities, such as simultaneous localization and mapping (SLAM), which is not investigated further in the scope of this project.

Lidar sensor systems for object detection, as inspected here, include further data processing steps for handling the raw scans. In most cases, at first, the raw scans captured at different moments in time by a single sensor are aligned with other sensors mounted on the same vehicle, meaning extrinsic calibration is performed. This addresses two basic issues:

Registering laser scans in time is necessary for lidar sensor systems that incorporate a motor unit and moving base. Registering laser scans in space is necessary for multi-sensor systems, which are typically used to achieve extensive coverage of the environment. Temporal alignment is also important here.

The data of the sensors are then transformed and fused into a single coordinate frame, which results in a point cloud at IF_2. Therefore, the intrinsic parameters of the lidar system (azimuth and elevation angle, ...) are used in order to map the raw scans from a spherical coordinate system into a metric, locally referenced Cartesian frame. This includes the alignment due to the movement of the car in between different measurements during one scan.

Based on the (fused) point cloud, different reflected points from the same object in the environment are then grouped together in clusters, while others are discarded during segmentation. This is realized based on the proximity of the points in the point cloud, as well as the historical data computed during previous measurement intervals. If available, those prior detections are indeed good starting points for the search for clusters at the current time step.

This results in the addition of two feedback loops in the sensor architecture in Fig. 3. In order to extract the positions of the objects surrounding the ego-vehicle (e.g. vehicles, pedestrians, cyclists, etc.) from the point cloud, the reflections returned by the road surface are usually removed during a ground plane segmentation step [12].

Similar to the raw signal in the time domain, the interface that could possibly be chosen at this stage, probably by the name "cluster list", is not considered by the working group to be defined as a simulation interface. It can serve as debug or inspection interface for function development, but is not considered to be served by actual simulation tools. Also, data processing steps often cannot be separated easily, due to their complexity and interweaving of and within further processing steps like tracking algorithms.

Tracking algorithms use state estimators in order to associate detections resulting from the same object over time. The combination of historical data and state estimation algorithms allows estimating quantities that are not directly measured by the lidar front-end, such as orientation, speed, acceleration or geometrical extents of a target [13, 14].

Finally, (tracked) clusters can be sorted into a finite set of possible classes (vehicle, bike, pedestrian, etc.), mostly done rule-based or by usage of machine learning [15]. Historical data can be helpful in order to improve the accuracy of the classification (e.g. the speed of a vehicle is generally higher than the speed of a pedestrian), but also the localization accuracy since different parts of a target object may be visible at different times. All described functional blocks result in the last interface defined here, namely the object list at IF_3.

4 Descriptions of Lidar Sensor Interfaces

The previously mentioned functional blocks, as visible in Fig. 3, are not described in more detail here, as they are well described in literature. This section now concentrates on the agreed interfaces, as already mentioned in the previous section and listed in the following Table 1. The exemplary data and header fields within the interface descriptions in the following tables are aligned with [1] and are seen as basic contents while not being limited to the listed fields. In the case of raw scans (IF_1), there are separate scans for each sensor in the overall simulated setup. Point clouds (IF_2) and object lists (IF_3), instead, can be derived by fusing multiple sensors.

4.1 Interface 1 (IF_1, Raw Scan)

The raw scans of a lidar sensor as defined here are usually described by vectors of measured ranges and intensities from measurements of a single lidar sensor. Each entry represents the processed signal for one specific scanning angle. This means the data results from a uniform sampling in a spherical coordinate space. The following definition of the raw scan interface is closely related to the message definition for single line lidars in the Robot Operating System (ROS) [16].

Table 1 Lidar sensor interfaces

Interface	Interface name
IF_1	Raw scan
IF_2	Point cloud
IF_3	Object list

The general data packets consist of a header and the following data. The header of the raw scan interface contains the sensor ID, which identifies the respective sensor that the data is perceived by, as well as its mounting position and orientation relative to the vehicle's coordinate frame, taken from ISO 8855:2011 [17]. This right-hand coordinate frame is centered in the middle of the rear axle of the vehicle. Its X-axis is aligned with the wheelbase, whereas its Y-axis is aligned with the rear axle and its Z-axis is determined with the right-hand rule. These extrinsic calibration parameters enable subsequent processing and interpretation of the data.

To deal with potential misinterpretation of scan data caused by the minimum or maximum ranges that the scanner can detect, these parameters can also be included in the header of the raw scan, as can be seen in Table 2.

The actual measurements are encoded in a data field in the raw scan, which contains consecutive blocks for the ranges followed by the intensities measured, which are sometimes replaced by the formerly described echo pulse width, as listed in Table 3. The related scanning angles can be assigned by knowing the order of the transmission of beams and saving the received reflections into the list. If the specific

Table 2 Interface 1—raw scan header

Variable	Description	SI-units	Type
sensor_id	Sensor ID	–	uint16
echo_id	Echo ID, if simulated sensor stores multiple echoes per transmitted beam	–	uint16
scan_id	Scan ID, unique for each scan during the whole simulation	–	uint64
time_stamp	Seconds of the timestamp of the first point in the current raw scan (e.g. simulation time or UTC/UNIX, etc.)	s	int64
time_stamp_ns	Additional nanoseconds of the timestamp	ns	uint32
mount_pos_x	Mounting position w.r.t. vehicle reference frame	m	float32
mount_pos_y	Mounting position w.r.t. vehicle reference frame	m	float32
mount_pos_z	Mounting position w.r.t. vehicle reference frame	m	float32
mount_ori_roll	Mounting orientation w.r.t vehicle reference frame	rad	float32
mount_ori_pitch	Mounting orientation w.r.t vehicle reference frame	rad	float32
mount_ori_yaw	Mounting orientation w.r.t vehicle reference frame	rad	float32
min_range	Minimum range of the sensor that can be detected	m	float32
max_range	Maximum range of the sensor that can be detected	m	float32

Table 3 Interface 1—raw scan data

Variable	Description	SI-units	Type
point_id	Point ID, unique for each point during one raw scan	–	uint64
range	Measured range for a certain beam and echo	m	float32
intensity	Measured intensity of a certain beam and echo	dB	float32

sensor is able to detect multiple echoes for one single transmitted pulse, the number of entries can simply be multiplied by that number of possible echoes.

4.2 *Interface 2 (IF_2, Point Cloud)*

Point clouds are very plausible, intuitive, but generic data structures consisting of a list of 3-dimensional vectors (points) [18]. Different definitions for point clouds and the interpretation of points of a cloud exist [18–20]. To save the points, Cartesian or angular coordinate systems can be chosen. Intensity values or echo pulse widths and laser scanner IDs can be added to the points' coordinates. To compute the point cloud, multiple lidar sensors can be fused together within the sensor system. Thus, it is important to have a semantically fixed interface. The lidar workgroup's definition is based on lidar interfaces from Ibeo Lux [21], Velodyne HDL-32E [22] and the interface definition for point cloud messages from ROS [19].

The general structure of the point cloud starts with a header that contains general information about the point cloud and the specific scan, as shown in Table 4. If only one sensor is used for point cloud determination, the header contains the extrinsic calibration of the laser scanner towards the vehicle's base coordinate system (Fig. 4) combined with the sensor identification number, similar to the previously defined raw scan. In any case, the header contains information about the extrinsic calibration of the point cloud towards the vehicle's base coordinate system (Fig. 4) and the amount of point data that follows the header.

Table 4 Interface 2—point cloud header

Variable	Description	SI-units	Type
sensor_id	Sensor ID	–	uint16
echo_id	Echo ID, if simulated sensor stores multiple echoes per transmitted beam	–	uint16
point_cloud_id	Point cloud ID, unique for each point cloud during the whole simulation	–	uint64
time_stamp	Seconds of the timestamp of the first point in the current point cloud (e.g. simulation time or UTC/UNIX, etc.)	s	int64
time_stamp_ns	Additional nanoseconds of the timestamp	ns	uint32
mount_pos_x	Mounting position w.r.t. vehicle reference frame	m	float32
mount_pos_y	Mounting position w.r.t. vehicle reference frame	m	float32
mount_pos_z	Mounting position w.r.t. vehicle reference frame	m	float32
mount_ori_roll	Mounting orientation w.r.t vehicle reference frame	rad	float32
mount_ori_pitch	Mounting orientation w.r.t vehicle reference frame	rad	float32
mount_ori_yaw	Mounting orientation w.r.t vehicle reference frame	rad	float32
number_of_points	The number of points in the cloud	–	uint64

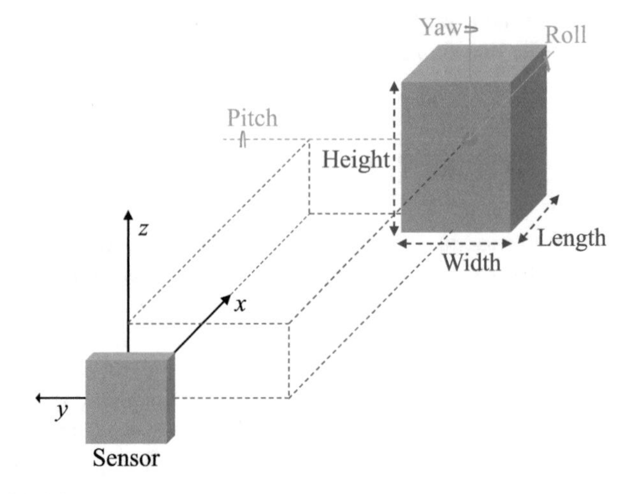

Fig. 4 Object list entries: geometrical and bounding box properties

Table 5 Interface 2—point cloud data

Variable	Description	SI-units	Type
point_id	Point ID, unique for each point in one point cloud	–	uint64
X	The x coordinate of the hit point w.r.t vehicle reference frame	m	float32
Y	The y coordinate of the hit point w.r.t vehicle reference frame	m	float32
Z	The z coordinate of the hit point w.r.t vehicle reference frame	m	float32
intensity	The intensity value related to the hit point	dB	float32

A point itself is described by three-dimensional coordinates, mostly Cartesian, in the vehicle reference frame described in the header and contains additional intensity and time information. Thus, points are five-dimensional vectors (see Table 5). The time information is needed, since in a single scan of a laser scan the different points may be related to different hit points in time. This information becomes more important with increasing angular velocity of the targets and the ego-vehicle.

4.3 *Interface 3 (IF_3, Object List)*

The points within a point cloud represent a measurement. Some sensor systems provide additional processing steps and outputs, e.g. for object detection, as described here. In order to find related structures and objects within these points, segmentation and clustering methods are applied [18, 23]. Therefore, assumptions about the environment (similar range or intensity) are made. The clusters generated in this way can then serve as object candidates. If an object is tracked over time, it will keep its ID in consecutive tracked object lists. The age of such an object, i.e. the number of measurements realized since its creation, is stored in the age property. The probability of existence gives the user information about the quality

Table 6 Interface 3—object list header

Variable	Description	SI-units	Type
sensor_id	Sensor ID	–	uint16
object_list_id	Object list ID, unique for each object list during whole simulation	–	uint64
time_stamp	Seconds of the timestamp of the object list (e.g. simulation time or UTC/UNIX, etc.)	s	int64
time_stamp_ns	Additional nanoseconds of the timestamp	ns	uint32
mount_pos_x	Mounting position w.r.t. vehicle reference frame	m	float32
mount_pos_y	Mounting position w.r.t. vehicle reference frame	m	float32
mount_pos_z	Mounting position w.r.t. vehicle reference frame	m	float32
mount_ori_roll	Mounting orientation w.r.t vehicle reference frame	rad	float32
mount_ori_pitch	Mounting orientation w.r.t vehicle reference frame	rad	float32
mount_ori_yaw	Mounting orientation w.r.t vehicle reference frame	rad	float32
no_of_objects	Number of objects in the following list	–	uint16

of the measurements and the tracking. It is usually computed directly inside the state estimation algorithm. Further output might be covariance measurements of the state estimation algorithm.

Typical state estimation algorithms like movement models based on constant velocity, if used during object tracking, allow estimating quantities that are not directly measured by the lidar sensor, such as the velocity and acceleration, but also the geometric extents of the objects. Both are provided in the sensor coordinate frame. Finally, state estimators usually provide a quality indicator per tracked state, which is stored in the corresponding measurement error fields.

An object list is divided into a header and the list of objects. If only one sensor is used for object list determination, the header contains the extrinsic calibration of the laser scanner towards the vehicle's base coordinate system (Fig. 4), combined with the sensor identification number, similar to the previously defined interfaces. If more than one sensor is used, these fields are obsolete. In all cases, the header contains the timestamp associated with the current measurement and the number of objects in the actual list at that time (Table 6).

Each detected object corresponds to an entry in the list and is defined by a set of properties. The structure of the proposed object list entries (Table 7) is inspired by the work proposed in the Open Simulation Interface [1] and object list descriptions used in the industry [24]. The ID of the object describes its position in the list. The object's class can be computed using classification algorithms, rule-based or by machine learning, on the cluster associated with the considered detection. The intensity of an object must be calculated using the corresponding points' intensities of the point cloud, e.g. by their mean value.

The geometrical information of the detection is provided in the coordinate frame of the vehicle that needs to be given. They correspond conventionally to the center of the bounding box, which is described by its length, width and height. The bounding box can subsequently be rotated around its center. Finally, the intensity

Table 7 Interface 3—object list data

Variable	Description	SI-units	Type
object_id	Object ID, unique for each object in one object list	–	uint16
object_id	Tracked object ID	–	Integer
class	The class associated with the detection, determined by a classification algorithm	–	uint16
class_prob	Probability of the chosen class	–	float32
exist_prob	Probability of existence of the object	–	float32
age	Object age/lifetime as detection	ms	uint32
x_rel	Relative long. distance w.r.t to vehicle reference frame	m	float32
y_rel	Relative lateral distance w.r.t to vehicle reference frame	m	float32
z_rel	Relative vertical distance w.r.t to vehicle reference frame	m	float32
v_rel_x	Relative long. velocity w.r.t to vehicle reference frame	m/s	float32
v_rel_y	Relative lateral velocity w.r.t to vehicle reference frame	m/s	float32
v_rel_z	Relative vertical velocity w.r.t to vehicle reference frame	m/s	float32
a_rel_x	Relative long. acceleration w.r.t to vehicle reference frame	m/s^2	float32
a_rel_y	Relative lat. acceleration w.r.t to vehicle reference frame	m/s^2	float32
a_rel_z	Relative vert. acceleration w.r.t to vehicle reference frame	m/s^2	float32
x_rel_std	Standard deviation on the relative distance in long. axis	m	float32
y_rel_std	Standard deviation on the relative distance in lateral axis	m	float32
z_rel_std	Standard deviation on the relative distance in vertical axis	m	float32
v_rel_x_std	Standard deviation on the relative velocity in long. axis	m/s	float32
v_rel_y_std	Standard deviation on the relative velocity in lateral axis	m/s	float32
v_rel_z_std	Standard deviation on the relative velocity in vertical axis	m/s	float32
a_rel_x_std	Standard deviation on the relative acc. in long. axis	m/s^2	float32
a_rel_y_std	Standard deviation on the relative acc. in lateral axis	m/s^2	float32
a_rel_z_std	Standard deviation on the relative acc. in vertical axis	m/s^2	float32
length	Length	m	float32
width	Width	m	float32
height	Height	m	float32
roll	Orientation of the target w.r.t to vehicle reference frame	rad	float32
pitch	Orientation of the target w.r.t to vehicle reference frame	rad	float32
yaw	Orientation of the target w.r.t to vehicle reference frame	rad	float32
intensity	Intensity	dB	float32

of light associated with the object can be added to the list. The principal properties associated with each object in the list are summarized in Fig. 4.

The bounding box is directly influenced by the selected tracking method. If it is computed using the actual size of the point cloud, it grows with lower distance and decreases with higher distance of objects. By using historical information, the size of the bounding box can e.g. be kept, once the point cloud is shrinking again.

5 Methods for Sensor Simulation

There are several methods to generate synthetic sensor data for the different interfaces. For raw scan and point cloud generation, ray tracing or ray casting are widely used in commercially available simulation tools. In addition, especially in the context of vehicle simulation, Open Source frameworks, such as GAZEBO, or 3D game engines, such as UNITY, can be used for lidar simulation. Ray tracing and ray casting are very efficient in computing multi-path propagation and reflections due to their ability to parallelize workloads.

Nevertheless, methods like beam tracing or Z-buffer are possible to use, as well. When it comes to beam divergence, other methods than ray tracing are possibly even more favorable. Beam tracing was invented for the simulation of diffraction. Z-buffer is very efficient if multi-path propagation or echoes are not of interest.

Object list simulation has the objective to accurately reproduce existence, state and class uncertainties of moving objects. In most cases, they are simulated directly via probabilistic models with data-driven stochastic for the three listed uncertainties and their subsequent object states and parameters. Besides, it is also possible to simulate data on prior interfaces like point clouds and apply the same algorithms as in the real sensor system, if available, until object lists are obtained or to use own algorithms and methods to derive them accordingly.

The authors propose the following strategy for modeling the functional blocks in order to serve the necessary interfaces in simulation:

At first, the requirements for the generation of synthetic lidar sensor data need to be defined. Therefore, physical effects that should be simulated need to be selected, like signal degradation, beam divergence etc. The selection must be derived from the intended usage of the synthetic data, which can be function development of further data processing steps, testing of such processing, or to serve as input for safety validation of a complete automated driving system.

Secondly, with the requirements at hand, the previously mentioned simulation approaches like ray casting, ray tracing or Z-buffer need to be evaluated and the best suitable with respect to the effects to simulate and the computational effort should be selected.

Since the simulated environment serves as a resource for the sensor simulation to gain information about the objects that get hit by the active sensor, requirements should be derived for them, as well. Of course, technical limitations come into play if a certain degree of fidelity is needed. As an example, in the case of fine structures like fences or grass, approximations and merging have to be used.

6 Conclusion and Outlook

The length of this chapter does not allow describing functional blocks, interfaces, or simulation methods in more detail. Still, the functional comprehension and the commonly agreed definitions of interfaces described here can be seen as a major achievement and milestone within the ENABLE-S3 project. Furthermore, the work performed by the different partners regarding lidar system simulation covers different modeling methods and interfaces, as they have been briefly described here.

A special task for all partners is the validation of the different models. One approach could be the validation by evaluating the effect of injecting partly augmented sensor data into black-box ECUs by monitoring the automated driving function's response to the stimulation, see [25]. There, the statement's validity originates from simulation and from independent examination of the black-box system. This includes the formulation of requirements and collection of parametrization and validation data. Besides, reference data needs to be collected during the measurements to be able to re-simulate the real driving maneuvers in a virtual environment that also needs requirements.

Another and more common approach for validation is the replay-to-sim method, where the real measurements are performed first while collecting reference data like GPS trajectories of all moving objects in high accuracy. Afterwards, these reference data together with the static scenery that needs to be modeled serve as input for the simulation. In the end, several metrics for the comparison of real and synthetic data are applied. If the sample experiments are chosen in a way that the overall parameter space of the model is covered properly, e.g. by applying a sensitivity analysis beforehand, the sample validities from single experiments gain overall trust in the model [26].

Therefore, to obtain valid sensor models, with the described interfaces at hand, the next logical steps are to find metrics for benchmarking of sensor models on different interfaces and for example to compute metrics on data of the same origin interface and further processing for correlation of metrics applied on those subsequent interfaces. In a first trial within ENABLE-S3, benchmarking of the implemented different approaches has been fulfilled on point cloud level and these results will be included in scientific publications and project reports.

References

1. Hanke, T., Hirsenkorn, N., van Driesten, C., Garcia-Ramos, P., Schiementz, M., Schneider, S.K.: Open simulation interface – a generic interface for the environment perception of automated driving functions in virtual scenarios. https://www.hot.ei.tum.de/forschung/automotive-veroeffentlichungen/ (2017). Accessed 19 Dec 2018
2. Pegasus Research Project: https://www.pegasusprojekt.de/en/home. Accessed 16 Dec 2018
3. Gotzig, H., Geduld, G.: Automotive LIDAR. In: Winner, H., et al. (eds.) Handbook of Driver Assistance Systems. Springer International Publishing, Cham (2016)
4. Weitkamp, C.: Lidar: Range-Resolved Optical Remote Sensing of the Atmosphere, p. 7. Springer Science & Business, New York (2006)
5. Ibeo Automotive Systems GmbH: ibeo LUX 2010, Laserscanner manual, version 1.6 (2014)
6. Pfotzer, L., Oberländer, J., Rönnau, A., Dillmann, R.: Development and calibration of KaRoLa, a compact, high-resolution 3D laser scanner, SSRR 2014, pp. 1–6

7. Davis, S., Rommel, S., Gann, D., Luey, B., Gamble, J., Ziemkiewcz, M., Anderson, M.: A lightweight, rugged, solid state laser radar system enabled by non-mechanical electro-optic beam steerers. In: LADAR Conference, 2016
8. Wang, Y., Kyoungsik, Y., Wu, M.: MEMS optical phased array for lidar. In: 21st Microoptics Conference, 2016
9. Suni, P., Bowers, J., Coldren, L., Ben Yoo, S.: Photonic integrated circuits for ceherent lidar. In: 18th Coherent Laser Radar Conference, 2016
10. Rosenberger, P., Holder, M., Zirulnik, M., Winner, H.: Analysis of real world sensor behavior for rising fidelity of physically based lidar sensor models. In: 2018 IEEE Intelligent Vehicles Symposium (IV), Changshu, pp. 611–616 (2018)
11. Amersbach, C., Winner, H.: Functional decomposition: an approach to reduce the approval effort for highly automated driving. In: 8. Tagung Fahrerassistenz, Einführung hochautomatisiertes Fahren, 2017
12. Nguyen, A., Le, B.: 3d point cloud segmentation: a survey. In: IEEE Conference on Robotics, Automation and Mechatronics, RAM – Proceedings, pp. 225–230 (2013)
13. Schreier, M.: Bayesian environment representation, prediction, and criticality assessment for driver assistance systems. PhD Thesis, Technische Universität Darmstadt, Darmstadt (2016)
14. Granström, K., Baum, M.: Extended object tracking: Introduction, overview and applications. CoRR, vol. abs/1604.00970 (2016)
15. Deshpande, S., Muron, W., Cai, Y.: Chapter 3: Vehicle classification. In: Loce, R.P., Bala, R., Trivedi, M. (eds.) Computer Vision and Imaging in Intelligent Transportation Systems, pp. 47–79. Wiley, Hoboken, NJ (2017)
16. Definition of the sensor_msgs::LaserScan message type in the Robot Operating System (ROS). http://docs.ros.org/api/sensor_msgs/html/msg/LaserScan.html. Accessed 08 Oct 2017
17. International Organization for Standardization, ISO 8855:2011: Road vehicles – Vehicle dynamics and road-holding ability – Vocabulary, Geneva, CH. Tech. Rep. ISO 8855:2011, 2011
18. Rusu, R.B., Cousins, S.: 3D is here: Point Cloud Library (PCL). In: IEEE International Conference on Robotics and Automation (ICRA), 2011
19. Definition of the sensor_msgs::Pointcloud2 message type in the Robot Operating System (ROS). http://docs.ros.org/api/sensor_msgs/html/msg/PointCloud2.html. Accessed 08 Oct 2017
20. Definition of the sensor_msgs::PointField message type in the Robot Operating System (ROS). http://docs.ros.org/jade/api/sensor_msgs/html/msg/PointField.html. Accessed 08 Oct 2017
21. Ibeo Automotive Systems GmbH: Ethernet data protocol ibeo LUX and ibeo LUX systems, version 1.36 (2017)
22. Velodyne Lidar, Inc.: User's Manual and Programing Guide – HDL-32E High definition Lidar Sensor
23. Douillard, B., et al.: On the segmentation of 3D LIDAR point clouds. In: 2011 IEEE International Conference on Robotics and Automation, Shanghai, pp. 2798–2805 (2011)
24. Continental Engineering Services GmbH: Standardized ARS Interface – Technical Documentation, 2012
25. Zofka, M.R., Essinger, M., Fleck, T., Kohlhaas, R., Zöllner, J.M.: The sleepwalker framework: verification and validation of autonomous vehicles by mixed reality lidar stimulation, SIMPAR 2018, pp. 151–157
26. Viehof, M.: Objektive Qualitätsbewertung von Fahrdynamiksimulationen durch statistische Validierung. PhD Thesis, Technische Universität Darmstadt, Darmstadt (2018)

Camera Sensor System Decomposition for Implementation and Comparison of Physical Sensor Models

Marcel Mohr, Gustavo Garcia Padilla, Kai-Uwe Däne, and Thomas D'hondt

In this chapter, as in the preceding chapters of lidar and radar modeling, we present results of the camera sensor modeling workgroup within ENABLE-S3. The main objectives are similar to those of the lidar and radar workgroups, namely to find a common language and understanding when describing camera sensors. These agreed results could now be used as a reference. The results of the functional decomposition can be used to identify and define standardized interfaces for camera (e.g. in the Open Simulation Interface (OSI) project [5]).

1 Camera Sensor Principle

This specification focuses on camera sensorics and its method of software simulation. Therefore, it is helpful to take a look at a real sensor implementation. Automotive camera sensors consist of camera Front-end, Signal Processing Unit and Control Unit. Figure 1 illustrates exemplarily the camera system block diagram.

Different interfaces (IF_n, $n = 0 \dots 6$) have to be defined in the camera sensor modelling approach to provide each project partner a detailed understanding what sub-models are available, what functions the submodels provide and how sub-models can be linked together. The goal of the approach should be to get between every interface sub-models to build up a model chain that provides the whole camera sensor functionality.

The following interfaces are considered and will be described in Sect. 2 (Table 1).

M. Mohr · G. Garcia Padilla (✉) · K.-U. Däne
Hella Aglaia Mobile Vision GmbH, Berlin, Germany
e-mail: marcel.mohr@hella.com; gustavo.garcia.padilla@hella.com; kai-uwe.daene@hella.com

T. D'hondt
Siemens Industry Software NV, Leuven, Belgium
e-mail: thomas.dhondt@siemens.com

A. Leitner et al. (eds.), *Validation and Verification of Automated Systems*,
https://doi.org/10.1007/978-3-030-14628-3_13

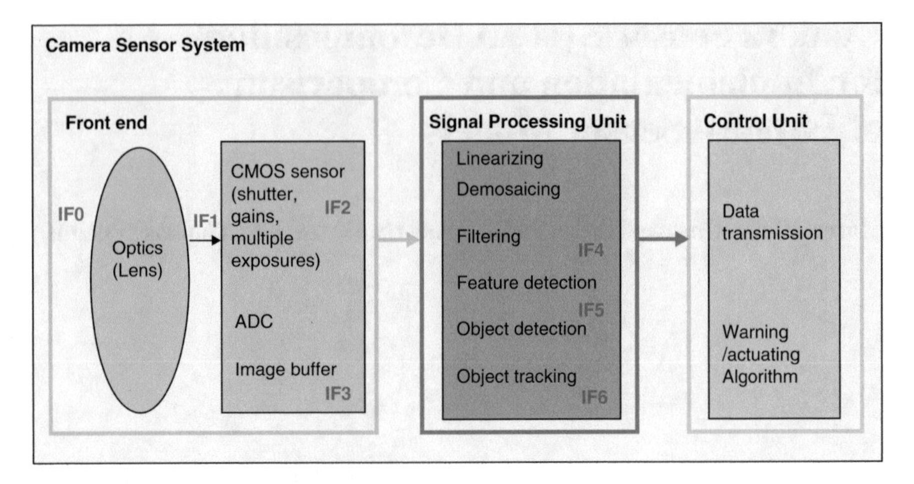

Fig. 1 Camera sensor system block diagram

Table 1 Camera sensor interface description

Interface	Interface name	Domain, signal-/data-type, parameter
IF_0	Visual path	Day, Night, Weather, Artificial light
IF_1	Optic	Camera model, Projection parameters, Distortion coefficients
IF_2	Image and grabber	Sensor format, Frametiming, Sensitivity
IF_3	Image buffer	Header and image data
IF_4	Filter output	Depends on ADAS function
IF_5	Feature list	Edge, Corner, Circle, Dynamic data, Color processing
IF_6	Object list	Time domain, Object location, Orientation

1.1 Front-End

The camera front-end consists of a lens, lens-mount and an imager, e.g. a CMOS sensor, with on-board electronics.

1.1.1 Visual Path

The virtual counterpart of a three-dimensional real visual scene is a three-dimensional modelled world [1]. Three-dimensional modelling today is performed with rendering. In general, rendering is the process of converting a description of a three-dimensional scene into an image. Algorithms for geometric modeling, texturing, light sources, and other areas influence the process of rendering. The 3D-models are usually composed of 2D meshes. These surface-meshes get assigned material properties, like color and reflectivity, e.g. BRDF—bidirectional reflectance distribution function and/or *BSSRDF*—Bidirectional scattering-surface reflectance distribution function. Light sources take care of the illumination of the scene: typically used light-sources are point-, area-, directional and goniometric light source. Then different light transport equations (LTE) take care of the illumination and sampling of the scene.

There are different approaches to the rendering tasks. Physically based techniques [1] try to use the principle of physics to model the interaction between light and matter. Realism is the primary goal, which usually means to sacrifice speed. In general, these renderers cannot produce real time images. However, they are able to produce realistic images, with multiple reflections, sub-surface scattering, absorption, depth of focus, etc. One text-book that describes and develops such a renderer is [1].

From here approximations may apply, that suffice the detail level needed. A few examples for such approximations:

- BRDFs are approximated with models and/or programmed in the GPU via shaders.
- Direct lighting may be used, which only computes light from source to object.

1.1.2 Lens

The purpose of the lens is to form an optical image. The full description of the image formation is a complicated process that depends on many parameters: lens curvatures, lens material, lens coatings, lens barrel material, etc. A simplified image formation description based on a few characteristic parameters like focal length and f-number of the lens can be found in [1].

The most general description of the lens model is the following: Take a 3D scene with illumination, place an optic into the scene and observe the image at the sensor surface, usually a plane. For a differential area dA on the sensor surface specify the number of photons passing this surface for a differential spectral range $d\lambda$ and for a differential solid angle $d\Omega$ [1]. If the light sources are not time-continuous, but time varying, e.g. pulsed, the number of photons also depends on the time. (A possible use case is that of flickering LED light sources described later.)

From here approximations and simplifications may be used:

- The differential area becomes the pixel area (the light sensitive area of a real pixel is smaller than the pixel area [which is defined as product of horizontal and vertical pixel pitch]).
- The continuous spectral range gets accumulated into bins of, e.g., 30 nm width.
- The solid angle gets fixed.
- The number of photons is constant in time, etc.

The lens in an automotive camera sensor is a static system: it has no zoom, autofocus or aperture control. The focus position is adjusted during production. The aperture is fixed. That implies that the interface between optic and imager is unidirectional, from optic to imager only. (The interface IF_1 becomes bidirectional if time-varying light sources are used, this will be explained later).

The interface IF_1 proposed here will contain the spectral power distribution (SPD) for each pixel element. The SPD describes the power per unit area per unit wavelength. Typically, it is defined in the visible spectral range with light wavelength between 400 and 800 nm. The spectrum gets binned, e.g. into 30 nm bins. Other ranges can be used, a tradeoff between physical reality and memory usage has to be made. The spectral power distribution can also be specified in RGB, however physical information gets lost.

From the SPD the number of photons can be calculated. Integrating the SPD gives the total power Ptot that arrives at the pixel element.

When time varying light sources are used the SPD has to be defined time-varying. This is the case, if the scene contains strongly pulsed light sources, like LED traffic signs, or LED head-/tail-/brake-lights.

The time-scales for these light sources (50–600 Hz) and those for the sensor (<50 kHz) are much smaller than simulation time-scales (~20–60 Hz), so special considerations/approximations have to take place to avoid the needed high sampling frequencies. An idea for an approximation is, to disclose the integration mechanism and timing of the sensor to the lens model and world, which calculates, if the light source is on or off (or a fraction of on/off) during the integration time for each pixel. This has to be calculated for each HDR exposure i, separately (The HDR mechanism will be explained in the next chapter). Then the interface IF_1 becomes bidirectional, as information from the sensor has to enter the lens model and/or the world.

The image formation on the sensor plane can be described using the projective camera model. (Please note that with this camera model only cameras with a field of view (FOV) <180° can be modelled. This is a theoretical limit, in real life this FOV reduces to realistic values of ~160°.)

1.1.3 Imager

The most general description of the imager model, usually a CMOS sensor is the following: The lens forms an image of the illuminated world on the sensor surface, usually a plane. This surface consists of many pixels. The image on the sensor surface is composed of photons hitting the sensor pixel. These photons may arrive from different solid angles and are focused by the microlenses of the sensor pixel. Also, the photons have different spectral wavelength. These photons are filtered by the color filters and/or the transmission of the microlenses. A number of photons hit the photodiode, where they create electrons. These electrons form a charge which is converted into a voltage by a capacitor. This capacitor is amplified and being readout by an analog-digital-converter (ADC). A well accepted model of the sensor pixel is presented in the EMVA1288-Standard [2]. Put in simple terms of a black box model, the imager eats photons and outputs digital numbers (DN).

The most distinct difference between an automotive CMOS sensor and consumer grade CMOS sensors is the ability to record multiple exposure high dynamic range (HDR) images. This currently implies a rolling electronic shutter, which is another distinction to the global shutter used in mobile devices or digital cameras [3]. The resolution of automotive CMOS sensors is much lower, currently in the 1–2 MP range. And also the color filter arrays (CFAs) may deviate from the standard bayer pattern (RGGB) used in multimedia devices.

During a multiple exposure a sensor row gets illuminated n-times (n usually between 2 and 4) with different sensor settings (at least different integration times, but other settings may be changeable, too). Then, these n measurements are combined to a single image row. This multi-exposure or HDR feature allows to record, real world scenarios with much more detail, especially in the high or low light areas.

1.1.4 Image Buffer

The image buffer contains the image data. In addition further information can be transmitted, e.g., using a header, which precedes each image. The header may contain metadata information like timestamp, camera id or frame information. Another way to transfer information is using the image data itself. There can be rows which contain information transmitted from the image sensor: this can be the current operating status of the sensor (the status rows) or statistical information on the current image content (the statistic rows).

The timestamp is needed for fusing different sensors. The camera id is needed in case that more cameras are used. Frame information is needed for the correct interpretation of the image data. Status rows are needed for detailed information of the sensor runtime parameters. Statistic rows can be used to gather statistical information of the image. Image data contain the pixel information. cImg (IF_3) defined below describes the image data.

1.2 Signal Processing Unit

This unit consists usually of a high performance digital processor and performs tasks like linearizing, color processing, filtering, feature detection, object detection and object tracking.

1.2.1 Linearizing

The dynamic range of the HDR sensor is usually much higher than the bit depth of the transmitted image. Usually, a companding mechanism is implemented in the sensor to reduce the bit depth of the internal raw image from $2^{20} \ldots 2^{24}$ to $2^{12} \ldots 2^{16}$ in the transmitted image. This may be done using a piecewise linear compression. The signal processing unit needs to reconstruct the original image intensities, which is called linearizing the image.

1.2.2 Demosaicing

The image sensor uses color filters in front of specific pixels to sense color information, e.g., a red color filter every second pixel. Demosaicing reconstructs a full color image.

1.2.3 Filtering

Different linear and non-linear filters, depending on the ADAS function, are used for preprocessing the image. For performance reasons, this is best be done using ASICS (application specific integrated circuits) or FPGAs (field-programmable gate arrays). Examples for filters are edge or corner extraction operators, low pass, band pass or high pass filters.

1.2.4 Feature Detection

Different features can be used, e.g.

- Color features like
  - color spaces like Grayspace, RGB, HSV and HLS, CMY, LAB,
  - color histograms,
  - spatial binning of color
- Gradient features like
  - Sobel operator
  - Histogram of Oriented Gradients (HOG)
- Thresholding

1.2.5 Object Detection

This consists in the detection and classification of an object based on their detected features. The following figure illustrates the object detection with an example (Fig. 2).

1.2.6 Object Tracking

Combining the measurements captured over a longer period of time, hence taking into account the history of the object list, allows improving the reliability of the measurements. This is realized through a tracking algorithm, which can be implemented in the signal processing unit or in the control unit. The techniques used for this, e.g. Kalman Filter, Data Association and Track Handling are similar to the methods for lidar and radar objects and not explained in more detail here.

1.3 Control Unit

The track list is finally transmitted to the advanced driver assistance system, where it can be used as input for a controller or a data fusion algorithm.

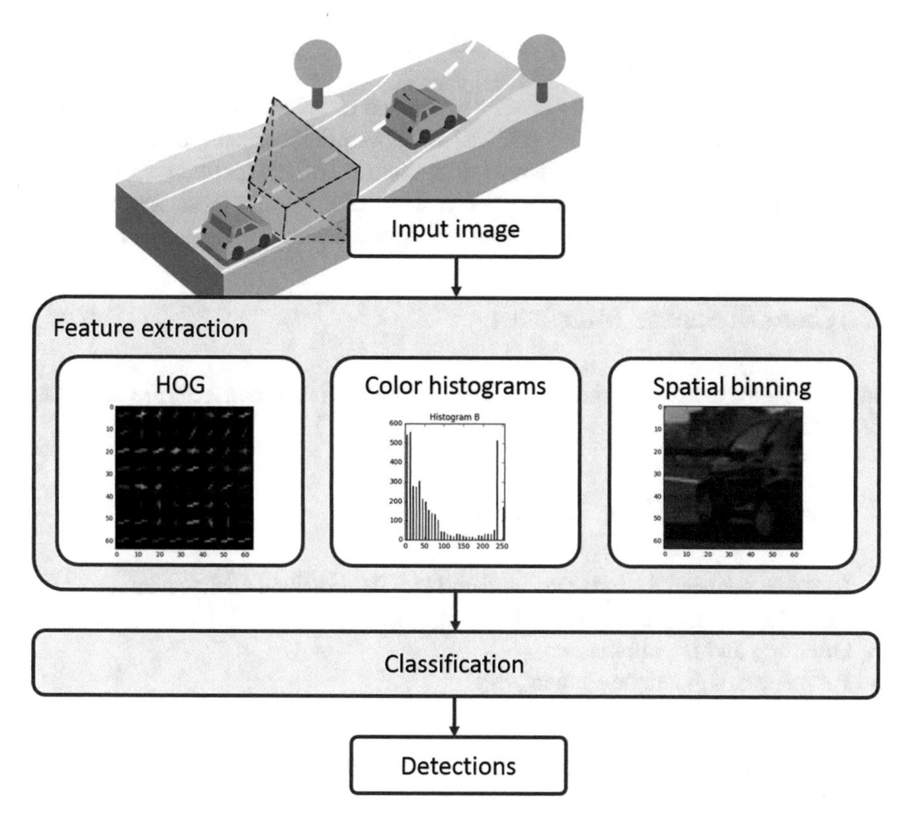

Fig. 2 Vehicle detection. Source: [SISW, T. D'Hondt]

1.3.1 Data Transmission

This transmission is usually done over a CAN-bus or a FlexRay-bus for improved performance. For this purpose, each track is split into several messages depending on the number of variables to be transmitted. Those messages are then sent serially over the bus, message per message and track per track [4].

In time critical applications, ordering those tracks before sending them can reduce the delay between a detection and the action of the controller. For instance, detections may be ordered by their range or their amplitude (Table 2).

Table 2 Possible output of the control unit

Type of signal	Transmission bus	HMI
Warning/Alarm signal	CAN/FlexRay/LIN/GPIO	Optical: Light/Symbol/Icon/ Haptic: Vibration Acoustic: Ton/Voice
Actuating signal	CAN/FlexRay/LIN/GPIO	–

1.3.2 Warning and Actuating Algorithm

This algorithm depends on the driving assistance or autonomous driving function (e.g. forward collision warning (FCW), autonomous emergency braking (AEB), blind spot detection (BSD), adaptive cruise control (ACC)) and may vary in their implementation and complexity (e.g. FCW can include or not presetting of brakes and seat belts and not only driver alerts or they can use or not sensor fusion).

2 Camera Sensor Interfaces

In this chapter the main parameters of the different sensor interfaces are described in table form.

2.1 *Interface 0 (IF_0, Visual Path)*

- Interface between the lens one hand and the 3D-modelled environment.
- The lens collects photons from different solid angles.
- Often IF_0 and IF_1 are merged.
- Parameters: solid angle element, SPD

2.2 *Interface 1 (IF_1, Lens) (Tables 3 and 4)*

Table 3 Description of interface 1 for time-constant light sources

Parameter	Abbreviation	Description
Horizontal/vertical pixel resolution	(H/V)PixRes	The horizontal/vertical number of pixels.
Horizontal/vertical pixel pitch	(H/V)PixPitch	The pixel pitch between two neighboring horizontal/vertical pixels.
Spectral power distribution	SPD[HPixRes*VPixRes]	Array of the spectral power distribution. The size of the array is the number of pixel elements. Possible candidates are the classes SPECTRUM or RGBspectrum from [1]. From the SPD the number of photons per spectral range and the total power can be calculated. For a simplified case the number of data points in the SPD can vary from one for monochrome camera to three or more for color or hyperspectral cameras.

Table 4 Extension of interface 1 in the case of time-varying (e.g. pulsed) light sources and/or HDR imagers

Parameter	Abbreviation	Description
Start of integration row 0	SOI_r0	The start of the integration of row 0 of exposure 1 in simulation time.
Line time	LINE_TIME	All pixels in a row are integrated at the same time. However, each row is integrated at different times. The starting time of the integration of the next row is LINE_TIME later.
Integration time	Itime_i $(i = 1 \ldots n)$	For each exposure $i = 1 \ldots n$. The integration time is measured in units of LINE_TIME. The exposures are sorted from long to short, i.e., the longest exposure is $i = 1$.
Spectral power distribution	SPD[HPixRes*VPixRes][i] $(i = 1 \ldots n)$	Array of the spectral power distribution for each exposure i. The size of the array is the number of pixel elements. Possible candidates are the classes SPECTRUM or RGBspectrum from [1]. From the SPD the number of photons per spectral range and the total power can be calculated.

2.3 *Interface 2 (IF_2, Imager)*

In the Table 5 the interface description of IF_2 between Sensor and Signal Processing Unit (Sensor parameters that can be controlled/Data transmitted via the interface) is given.

The digital gain can be set individually for all four pixels of a 2×2 group of pixels (R, GR, GB, B pixels in case of Bayer color pattern). These names have been used for clarity.

The following parameters are considered static and therefore they are handled separately (Table 6).

2.4 *Interface 3 (IF_3, Image)*

In the following, we will describe the cImg class, which provides support for images. Primarily an image consists of a cImgType (an enum, which will be defined lateron) as decorator, the pixel buffer and a timestamp. Optionally user-defined meta data, such as source-ids or framerates, can be contained. Several images can be related to the same pixel buffer. This can be achieved by the shallow copy concept. The idea behind this is, to concentrate on specific area of interest (aoi) or on sub-

Table 5 Description of interface 2

Parameter	Abbreviation	Description	CS	In/Out
Context switching	CS	A set of parameters between which the sensor switches during each frame (even and odd). Actually all parameters tagged in column CS are defined in pairs, i.e., in a Context A and a Context B set. (This feature is implicitly contained in the framewise control of the exposure. However, the exposure control usually is very slow, remaining constant during hundreds of frames, whereas the context switching occurs at every second frame. So using this mechanism reduces control overhead.) Context switching is also for more than two contexts possible. The static parameter Number of contexts may be used to describe more contexts.		
Integration time	Itime_i $(i = 1 \ldots n)$	For each exposure $i = 1 \ldots n$. The integration time is measured in units of LINE_TIME. The exposures are sorted from long to short, i.e., the longest exposure is $i = 1$.		
Analog gain amplification	AGA_i $(i = 1 \ldots n)$	The analog amplification of the measured pixel voltage.		
Pixel conversion gain	CG_i $(i = 1 \ldots n)$	The pixel can be operated in two different conversion gain modes.		
Global digital gain	GDG_i $(i = 1 \ldots n)$	A simple factor multiplication of the digitized value, applies to all pixels.	X	In
R(ed) digital gain	RDG_i $(i = 1 \ldots n)$	A simple factor multiplication of the digitized value, applies to red pixels R.	X	In
GR (Green1) digital gain	GRDG_i $(i = 1 \ldots n)$	A simple factor multiplication of the digitized value, applies to the green pixels GR.	X	In
GB (Green2) digital gain	GBDG_i $(i = 1 \ldots n)$	A simple factor multiplication of the digitized value, applies to the green pixels GB.	X	In
B(lue) digital gain	BDG_i $(i = 1 \ldots \text{n})$	A simple factor multiplication of the digitized value, applies to the blue pixels B.	X	In
Compression parameters	–	Parameters for the compression scheme.	–	Out
Image	img	The transferred image. The height is the optical active height plus additional status and/or statistics rows. A possible specification of the img is the CImg class, defined in next chapter for the interface 3 with type UInt16 or UInt32.	–	Out
Status rows	(embedded in img)	Sensor specific information, like framecounter, sensor settings, etc. can be found encoded in the image, usually at the top or the bottom of the transferred image. Please note: When using the images in a HiL-environment, these rows are needed, in general.	–	Out
Statistic rows	(embedded in img)	Histogramm and statistic information of the current image can be found encoded in the image, usually at the top or the bottom of the transferred image. Please note: When using the images in a HiL-environment, these rows are needed, in general.	–	Out

Table 6 Static parameters of interface 2

Parameter	Abbreviation	Description
Frame rate	fps	The number of frames per second.
Bit depth	bitdepth	The bitdepth of the image.
Compression scheme	–	The HDR information depth is usually not transferred directly, but compressed, resulting in transferred bit depth of 12, 14, or 16. This compression scheme has to be known for linearizing the image again in the signal processing unit.
Number of contexts	numContexts	Number of contexts.

Table 7 Class cImg

Type	Name	Description
cImgType::ImageTypes	type	The image type.
uint32	width	The width of the image.
uint32	height	The height of the image.
uint32	xPitch	The number of bytes in the buffer between two consecutive pixels in the same row.
uint32	yPitch	The number of bytes in the buffer between two consecutive rows.
uint8	*buf	A pointer of the image buffer.
uint32	numChannels	The number of channels (default is 1). A real image sensor usually has only single channel pixels. In front of this pixels are color filters that are used for extracting color information. In computer images, a colored image contains three channels: red, green, blue. (Sometimes four channels are used, the fourth channel then is transparency. However this does not apply to camera images.)

matrices. These sub-matrices can contain information from a pixel class, consider, for example, the red pixel of a RGGB bayer pattern, which occurs only at every second pixel in every second row.

2.4.1 Class cImg

Metadata information, such as frame rates or sensor ID-s etc. can be prepended in the buffer $_*$buf by using map objects. If this is the case, a buffer offset has to be specified, which defines the starting of the image data (Table 7).

2.4.2 Class cImgType

The cImgType describes an image type, that means how to interpret a pixel.

The following main features are supported:

- A textual name of the type: cImgType::getName
- Conversion between image types: cImgType::convertTo
- Visualization style: cImgType::getStyle. Each image type supports a list of styles, which describe how to display the image of this type.

2.5 Interface 4 (IF_4, Filter Output)

This interface will not be described because of intellectual property protection reasons.

2.6 Interface 5 (IF_5, Feature List)

This interface will not be described because of intellectual property protection reasons.

2.7 Interface 6 (IF_6, Object List)

Further processing of the feature list allows to associate an object class to the detection. Therefore, the data of the feature list can be reduced to a list of relevant object detections, where each one is described by a set of properties. Such an object list is divided into a header and the list of object detections. The header contains information that is relevant for the entire list, including the sensor identification number, as well as the timestamp associated to the current measurement [4]. Additionally, the position and orientation of the camera sensor in the car coordinate frame are added. The addition of these fields in the header allows to take into account drift of the sensor mounting position over time.

The object list header is already specified for the lidar sensor. The same table applies to the camera sensor.

The mounting position and orientation are referred in the vehicle coordinate frame. This right-hand coordinate frame is centered in the middle of the rear axle of the vehicle. Its X-axis is aligned with the wheelbase, whereas its Y-axis is aligned with the rear axle and its Z-axis is determined with the right-hand rule.

Each detected object corresponds to an entry in the list and is defined by a set of properties.

The structure of the proposed object list entries is inspired by industry-used object lists [4, 5]. The ID of the object describes its position in the list. Its class can be computed using classification algorithms on the features associated to the considered detection.

An object list entry is given for the lidar sensor, as well. The same table applies to the camera sensor with the only difference that the variable intensity is not used for the camera sensor. The detection and use of color features is specific of the camera sensor but they are already considered by the object detection not being necessary to include them in this interface.

The geometrical information of the detection is then provided in the coordinate frame of the sensor. These coordinates correspond conventionally to the center of the bounding box associated to the detection, which is described by its length, width and height. The bounding box can subsequently be rotated around its center. Finally, the color associated to the object is added to the list through the object color. The principal properties associated to each object in the list are already summarized for lidar sensors.

Tracking methods combine object lists collected over time with models of the expected target. This reduces the effect of measurement noise, allows to estimate unmeasured variables and filters out part of the false positives.

The age of a track, or the number of measurements realized since its creation, is stored in the age property. The hypothesis probability of existence gives the user information about the quality of the measurements and the tracking. It is usually directly computed inside the state estimation algorithm.

Typical state estimation algorithms used during object tracking allow to estimate quantities that are not directly measured by the camera sensor, such as the velocity and acceleration of the objects. Both are provided in the sensor coordinate frame. Finally, state estimators usually provide a quality indicator per tracked state, which is stored in the corresponding measurement error fields. A conversion from polar coordinates to Cartesian coordinates is not needed for the camera sensor.

The detection and use of color features is specific of the camera sensor but they are already considered by the object detection not being necessary to include them in this interface.

References

1. Pharr, M., Jakob, W., Humphreys, G.: Physically Based Rendering: From Theory to Implementation. Morgan Kaufmann, Cambridge, MA (2016)
2. The EMVA1288 Standard version 3.1. http://www.emva.org/standards-technology/emva-1288/emva-standard-1288-downloads/
3. Maddalena, S., Darmon, A., Diels, R.: Automotive CMOS image sensors. In: Valldorf, J., Gessner, W. (eds.) Advanced Microsystems for Automotive Applications 2005. Springer, Heidelberg (2005)
4. Continental Engineering Services GmbH: Standardized ARS Interface – Technical Documentation (2012)
5. Open Simulation Interface: https://github.com/OpenSimulationInterface

Seamless Tool Chain for the Verification, Validation and Homologation of Automated Driving

Andrea Leitner, Jürgen Holzinger, Hannes Schneider, Michael Paulweber, and Nadja Marko

1 Introduction

Autonomous vehicles are becoming increasingly capable and frequent. Any automated application, whether Advanced Driver Assistance Systems (ADAS) or Highly Automated Driving (HAD) functions, needs to go through rigorous testing procedures due to the traditionally strict safety requirements of the automotive industry. However, purely physical testing is cost- and time consuming. Varying road infrastructure, changing traffic and weather conditions, different driver behavior and country-specifics need be considered. Furthermore, there is a heavy tail distribution of surprises from the real world and an automated vehicle needs to be able to detect and react to these unforeseen situations [1]. All this makes a so-called proven-in-use certification solely based on physical test driving almost impossible [2]. Software-based (virtual) validation approaches are therefore promoted as a viable alternative to support the testing process by enabling a higher scenario coverage at lower costs and in shorter time. Simulations can further be used to identify the sub-set of test cases that should be executed in a more accurate test environment (Vehicle-in-the-Loop or proving ground). At the end, the whole validation process will be a mixture of different test environments and the results of this overall process will be used as an input for homologation. Especially for virtual test environments it is also important to show the correctness and accuracy of the simulation results. This means that also all models used in the virtual environment must be compared to real world behavior to ensure trustable results [3]. For all these different tasks it is important

A. Leitner (✉) · J. Holzinger · H. Schneider · M. Paulweber
AVL List GmbH, Graz, Austria
e-mail: andrea.leitner@avl.com

N. Marko
Virtual Vehicle Research Center, Graz, Austria

A. Leitner et al. (eds.), *Validation and Verification of Automated Systems*,
https://doi.org/10.1007/978-3-030-14628-3_14

to have a seamless toolchain, where models as well as test cases can be reused independent of the test environment.

This paper discusses the requirements and potential solutions for the realization of a seamless toolchain for the validation of automated driving functions.

2 Test Execution Platforms in the Context of the Generic Test Architecture

Figure 1 shows an exemplary instantiation of the generic test architecture [15] proposed by the ENABLE-S3 project and introduced in chapter "ENABLE-S3: Project Introduction". The focus of this paper is on the lower layer called Test Execution Platform. This layer subsumes the different test environments from purely virtual testing to real-world testing.

In a seamless toolchain, we do not only assume that models can be reused throughout the different test environments, but also that there is a common interface for test cases as well as for collecting test results. This enables the reuse of test cases and test plans on different test execution environments. Of course, the Test Planning tool has to take care about preparing the test cases in a way that they are executable in a specific test environment (i.e. considering the constraints and formats of the execution environment).

One dimension in Fig. 1 distinguishes between three phases: *Preparation*, *Execution* and *Evaluation & Reporting*. The focus of this paper is on the Execution phase, but for completeness, we provide a brief introduction into the other phases as well.

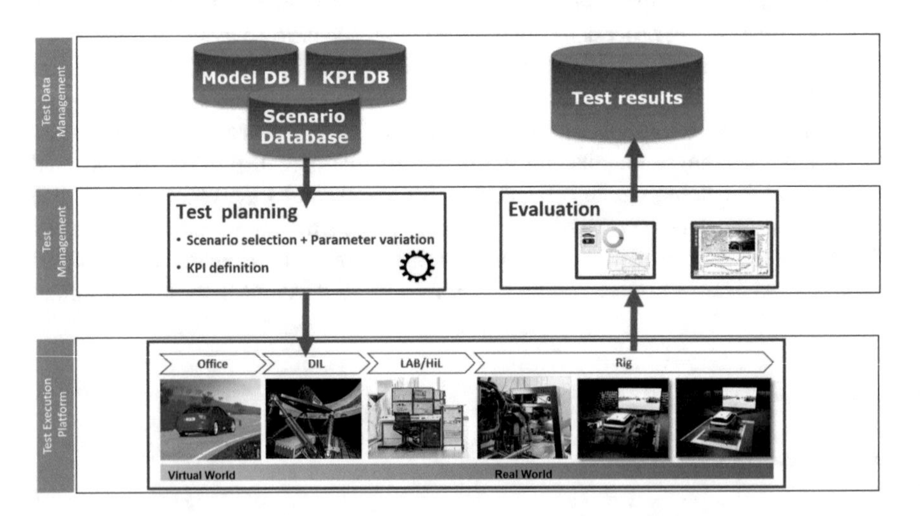

Fig. 1 Exemplary instantiation of the generic test architecture with different Test Execution Platforms

The *Preparation* phase includes all tasks, which are required for preparing the test execution. This basically means that the test input data needs to be selected and prepared in a way that makes it executable. An important asset for scenario-based validation are of course scenarios. Scenarios can come from different sources (e.g. engineered or extracted from measurement data) and consists of different aspects, such as dynamic aspects (traffic participants), static aspects (road and traffic signs) and the scenery (3D environment as input for sensor models). All these aspects need to be represented in a scenario database. For test case generation, these aspects are combined and parametrized in a certain way. Of course, the level of detail and the format is dependent on the test environment. For virtual environments, OpenDrive[1] and OpenScenario[2] can be used as standardized formats. For proving ground testing, the scenarios need to be transformed in a format, which can be used for test equipment automation. In addition, KPIs need to be assigned to the test cases dependent on the test purpose to configure the required measurements. As a last step, the test execution platform configuration needs to be prepared. This includes the selection of the required models and hardware components and their communication. In general, there are two ways of test case generation. In offline test case generation, a complete test plan is prepared beforehand. This test plan consists of a series of test cases, which can be executed either in a virtual test environment or on the proving ground (see e.g. [4]). For online test case generation, an initial test plan is prepared and based on the results of these test cases further test cases are generated and executed on-the-fly (see e.g. [5]).

The *Evaluation & Reporting* phase takes the results of the test execution and assesses them. This can again be done online or offline. In the case, where test plans are updated on-the-fly, the respective KPIs also need to be calculated during the execution or immediately afterwards. Especially on the proving ground, measurements need to be taken and uploaded for further processing. Either way, we propose that the same data format is used in order to be able to reuse the KPI evaluation scripts. In our case, the Open Simulation Interface (OSI) [6] specification is used to describe sensor measurement data, no matter if it is coming from simulation or from real sensors.

The *Execution* phase includes the test execution environments. They and their potential usages are described in more detail below.

Cloud/HPC Simulation Especially in early phase testing, the huge test space needs to be scanned quickly. This requires a scalable and fast simulation environment, where many different parameter combinations can be tested in parallel. Usually, simulation environments are built for desktop simulation and are not prepared for scalability. This is especially true for co-simulations, which include various simulation tools. To support fast and scalable simulation, different cloud-based

[1] https://www.asam.net/standards/detail/opendrive/

[2] https://www.asam.net/standards/detail/openscenario/

simulation environments have popped-up recently (e.g. Metamoto,[3] Cognata,[4] etc.). These companies promise Simulation as a Service to train, test, debug, and validate automated vehicle software. Model.CONNECT provides a solution based on Docker Containers,[5] which can be used for setting up scalable co-simulations. The main advantage is the ability to use the same simulation infrastructure and models as for other test environments. Additionally, the interfaces to test case generation and test evaluation tools stay the same and can be used in the same way as for other test environments.

This setup can also be used for continuous integration and running test cases automatically for each new software version.

Model (Software) in the Loop This non-real-time test environment features all system aspects as models or software components. It is mainly used by function developers, who are testing their function (sensor fusion, trajectory planning, etc.) or reproducing erroneous situations identified in real-world testing. It further allows to compare different sensor concepts and algorithm parameterizations in a safe und reproducible way even before the availability of actual hardware prototypes. There are many providers for simulation environments (e.g. Vires VTD,[6] TASS PreScan,[7] Oktal Scaner,[8] etc.). All have their strength and weaknesses and depending on the testing purpose the one or the other might be suited better. There are also several in-house solutions to meet the specific requirements of companies. With respect to the overall toolchain, an important requirement of a simulation environment is the possibility to integrate it with real-time components to be able to reuse the same environment also in combination with hardware. This requires a modular structure and certain interface capabilities.

Hardware in the Loop Hardware in the loop is a very common environment to test different kinds of control units standalone or in combination. The main purpose is to identify timing and communication issues.

Driver (Human) in the Loop Driver/human-in-the-loop simulation provides a highly realistic driving experience to the human/driver which is indispensable for investigating the influence of new products on human [7].

Vehicle in the Loop This test environment bridges the gap between conventional HiL testing and real road testing for efficient and reproducible validation of fully integrated vehicles. One concept for vehicle in the loop testing is the DrivingCube concept [8]. The advantage is the transfer of tests to a controlled environment where

[3]https://www.metamoto.com

[4]https://www.cognata.com

[5]https://www.docker.com

[6]https://vires.com/vtd-vires-virtual-test-drive

[7]https://tass.plm.automation.siemens.com/prescan

[8]https://www.avsimulation.fr

the risk for man and machine is reduced and the effectiveness is increased. It can be either a chassis dynamometer or a powertrain test bed for complete vehicles. Since the interaction with the steering system was not considered until now, the test of lateral controllers was not feasible. Nevertheless, new developments enable the stimulation of the steering system of the vehicle while keeping the interference with the vehicle at a minimum. This is achieved by a mechanical decoupling of tires and the steering system. Instead, a compact and universal steering force module is used to induce forces to the tie rod [9]. An additional requirement is the need to stimulate the environment sensors. There are several approaches for sensor stimulation (e.g. over-the-air stimulation of radar sensors [10] or the use of moving bases [11]).

Proving Ground Currently, tests on proving grounds are mainly focusing on standardized testing of active safety systems [12] and evolving gradually towards driver assistant systems. These testing approaches will not scale for automated driving, which means that test automation and new techniques (e.g. augmented reality on proving grounds) are required. One challenge is the increased complexity of orchestrating a growing number of robotic test equipment on proving grounds. Another one is the ability to collect, exchange, and compare data from test tracks as well as from virtual test environments. This requires a certain standardized data format [13]. Currently, for testing of active safety systems, the proving ground is the last instance of testing activities. Nevertheless, with the growing importance of virtual validation, proving grounds will also be an important environment for model calibration and validation.

Road testing is of course still a required test environment. Nevertheless, because of the different scope and requirements (no test planning, no models, etc.) we will not discuss this test environment in detail here.

In our setup, the co-simulation platform Model.CONNECT[9] and real-time integration platform Testbed.CONNECT[10] are used to couple the different elements (models and hardware components) and to take care about the communication and timing issues between these elements. Model.CONNECT interlinks simulation models into a consistent virtual prototype, regardless of the tool they were created with. Simulation and hardware components can be easily integrated into a complete virtual/real system. This facilitates the continuous, model-based development in a wide range of powertrain and vehicle applications (e.g. driver assistance systems). Testbed.CONNECT connects simulation models with the testbed. The testbed engineers do not have to wait for all the hardware components to be available but can simply replace them with the corresponding simulation models. Even complex models from the concept phases can be easily and robustly integrated on any kind of testbed. The system developers gain a deeper understanding of the complex interactions of their systems by using their models at the testbed. The acquired findings are continuously used to the further improve the simulation models. The

[9]https://www.avl.com/-/model-connect-

[10]https://www.avl.com/web/guest/-/testbed-connect-

concrete architecture used within the co-simulation and integration platform is described in more detail in Sect. 3.

Model.CONNECT is furthermore used as the interface for test planning and evaluation. This means that Model.CONNECT is taking care about the execution of the test cases in the respective test environment. There is also one interface to collect the simulation results for evaluation and post-processing.

Another important building block is a model repository, which can be used to manage and access different versions of different simulation models (e.g. various levels of detail for different testing purposes). The tight integration in Model.CONNECT enables the definition of different simulation configurations, which can be switched on the fly as needed.

3 Test Execution Environment: Architecture

The Test Execution Platform covers all relevant aspects of an automated cyber physical system as described in chapter "ENABLE-S3: Project Introduction" and shown in Fig. 2. An essential aspect of automated driving functions is their tight interaction with the environment (i.e. other traffic participants). This means that the environment needs to be simulated in an appropriate manner. For the perception of the environment, the automated vehicle uses different kinds of sensors. Therefore, it is also important to represent the characteristics and potential limitations of the sensors in terms of sensor models in the simulation. The required level of detail and accuracy of these models depend on the testing purpose. For testing the trajectory planning, ideal object level sensor models might be sufficient. For testing sensor fusion algorithms or even for system validation, more accurate and detailed sensor models (either phenomenological or physical) reflecting the properties of the actual sensor types are required. If the automated vehicle uses communication with the infrastructure or other vehicles, this aspect needs to be considered in simulation as well.

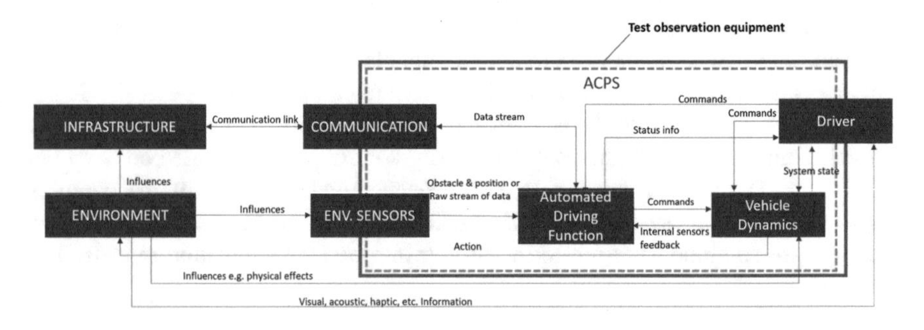

Fig. 2 Test execution platform—detailed architecture

Depending on the function under test, the dynamics of the vehicle need to be modeled as well. It gets evident that a modular structure is important here as well, since the required level of detail of the models differs for different tests. A modular structure enables an efficient exchange of models. Usually these models are developed using different simulation tools, each specialized on a specific aspect.

A co-simulation platform is required to couple the different simulation models to a holistic closed-loop simulation. The co-simulation platform is responsible for establishing the communication and minimizing latency effects during simulation. Depending on the development stage, there will be different instances of the test execution platform. For example, in a MiL environment all components will be available as simulation models. Later simulated components will be step by step substituted by real physical components resulting in a mixed environment of real-time and non-real-time components. Especially this latter case presents additional challenges for the test execution platform as there are various real-time systems with different properties that have to be integrated. Additional requirements have to be fulfilled; there are hard real-time conditions that have to be considered, communication is needed in real-time, synchronization of real-time and wall clock time as well as communication delays have to be handled by the platform. Real hardware has to be operated safely which means that safety mechanisms have to be implemented in order to avoid damage of hardware. To facilitate the integration of different models and real-time components, standardized interfaces can be used (FMI[11] for non-real-time communication and DCP[12] for real-time communication).

Another important aspect, which influences the performance of the simulation is the system decomposition. The main questions are how to split aspects in different models, how to distribute the execution of the models and how to define the model interfaces. Splitting and distributing models can be an important mean to improve the performance (e.g. distribution to different cores). The definition of model interfaces is often constraint by I/O capabilities of simulation tools but can also include a lot of integration experience and know how (e.g. taking constraints by physics as low inertia vs. high inertia into consideration). This knowledge could be made explicit and reusable by providing best practice templates and standardized interfaces.

4 Open Simulation Interface

Traditional automotive testing is mainly based on time-series signals and most tools are designed and optimized for this kind of data. For testing automated cyber-physical systems, large data sets with complex data types/data structures (object lists, images, point clouds, etc.) have to be exchanged. This also includes the

[11] https://fmi-standard.org

[12] https://dcp-standard.org/

challenge of how to interpret the data. Regarding the generic test architecture, OSI provides a standardized interface for environment and environment sensor data, which can be used by automated driving functions. Hence, this interface enables the connection between function development frameworks and the simulation environment.

The OSI implementation is based on google protocol buffers[13] that provide a mechanism for serializing data which is defined in language-neutral and platform-neutral messages. OSI specifies an object-based environment description by defining messages describing the ground truth as well as the sensor data for testing in simulation environments. Ground truth data contains unmodified object data describing the environment of the ego vehicle that is the output of the simulation framework. It is based on a global reference frame. In contrast to the ground truth, sensor data describes object data in the environment relative to one specific sensor which is thus based on the sensor reference frame. This data structure contains input as well as output of statistical sensor models.

OSI is in an early development stage and thus the specification can be subject to change. Further, there are large data sets necessary to describe environment data and experiences are needed in using OSI for simulations with such large data sets.

In addition to the data structure, OSI sensor model packaging is specified that defines how the OSI sensor models have to be packaged as FMU for use in simulation environments.

During the project a simple OSI demonstrator has been set up, which shows the application of the standardized interfaces to test an Adaptive Cruise Control (ACC) function. OSI is used for the communication between the environment simulation and the function under test. The main task of the environment simulation is the generation of realistic ground truth data based on the selected scenario. The output of the environment simulation varies from general simulation data to simple and complex object lists and beyond to realistic raw sensor data. For the demonstrator "VIRES Virtual Test Drive" (VTD) is used as environment simulation software, which covers the full range from the generation of 3D content to the simulation of complex traffic scenarios. In the demonstrator mainly object lists are generated in an OSI compliant form (osi3::GroundTruth).

Model.CONNECT™ uses a TCP connection to receive the OSI ground truth data from VTD and to hand it over to a sensor model. The sensor model has been implemented as an FMU running in Model.CONNECT. For the demonstrator a simple phenomenological sensor model was implemented, which transforms the global coordinates to relative coordinates with respect to the ego car and applies a filter function to the detected objects list from the ground truth data. The filter reduces the detection range of objects based on precipitation, fog and illumination. Based on this 'real' sensor data the ACC function is tested.

To compare ground truth and the sensor data, the object lists are visualized in ROS (Robot Operating System). Therefore, OSI data is exchanged between

[13]https://developers.google.com/protocol-buffers

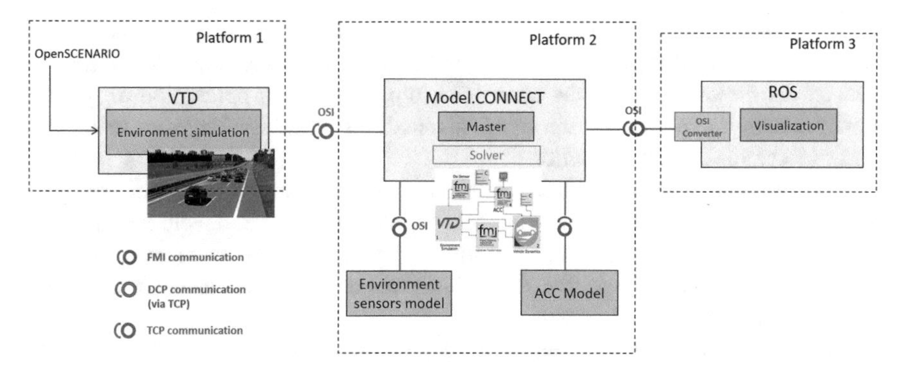

Fig. 3 Setup of the Open Simulation Interface demonstrator

Model.CONNECT and the ROS environment via DCP (Distributed Co-Simulation Protocol) over UDP. DCP is applied to evaluate this upcoming standard for distributed simulations. This standard should facilitate managing co-simulations and enables a standardized way of integrating various tools as well as real physical components. Figure 3 illustrates the demonstrator setup.

The demonstrator has mainly been used to gather experience with the specification and to assess the practical applicability. Results have been fed back into the OSI working group.

5 Simulation Model Preparation

An important ingredient for most of the test execution environments are simulation models. Simply simulating millions of test kilometers is of no value, if the simulation does not reflect the reality, at least to a certain degree. This means, the simulation is only as good as the match between the simulated signals from the sensor and vehicle model and corresponding values in the real vehicle. A model should be developed for a specific purpose (or application) and its validity determined with respect to that purpose. There are various validation techniques described in [14].

With respect to environment simulations, there are several factors, which can lead to significantly different results between tests in the real and virtual world. It starts with the modelling of the scenarios and 3D-environment, continues with material parameters of the modelled simulated objects, goes on with measuring and transferring weather conditions from real world to environment simulation and ends in the accuracy of the simulation models of sensors and ego-vehicle.

Here, we are mainly focusing on sensor models. We distinguish between generic sensor models, capable of simulating the main features of different sensor types (such as ultrasonic sensors, cameras sensors, radar sensors or LiDAR sensors), and specific sensor models used to replicate the behavior of a specific version of a sensor from a specific manufacturer. Specific sensor models have to model the normal behavior of a sensor type as well as the peculiarities and imperfections of a specific sensor. As these imperfections often lead to a non-perfect perception of the real environment, it can lead to safety-critical situations which are the most interesting cases in vehicle system validation.

The required level of realism depends on the development phase. For early function verification, ideal sensor models are sufficient as the developed functions will have to work with sensors from different manufactures. In many cases, it is even not decided, which specific sensor instances or makes will be used in the final automated cyber-physical system.

Therefore, we need two sensor model preparation activities: first, tune the parameters of an ideal sensor model to create a specific sensor model for a specific instance of a sensor; second, validate that the specific sensor model replicates the behavior of a specific sensor in all relevant scenarios and weather conditions.

More information about sensor models and sensor model architectures is given in the respective chapters ("Radar Signal Processing Chain for Sensor Model Development", "Camera Sensor System Decomposition for Implementation and Comparison of Physical Sensor Models", "Functional Decomposition of Lidar Sensor Systems for Model Development").

Here, we will spend a few more words on the challenges of sensor model parametrization. Sensor model parametrization describes the procedure to tune a generic sensor model in a way that it reflects the properties of a specific real sensor. Therefore, a lot of measurements have to be taken. First, the behavior of the sensor has to be measured. This can be done best on a proving ground, because the second important part is the measurement of the ground truth. This means that all involved traffic participants need to be instrumented with very accurate measurement devices. Only then we can determine the detailed position of each traffic participant at any time. Furthermore, all the participants of the scenario are under our control. This means that we can exactly determine their dimensions, materials, and behavior. All this information is required to reconstruct the ground truth information with the required accuracy. Another aspect which needs to be considered are environment conditions (such as weather conditions). This data is needed to calibrate the environment simulation with weather data to ensure, that different environment simulations deliver the same information to sensor models at same weather conditions. Unfortunately, no standardized metrics for environment conditions exists and it is not completely clear which metrics are required.

The basic idea of the parametrization procedure is to transfer the ground truth information into the environment simulation. The sensor models are then fed with ground truth and weather data from the simulation. At the same time the sensor model parameters are adjusted until the simulated output of the sensor model matches the measurement taken with the real sensor.

There are still a lot of unknowns for this parametrization procedure and for the final validation that the model is reflecting the reality with sufficient accuracy for the specific validation task. Nevertheless, this is a major prerequisite to make virtual validation a useful alternative to real-world testing of automated cyber-physical systems and thus to reduce the required test effort.

6 Conclusion

In this paper we propose a seamless validation toolchain, which aims to overcome the challenges of testing automated cyber-physical systems in general and automated driving functions in concrete. The validation toolchain promotes an open and modular architecture to use different simulation models as well as to support reuse of test cases throughout various test execution environments. We highlighted that there are already first standardization activities for interfaces, which support the modular structure. Nevertheless, there are still some challenges to ensure that the simulation reflects the reality.

References

1. Koopman, P.: The Heavy Tail Safety Ceiling; Automated and Connected Vehicle Systems Testing Symposium 2018
2. Winner, H., Wachenfeld, W.: Effects of autonomous driving on the vehicle concept. In: Autonomous Driving: Technical, Legal and Social Aspects, pp. 255–275 (2016)
3. Ahmetcan, E., Kaplan, E., Leitner, A., Nager, M.: Parametrized end-to-end scenario generation architecture for autonomous vehicles. In: CEIT Conference 2018, Istanbul (2018)
4. Schneider, H., Weck, T.: Efficient active-safety-testing using advanced on-road data analysis and simulation. In: SIMVEC Conference (2018)
5. Beglerovic, H., Ravi, A., Wikström, N., Koegeler, H.M., Leitner, A., Holzinger, J.: Model-based safety validation of the automated driving function highway pilot. In: Pfeffer, P. (ed.) Proceedings of 8th International Munich Chassis Symposium 2017. Springer, Wiesbaden (2017)
6. Hanke, T., Hirsenkorn, N., Van-Driesten, C., Gracia-Ramos, P., Schiementz, M., Schneider, S.: Open simulation interface: a generic interface for the environment perception of automated driving functions in virtual scenarios: research report. http://www.hot.ei.tum.de/forschung/automotive-veroeffentlichungen/ (2017). Accessed 28 Aug 2017
7. Moten, S., Celiberti, F., Grottoli, M., van der Heide, A., Lemmens, Y.: X-in-the-loop advanced driving simulation platform for the design, development, testing and validation of ADAS. In: Intelligent Vehicle Conference (2018)
8. Schyr, C., Brissard, A.: Driving Cube—a novel concept for validation of powertrain and steering systems with automated driving. In: Advanced Vehicle Control: Proceedings of the 13th International Symposium on Advanced Vehicle Control (AVEC'16), Munich, Germany, 13–16 September 2016, p. 79. CRC Press (2016)
9. Förster, M., Hettel, R., Schyr, C., Pfeffer, P.E.: Lateral dynamics on the vehicle test bed – a steering force module as a validation tool for autonomous driving functions. In: Pfeffer, P. (ed.) Proceedings of 9th International Munich Chassis Symposium 2018. Springer, Wiesbaden (2019)
10. Gruber, A., et al.: Highly scalable radar target simulator for autonomous driving test beds. In: 2017 European Radar Conference (EURAD), Nuremberg, pp. 147–150. https://doi.org/10.23919/EURAD.2017.8249168

11. Gietelink, O.J., Ploeg, J., De Schutter, B., Verhaegen, M.: VEHIL: test facility for fault management testing of advanced driver assistance systems. In: Proceedings of the 10th World Congress on Intelligent Transport Systems and Services(ITS), Madrid, Spain, 16–20 November 2003. Paper 2639
12. T. European New Car Assessment Programme. Test protocol – AEB systems, EuroNCAP, Test Protocol 1.0, July 2013
13. Knauss, A., Berger, C., Eriksson, H., Lundin, N., Schroeder, J.: Proving ground support for automation of testing of active safety systems and automated vehicles. In: Fourth International Symposium on Future Active Safety Technology Toward Zero Traffic Accidents (FASTzero) (2017)
14. Robert, G.: Sargent, verification and validation of simulation models. In: Jain, S., Creasey, R.R., Himmelspach, J., White, K.P., Fu, M. (eds.) Proceedings of the 2011 Winter Simulation Conference. IEEE Press, Piscataway, NJ (2011)
15. Leitner, A.: Generic test architecture ENABLE-S3. Technical Report. https://www.enable-s3.eu/ (2018)

Part III
Applications

Validation Framework Applied to the Decision-Making Process for an Autonomous Vehicle Crossing Road Intersections

Mathieu Barbier, Javier Ibanez-Guzman, Christian Laugier, and Olivier Simonin

1 Introduction

Through the media almost on a daily basis, we hear of the benefits and potential brought by Autonomous Vehicles to society. This is referred in terms of accessibility to land transportation for people unable to drive, improving driver productivity by reducing or eliminating altogether the driving load and improving safety by minimising driver errors [1].

Navigation in road networks is a complex task due to the randomness that exists on the behaviour of the different entities sharing the same space. Multiple situations emerge, requiring different behaviours of vehicles as they navigate autonomously. For this purpose, it is necessary to have a digital representation of the world through the acquisition of data from multiple sensors. This is then used to reason and determine the vehicle behaviour before a given situation. Finally, the command signals are generated to actuate the vehicle to obtain the vehicle control. These cyber-physical systems (CPS), if they are to operate autonomously, (under full computer control) need to address complex environments, need to be acceptable and they must be safe and operate according to the expectations of all stakeholders.

For the purposes of this chapter, Autonomous Vehicles (AVs) are regarded as those where the computer has overall control of the vehicle navigation function.

M. Barbier · J. Ibanez-Guzman (✉)
R&D Division, Renault S.A., Boulogne-Billancourt, France
e-mail: javier.ibanez-guzman@renault.com

C. Laugier
Inria Chroma Team, Villeurbanne Cedex, France

O. Simonin
INSA Lyon, CITI Lab, Villeurbanne Cedex, France

Inria Chroma Team, Villeurbanne Cedex, France

A. Leitner et al. (eds.), *Validation and Verification of Automated Systems*,
https://doi.org/10.1007/978-3-030-14628-3_15

Thus, according to the SAE definition of vehicle automation levels, this chapter addresses AVs Level 4 SAE of automation[1] which can navigate autonomously without any driver intervention in a designated area.

Within a road network, intersections are likely the most complex areas, most entities (i.e. vehicles, motorcycles, pedestrians, emergency service vehicles, etc.) converge to them. Thus, the crossing of an intersection by an autonomous vehicle is difficult. Further, when intersections most of the time vehicles negotiate their passage with other road users, making such a task very complex for the machine intelligence controlling the vehicle behaviour. From a road accident statistics perspective, intersections represent areas where most accidents occur due to human error, with those involved being either young or elderly drivers [2].

Whilst it is imperative to test all vehicle functions as much as it is statistically credible, this is a major endeavour and subject of much research. This chapter addresses the evaluation of the decision-making & navigation process as an application of the ENABLE-S3 V&V framework to the crossing of road intersections. To narrow the scope, other autonomous driving functions such as localisation, perception, etc. are included in a functional level, that is, they operate nominally. The chapter is structured as follows: Sect. 2 provides a systems perspective background on autonomous vehicle functions, their complexity and testing, and validation challenges. Section 3 introduces the system under test, it centres on the Test Platform, Test Management and Validation & Verification management. It presents the ENABLE-S3 framework from a vehicle automation perspective and formulates the notion Key Performance Indicators (KPIs). Section 4 introduces the test platform by initially providing details of the system under test—the behaviour planner. It then describes the simulation environment configured for its evaluation. Section 5 introduces the Key Performance Indicators defined and applied for the evaluation of the system under test. This is equivalent to what is known as test management. The chapter concludes by providing some perspectives with regards to the evaluated system, the evaluation of the testing system deployed as well as the importance of statistical model checking algorithms.

2 Background: System Perspective

Autonomous vehicles represent one of the most complex CPS. To navigate in public roads, they must demonstrate the same capabilities as human drivers and must be able to respond to the random situations that can be encountered when driving in public road networks. In a very simplistic manner, they can be regarded as a computer-controlled system interacting with its immediate environment. That is, the system will need to sense the environment, reason and decide on what it has sensed plus its own state and ultimately act on the environment.

[1]SAE J 3016-2018: Taxonomy and Definitions for Terms Related to Driving Automation Systems for On-Road Motor Vehicles.

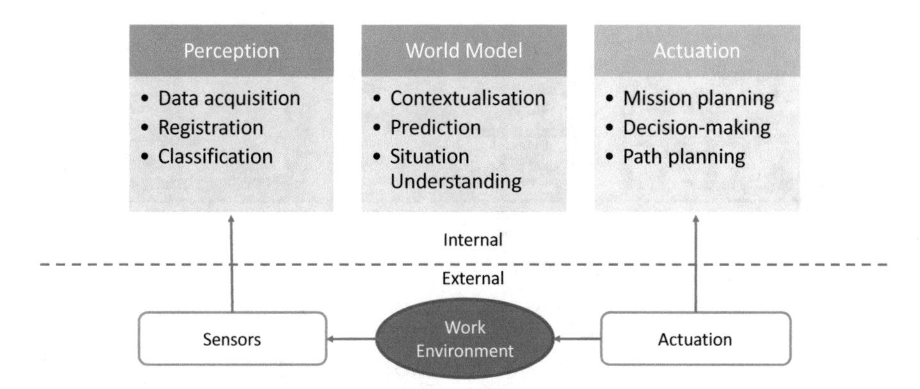

Fig. 1 Simplified block diagram representation of a computer-actuated system (a CPS)

Figure 1 shows an oversimplified view of a computer-controlled system when interacting with its work environment. It can be partitioned into Perception, World Model and Actuation. The first is concerned with building a digital representation of the perceived environment. The second contextualises this representation by incorporating stored knowledge, for example in the form of maps. The third provides the reasoning including tasks such as situation understanding and behaviour generation. The latter generates commands to the physical system so that it interacts with its environment. That is, it generates the control signals that are used to control the vehicle.

In this section, a summary description is provided on the functions needed for a passenger vehicle to demonstrate autonomous capabilities. It shall provide the reader with comprehensive insights into the complexities of such CPS. This is followed by an examination of the complexities that exist when driving a vehicle under computer control. The section concludes with the challenges related to the testing and validation of autonomous vehicles in terms of driving scenarios and safety.

2.1 Functional Description of an Autonomous Vehicle

Autonomous vehicles have been developed since the mid-90s. There have been two milestone projects: Prometheus in the European Union, enabling a vehicle to cross Europe through different motorways mainly using computer vision [3] and NavLab from researchers at Carnegie Mellon University who crossed the USA on a modified vehicle using computer vision and active sensing as well as machine learning methods [4, 5]. Since then multiple efforts across the automotive industry and the technological giants such as Google (parent company of Waymo), as well as ride-hailing companies such as Uber have emerged. The major milestone is

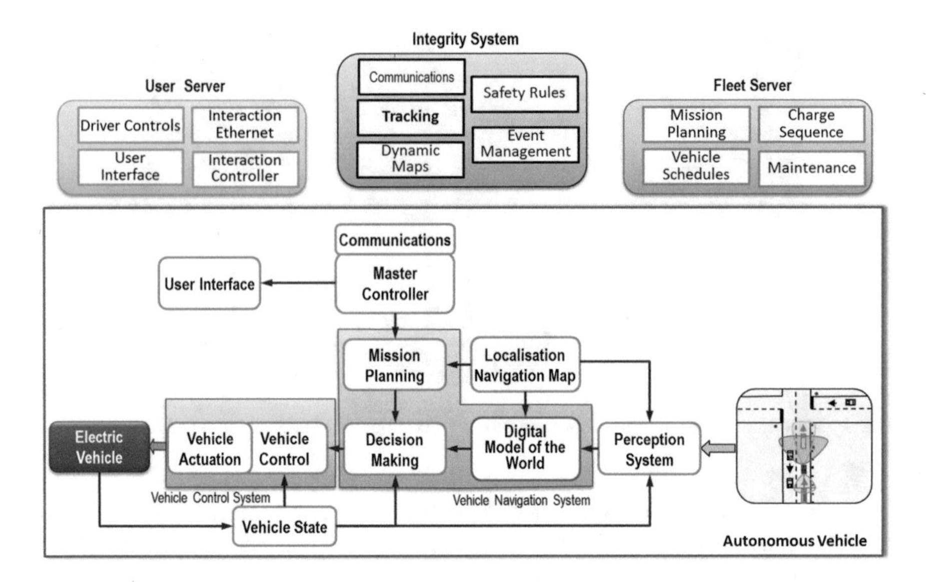

Fig. 2 Functional architecture for an autonomous vehicle and its ecosystem

Waymo's report of having driven 10 million miles in autonomous mode (with a safety driver) towards the end of 2018 [6].

A summary description on the early evolution of architectures for autonomous vehicles can be found in [7]. It provides information on the manner these vehicles have evolved.

Figure 2 shows the functional architecture for a vehicle deployed by Renault as an unmanned Valet parking within a private compound. It reflects the basic components that can be found in most AV architectures. It includes the ecosystem needed for the operation of such vehicles when deployed as large fleets. They are only included for completeness, as in this chapter we are concerned with the AV as a standalone unit.

The major functions within an autonomous vehicle are shown in Fig. 2. Each of them can be considered as a system due to their complexity. The functions used to test the system object of the study are described with some detail.

Perception System It acquires data representing the immediate environment. This data is processed via a combination of machine learning methods and computer vision techniques. There are no sensors capable of perceiving everything, thus multisensory fusion approaches are preferred using video cameras and LiDAR (laser rangers). Much progress has been achieved, however, to date there is no complete solution. Sensor layout, occlusion, sensor resolution, etc. imply that there are conditions where false positives and false negatives exist, resulting in hazardous situations. These systems need to be tested in difficult conditions, which currently cannot be modeled or set in a representative way. The simulation of sensors is difficult. Whilst models exist, the environment needs to be simulated so it reacts to the sensor physics. Each observed item needs to have such features, further irradiation processes need to be considered, the gap between the sensor and the perceived object filled by light conditions, etc. For the testing of full CPS, it represents a major constraint.

Localisation System The vehicle pose (position and attitude) is a fundamental function. By knowing where the vehicle is, it is possible to determine the road the vehicle needs to take to reach its destination. Current solutions combine global navigation satellite systems (GNSS) with vehicle odometry, inertial information as well as external GNSS corrections (RTK messages via cellular networks) to determine the absolute vehicle location [8]. However, in case of tunnels, strong multipath and tree canopies, etc., preclude the use of GNSS signals, other techniques are used to estimate the relative localisation of the vehicle. These are based on the Simultaneous Localisation and Map Building (SLAM) as well as optical odometry [9]. They enable the estimation of the vehicle position when no GNSS signals are available. Currently, solutions are combined into what is known as high integrity navigation. Testing these systems is also complex. GNSS simulators exist for GPS and other constellations, however, the physics of the operating environment needs to be considered, which is more complex. The same applies to the simulation of SLAM-based methods.

Maps They can be considered as the storage of a priori knowledge about the road network. AVs use lane level navigation maps (HD-MAPS). They include good geometric descriptions on road level, and navigation information like the description of paths at complex intersections, the attributes associated to the lanes, the presence of features like pedestrian crossings, etc. [10]. They differ from commercial navigation maps, like those used in navigation systems of passenger vehicles by the precision of the geometric road descriptions and the richness of attributes. Information from maps is shared with other systems such as the localisation system in the form of lane marks. For the case of road intersections, without maps the vehicle must determine that it is arriving to an intersection and its options open once it reaches. This is an almost impossible demand to the perception system. Nevertheless, maps can be outdated or simply contain errors, thus mechanism to detect their integrity are needed [11].

World Model This is the digital model of the world that combines the perceived environment with the navigation maps, to provide a contextualised representation of 'what is going on' within the immediate navigation environment of the AV. Each of the perceived entities together with the vehicle state is represented in the World Model [12]. This is equivalent to the human mental model of our surroundings that we use to make decisions. By associating the perceived with maps, errors in localisation and perception can be unmasked. For example, if the vehicle localization is wrong, perceived vehicles might be associated with the wrong lane. Such error can induce hazardous conditions. The world model allows also to project into the future, e.g. the likely trajectories or intentions of the perceived entities, to facilitate understanding and infer what could happen next.

Decision-Making and Navigation The behaviour of the vehicle is determined in this system. It generates the trajectory the vehicle needs to follow the trajectories are generated after a complex process that implies understanding the current situation of the AV, evaluating what the perceived entities might be doing next, incorporating the destination where the vehicle will be going and the situation of the road network (next junctions, intersections, pedestrian crossings, etc.). The challenge here is to incorporate uncertainty into the whole process, as it exists at different levels: perceived world, localisation estimates, map precision, behaviour of the perceived entities, etc. Different techniques can be used to solve this task. These include different optimisation methods to assess a cost function, Markovian processes to incorporate hidden variables like intention, etc. [13, 14]. Further details on this system will be included in later sections.

Vehicle Control System The command signals in terms of heading and speed or a collection of points that describe the lateral and longitudinal displacement of the vehicle (could even be torque commands) are transformed into control signals to the controlling actuators like the accelerator, steering wheel and brake. The system ensures the stability and response of the vehicle to the commands generated by the vehicle intelligence. Different safety mechanisms are included to avoid hazardous situations. The control algorithms vary according to their level of integration, that is vehicle OEMs having direct access to the actuations mechanisms whilst other need to find ways to adapt them [15, 16].

Actuation System This is the mechanism that acts directly on the vehicle actuators when controlled by a computer. Two different approaches exist: Adaptation of the actuators through interference with the CAN-bus messages to command brake, accelerator and steering wheel. It is basically reverse engineering & hacking to an extend [17]. Controlling directly the actuators via CAN-bus message with the support of tier 1 suppliers is privileged by the vehicle OEMs. Likely these represent the safest in vehicle actuation. Changes within the vehicle architecture might not be easily perceived by the reverse engineering approaches, resulting in hazardous situations. Other systems and modules exist as described in Fig. 2. However, for the purpose of this chapter, those described are part of the System Under Test (SUT).

2.2 Operational Complexity

Autonomous Vehicles operate in specific areas, these are mainly geofenced either by the availability of HD maps or the permissions provided by public authorities for trials. Recently, the National Highway Traffic Administration (NHTSA) has introduced the notion of Operational Design Domain (ODD) that describes in a detailed manner the conditions in space, traffic and safety under which AVs ought to operate [18]. This is very important as the ODD defines also the conditions of use of such vehicles. An ill-defined ODD results in user confusion with consequences

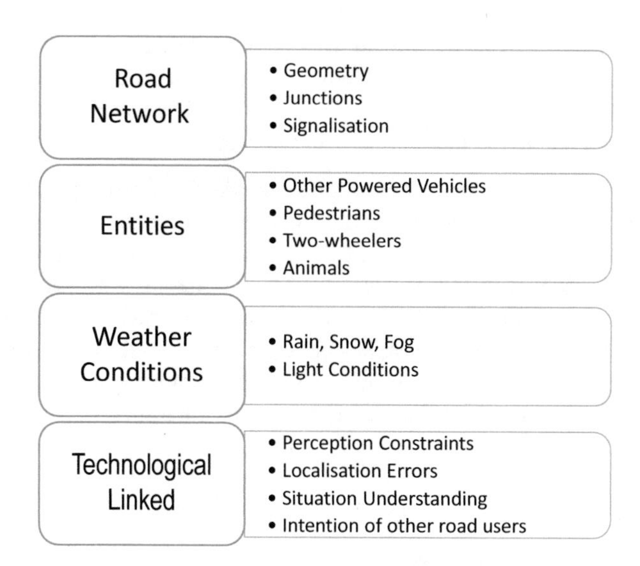

Fig. 3 Sources of complexity related to autonomous driving

that might be fatal. Thus, an ODD defines precisely the working conditions of the AVs and thus attempts to reduce the perceived operational complexity.

For this work, the sources of complexity can be classified as: the road network, the entities sharing the same road network, weather conditions, and those related to the technological solutions. Figure 3 presents a summary of these sources.

Road Networks In many cases, these have evolved from linking medieval cities to linking large metropolis with core urban centres—unlike the matrix structure of US cities or new developments in China. Thus, road geometry is an important factor. Junctions are likely the most complex part of road networks. Traffic flow converges to intersections, all entities converge to them, so they can traverse across streets, etc. Signalisation can be complex, in the form of traffic signs and writings on the road surface or active signals in the case of traffic lights. A high degree of complexity exists just by observing how road networks evolve as they cross rural, peri-urban, rural, urban or core urban areas.

Entities There are different types of vehicles sharing the same road network, each with their own features. All drivers behave somehow differently. Pedestrian are a major source of complexity; their behaviours can be considered almost random. Motorcycles have always represented a source of complexity due to their manoeuvrability and speed. However, these days the emergence of motorised bicycles and other two-wheelers attaining fast speeds has compounded the problems. The presence of animals in the road network or next to it represents an area of concern, as perception systems might have difficulties. They are a potential source of outliers during testing.

Weather Conditions As with human drivers, they are disturbances to perception systems and to the vehicle response. Many of the sensors used have difficulties with light conditions and with the presence of fog or snow. The weather also disturbs the vehicle response due to the interaction with the ground.

Technological Linked The different systems are far from perfect. Perception systems have their limits, there are errors from localisation systems, multiple times situations are encountered that are not fully understood by the computers, it is very difficult to infer the intention of other traffic participants, etc. All that makes it impossible to plan with certainty. Further, many of the tasks when driving a car are related to the negotiation that exists between the AV and the other entities. All these constraints imply that the systems we test are complex and different functions have different performance levels. The whole makes the testing of autonomous CPS very complex.

2.3 Validation Challenges

Individual systems encountered in an AV are validated in a stand-alone manner. However, their interdependencies like in the case of the world model (convergence of localisation, map and perception) implies that full tests are needed as systems are integrated into the final vehicle. Whilst simulation and emulation can facilitate this process, it is insufficient as there is the need for tests to validate the simulation models.

The AV system architecture needs to have a modular structure that allows the stand-alone operation of the various systems or their partial integration, so that they can be fully tested. The Supervisory Control System shown in Fig. 2 has that function. It allows for the individual control of each of the system components and thus their testing.

Validation can be grouped in three phases, each has a different purpose and challenges. The phases can be described as follows:

System Verification Each system is verified with regards to a particular ability. Test procedures exist for example for parking manoeuvres [19]. The performance of the system is compared with the ground truth that enable a quantified measure of performance [20]. However, for some systems like the perception of the world model this is more difficult, as there is at times no ground truth. Further, exhaustiveness is lacking, it is difficult to have the different conditions under which the systems would operate.

The notion of ODD is very powerful in this sense, it allows for narrowing the test as a function of where the vehicles are to be deployed.

Simulation plays an important role, it allows for the testing of extreme cases, to do the preliminary debugging, etc. However, in many cases validating simulation models is also a challenge, to provide confidence upon them, as explained in more detail in the chapter "Seamless Tool Chain for the Verification, Validation and Homologation of Automated Driving".

The exhaustivity of tests is a major concern, if a parallel is made with the validation of driving assistance systems, for production vehicles, there is a major gap.

Test Track Validation Initial full tests are done in test tracks or modified compounds (e.g. old industrial centres or military compounds). Different scenarios are constructed, and evaluations made. Emphasis is made on the safety content of such systems, as this allows safety officers of companies as well as regulators to test the capabilities of the systems prior to their deployment in public roads.

One notable example is the Centre of Excellence for Testing & Research of AVs (CETRAN) at the Nanyang Technological University in Singapore. The focus is not on the development of AV technologies but on how these systems should operate, developing testing requirements, and establish an international standard for AVs. The CETRAN Test Circuit provides a simulated road environment for the testing of AVs prior to their deployment on public roads, and it complements the AV testbed in an area of Singapore known as 'one-north'.[2] The track testing will have the dual function of testing the most critical functionality in the real world, while at the same time validating that the simulation in the virtual world is correct.

Whilst the above facilities are a good contribution, identifying all edge-cases remains a major issue. Further, to experiment in situations where hazardous conditions exist is very difficult. Hence, there is interest on additionally developing a digital world that allows for the testing of the vehicles in extreme conditions. This has been done on a project by Renault that aims at developing a 'digital twin' of this site in cooperation with the ASV project together with NTU, SNCF, Systra, System-X and AV Simulation.[3]

Capability Validation The validation of capabilities of AVs is amply documented in the literature and technical press. Different test sites exist notably in California, Texas, Nevada and Arizona, where different companies like Waymo, Cruise-GM, Aurora, Aptiv, etc. test their systems. The only parameter used is the Miles per Disengagement. This metric reports the average distance vehicles operate without human intervention as reported by law to the Department of Motor Vehicles in California (DMV).[4] This can be used only as an indicator of the progress made.

Testing AVs is very costly, the prototypes are still expensive likely above 150,000 Euros. The logistics needed is also very high, there is the need for a skilled safety driver, support engineers, safety officers, an infrastructure to store, and analyse performance data, etc. Despite different procedures accidents have occurred. A fatal one in 2018 has been widely documented in the press. Another challenge resides on the edge-cases, despite the millions of kilometres driven, there are still situations that systems are unable to handle. Thus, as of March 2019 there are

---

[2]http://erian.ntu.edu.sg/Programmes/IRP/FMSs/Pages/Centre-of-Excellence-for-Testing-Research-of-AVs-NTU-CETRAN.aspx

[3]https://www.irt-systemx.fr/en/project/asv/

[4]https://www.dmv.ca.gov/portal/dmv/detail/vr/autonomous/auto

still no vehicles without drivers being tested in public roads despite some early attempts. In October 2018, Waymo accomplished 10 million miles of autonomous driving in public roads.[5] In addition to these, millions of miles were driven in virtual environments. This alone shows the complexity of testing and the founded fear to edge-cases.

Validating AVs remains a major challenge.

3 System Under Test and the Test Platform

This System Under Test (SUT) addresses the decision-making & navigation system of an autonomous vehicle, crossing a road intersection, mainly the cross-cutting manoeuvre where most accidents occur, cutting collisions. Figure 4 represents such scenarios, it includes the subject vehicle (SV) and the other vehicle (OV). Three types of scenarios are examined: Scenario A, where the SV is required to yield and Scenario B and C, where the SV has the right of way.

The decision system will control the subject vehicle acceleration as it approaches the intersection according to the situation and associated uncertainties. A partially observable Markov decision process and an online solver are used to find the action to be performed by the vehicle to react to the situation and future evolution. The state and observation spaces include behavioural variables that represent three possible manoeuvres at an intersection: Pass, Stop and Yield. The reward function considers problems related to the driver's behaviour, comfort and risk. This is what this chapter considers as the SUT. It is to note that the other AV system components as described in Sect. 2 are not being tested, they are assumed to operate at a nominal level and thus there are no source of uncertainty or errors.

In this section a summary description of the decision-making system is presented, and the approach taken to define the Test Platform used for evaluation purposes is presented.

3.1 System Under Test: Decision-Making System

There are different approaches to determine the behaviour of the vehicle as it navigates autonomously. These can be summarised as generating control commands from sensory data, namely traditional planning and control, behaviour-aware planning and end-to-end planning. For this purpose, different algorithm and modes are used including Interactive Risk-based Motion Planning (e.g. Risk-RRT), Decision-

[5]https://www.theverge.com/2018/10/10/17958276/waymo-self-driving-cars-10-million-miles-challenges

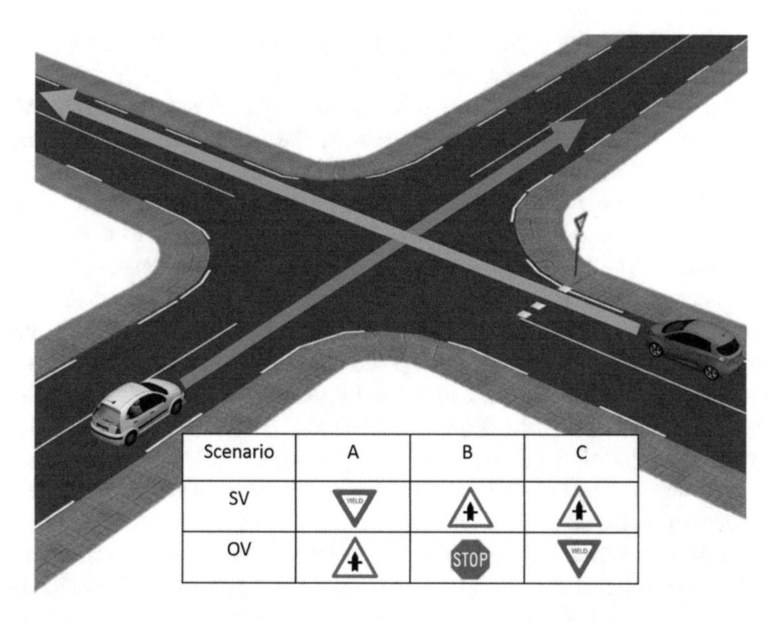

Fig. 4 Crossroad intersection representation including possible scenarios and the main actors: subject vehicle (SV) on the right side, other vehicle (OV) on the left side

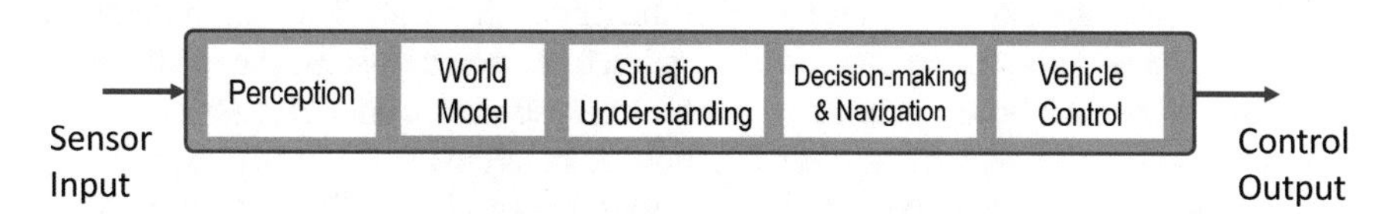

Fig. 5 Context of the SUT with respect to different AD functions

making using Contextual data & Bayesian Networks. Details on the different methods can be found in [14, 21].

Figure 5 shows the sequence of functions typically found to control a vehicle from the sensor input component to the actuation of the vehicle. The System Under Test relates only to the Decision-Making & Navigation component. It assumes the vehicle understands that it is to enter to a road intersection, its output is sent to the vehicle control that acts on the vehicle controllers. The perception and localisation functions are assumed to operated nominally and are outside the SUT.

Decision-making can include Probabilistic and Bayesian approaches that consider uncertainty into their representation. Partially Observable Markov Decision Process (POMDP) is a model that relies on a Bayesian network to represent dynamics of the system (with a special action variable) and a reward model to keep track of action values. Planning with POMDP models is well known and has been experimented for various robotic applications [21]. The complexity of problems, like autonomous driving, with large number of possible states, observations and actions affect the possibility to find the optimal policy in an acceptable time. Efforts on methods to keep track of the belief [22] and to solve POMDPs online [23], have made their usage effective in the last couple of years.

The state space of a POMDP represents all the variables required to characterise the situation. At road intersection crossings, the state space is composed of variables for the vehicle pose, speed and behaviour. The behavioural variables can represent, for example, the motion intention of drivers partitioned into stopping, hesitating, normal and aggressive [24]. These were inferred from a previously learned context. The behaviour variable is modelled to either constant velocity or constant acceleration that are then used to update the pose of vehicles. However, these behavioural variables are not used to calculate the reward. The driver intention enforced by regulation (traffic signs) in place is not considered in referred works. A coarse discretization is often used to represent the pose and velocity of vehicles as it simplifies the problem without performance loss. The situation understanding is done within the POMDP, but in a more complex system this function can be done by other systems. The POMDP reward model promotes states that are desirable for the system. Thus, states after the intersection are highly rewarded and penalties given to collision states. However, these long-term reward systems are insufficient for decision-making. It can be enhanced by penalizing changes of acceleration to guarantee the comfort of passengers or deviations from a reference speed [25].

The underlying components of the model used for the decision-making are detailed in this section.

Uncertainties are present at all levels of the navigation task. This can be present as lack of precision on the spatial representation of detected objects, the localisation estimates, maps, etc. Ignoring uncertainties is likely to lead to an under-estimation of risk in some situation whilst for other situations to result in a conservative behaviour. If these uncertainties are fixed to reduce complexity, this is likely to culminate into deadlock situations as well with vehicles taking hazardous decisions. In the proposed solutions a probabilistic reasoning approach is used, it models risk as an expected penalty. A POMDP models the decision process of an agent acting in an uncertain environment. Figure 6 represents the interaction between the vehicle (agent) and the environment that includes the other vehicle.

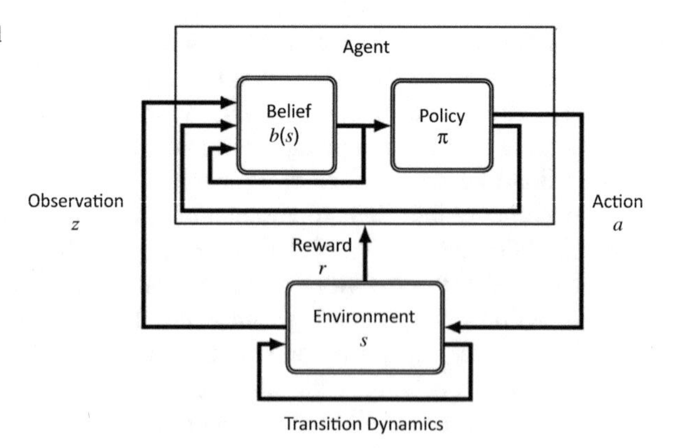

Fig. 6 A POMDP model and interactions between the agent (the ego vehicle) and the environment (other vehicle)

The POMD is formally made of {X, A, Z, T, R, O}, where X is the state space, A a set of actions that the agent can take, Z a set of observations that can be obtained by the agent. T is the transition function—T(x′, x, a) = P(x′|x, a) describes how the system changes when the agent takes action a when in state x with x′ the state at the next time step. The reward function R(x, a) indicates the value obtained after performing an action in a given state. O(z, x, a) = P(z|x, a) represent the probability to obtain an observation z ∈ Z given a state and an action. In an uncertain environment, the agent does not know the real state after an observation, it reasons with a belief b ∈ $\mathcal{B}$ with b:$\mathcal{B} \rightarrow \mathbb{R} \geq 0$ and $\int_{s \in S} b(s)ds = 1$. Thus, the goal of the agent is to maximize the value V:$\mathcal{B} \rightarrow \mathbb{R}$ for an initial belief. In a POMDP there exists an optimal policy $\pi_*:\mathcal{B} \rightarrow$ A that maximizes V. The value of a policy can be estimated as the expected future sum of rewards for an initial belief

$$V^{\pi}(b_0) = \sum_{t=0}^{\infty} \gamma^t R(b_t, a_t) = \sum_{t=0}^{\infty} \gamma^t E(R(s_t, a_t) | b_o, \pi)$$

γ the discount factor penalizes rewards more and more after each time step t.

This search for the optimal policy is intractable in time as the solver must explore the entire combination of state, belief, observation and actions. An online solver has been chosen with our approach to overcome this problem [25]. This solver samples the policy space to generate policies that evaluated using models of the problem from the current belief. The policy with the highest value at the time is chosen to be executed. Under computation constraint this process is likely to be non-deterministic as sampled actions can be different at each execution time. This is to be compounded with the non-deterministic process, i.e. to build the belief from the measurement. It is these phenomena that makes the system complex to validate as each execution can be different even in simulation.

3.2 POMDP Applied to Road Intersection Crossing

The scenario for which the system is designed is as follows: The subject vehicle (SV) must cross an intersection with or without having right of way (priority). Other Vehicle (OV) is approaching to the left of the SV at different velocities and manoeuvres. This scenario and the associated state variables are illustrated in Fig. 7.

The state space is defined by:

d_{sv}, d_{ov}	Distances towards the intersection entrance
s_{sv}, s_{ov}	Speeds of each vehicle
$\boldsymbol{e_{sv}}$	the expected manoeuvre of SV
$\boldsymbol{i_{ov} e_{ov}}$	the intended and expected manoeuvre of the OV

The intended manoeuvre models how the OV is currently behaving whilst the expected manoeuvre indicates how the vehicle should behave given the current

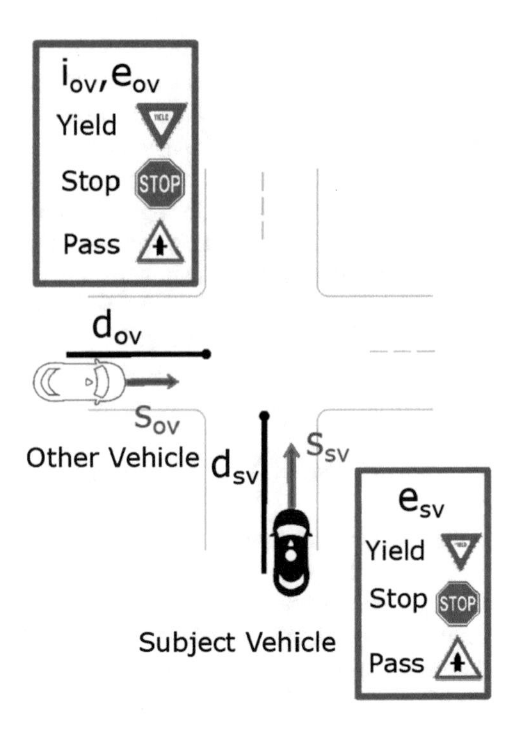

Fig. 7 Scenarios addressed by the system under test

situation. By using these variables in the transition and reward function, the SV can interact with the other agent. This is essential when crossing intersections in the presence of another vehicle.

The reward function of the model considers: comfort, velocity, time to collision, traffic rules and differences between intention and expectation. Each of the component is weighted by an affine function, consequently 10 parameters can be tuned to change the system behaviour. Full details on the decision-making framework can be found in [25].

3.3 The Test Platform

The Test Platform should allow to consider the measurement of as many variables as possible, especially those measurement variables that represent the key performance indicators (KPIs) of the SUT. The approach is based on the capability to cover as many different scenarios as possible by testing using physical components to validate the models used for the simulation part. This is based on the premise that testing for edge-cases and hazardous conditions is virtually impossible to do on public roads. Simulation is considered effective though the modelling of sensors and disturbances could be complex.

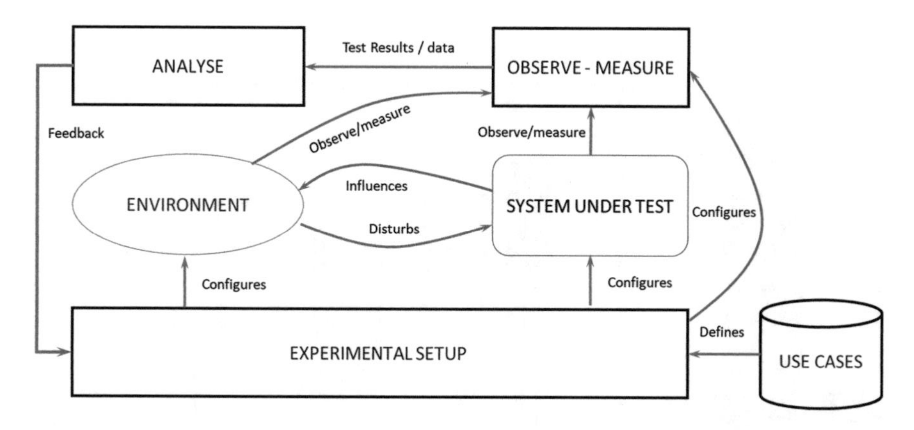

Fig. 8 Functional representation of the Testing System and SUT

The approach taken is based on system engineering principles and the tacit separation between the SUT and the Test System whist formulating the Key Performance Indicators measuring the capability of the SUT. Major constraints include to be as exhaustive as possible (how can we demonstrate this?), to be able to demonstrate that the test performed can be repeatable and to have the ability to demonstrate confidence with the test results.

Figure 8 shows a functional system representation of the Testing System. The SUT can be regarded as embedded within a series of functions. In order to perform the tests, it is important to design experiments, that is to find the means to configure its Test Environment. This is governed by the scenarios (use cases) within which the SUT is expected to operate. Interactions are created between the SUT and the environment that represents the work space. Different means are then needed to measure and to create observations of the relevant variables. All the observations and system parameters are logged. These are then analysed with respect to a pre-established evaluation criterion (the KPIs). In general, after analysis further tests are made with the experimental setup modified as necessary.

Tasks within the ENABLE-S3 project led to the formulation of what is known as the ENABLE-S3 architecture or framework. The purpose was to partition the test system in a way to provide the functions needed to ensure that the SUT is embedded within the test architecture. Figure 9 shows the ENABLE-S3 architecture.

The mapping of the SUT (the Decision-Making component) within the ENABLE-S3 framework (Fig. 9), for testing purposes can be made as follows: The SUT is equivalent to the ACPS Control System. As it is, an autonomous vehicle, there is no operator, the system dynamics refer to a vehicle dynamic model including the computer actuation part. Environment sensors for our experiment operate in a nominal manner, whilst part of the system they are not considered under test. There are no communications with the infrastructure. Within the environment, the road intersection is modelled, the lane level maps as well as the presence of other vehicles. For testing purposes, the velocity at which the ego vehicle arrives at

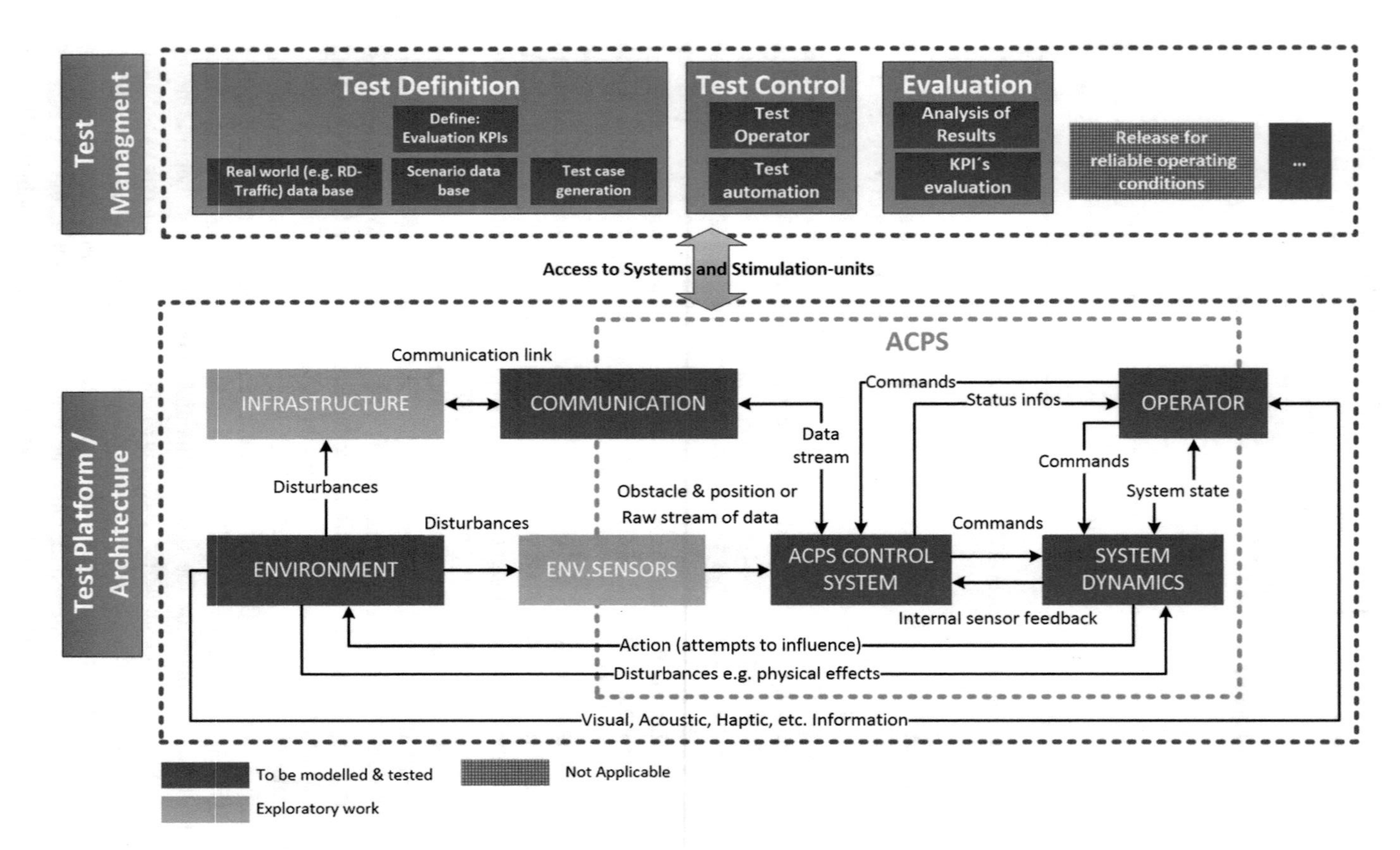

Fig. 9 The Enable-S3 testing framework

the intersection is a variable, as also the velocity of the other vehicle arriving at the intersection (c.f. Fig. 1). This configuration is set as part of the Test Definition that allows for the generation of the tests to be executed. It includes the definition on the manner how the SUT is to be evaluated, that is the Key Performance Indicators (KPIs) described in the next section. The Test operation is done in a manual manner through different setups by the simulation operator. The Evaluation is made through the application of the KPIs, to the tests as well as domain knowledge.

3.3.1 Implementation

A configuration of a typical implementation of the SUT on a prototype vehicle is shown in Fig. 10. The purpose is to provide a link with the physics of the implementation and the manner how this was simulated as the SUT and the Testing Framework.

The system will include a perception system to detect the other vehicle (OV), that is a LiDAR and a Camera in many cases. A localisation system that consist of GNSS receivers, vehicle odometry and a camera to detect lane markings. A lane-level map (navigation map) that when the output of the perception system is projected to the map results in contextualised data. As shown in Fig. 10, the testing system will include then the situation understanding part, the POMDP acting as the decision-making process and the conversion of the decisions into actuation.

A functional representation of the experimental setup is shown in Fig. 11. The testing system is made of the simulation of the vehicle and sensors, the other vehicle arriving at the intersection, the scenario creator, mainly controlling the arrival of the OV to the intersection and the evaluation component. The SUT is configured according to the scenarios, it is stimulated by the observations provided from both the Ego Vehicle and Other Vehicle. The result are actions that control the Ego Vehicle.

The actual implementation was made using the SCANer[6] simulation software, which is scenario driven. The ego vehicle model under computer control, plus sensors, the other vehicle and the environment are modelled within the SCANer software. The decision-making framework was coded in C++ in a Linux environment. Figure 12 shows the overall implementation. The software in the vehicle is built using ROS libraries for interacting with the various sensors and integrating maps, localisation and the decision-making framework. Therefore, a ROS node is created to serve as an interface between the vehicle intelligence and the simulator. Through this node, simulated sensor data and the vehicle state is sent to the decision-making system as observations. This generates the actions that are converted into commands that control the ego vehicle. Scenarios are controlled within SCANer by changing the other vehicle behaviour, that will change the scenario in the simulated world.

[6]https://www.avsimulation.fr/

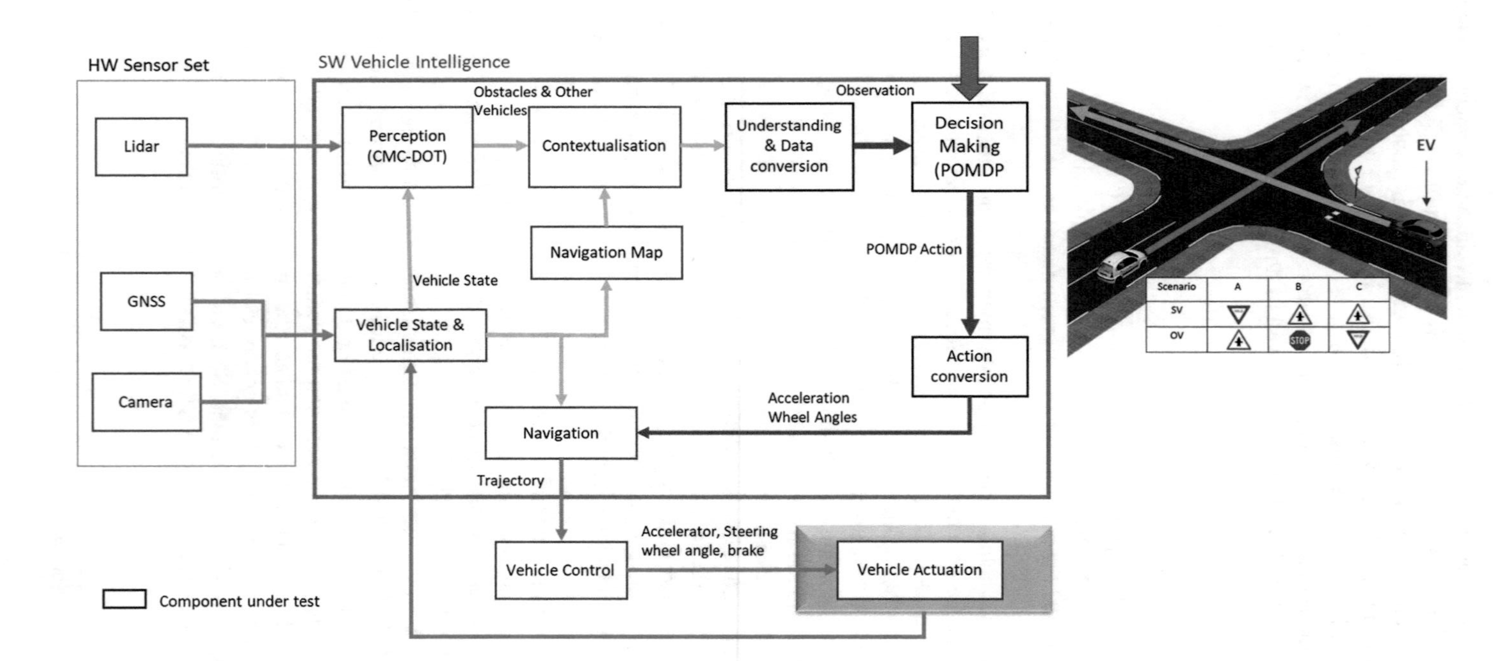

Fig. 10 A typical implementation of the decision-making process within an AV architecture. Includes, the test environment

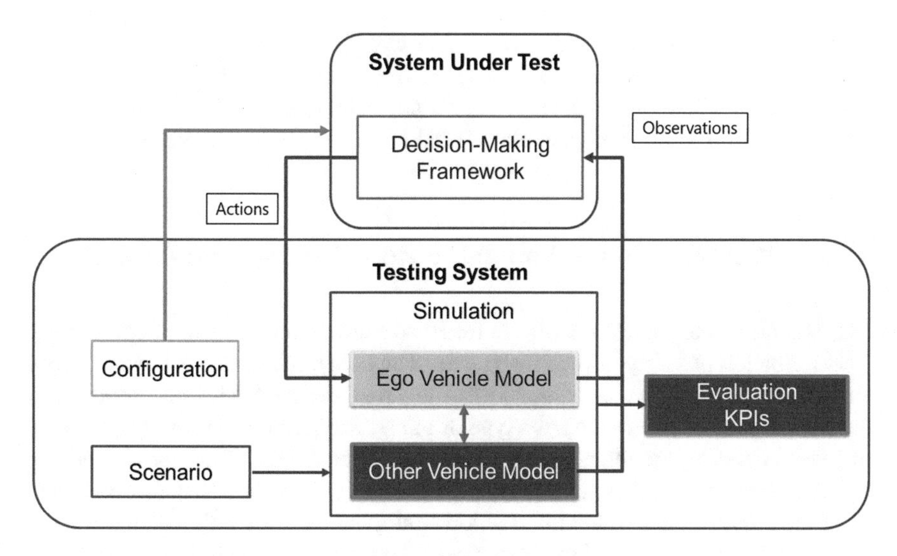

Fig. 11 Functional description of the experimental setup

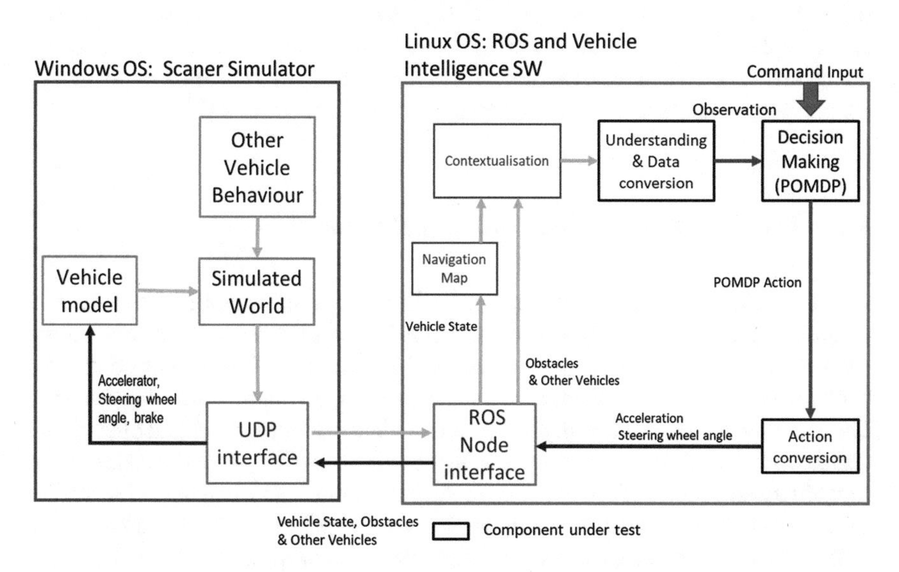

Fig. 12 Software in the loop implementation of the SUT and the Testing Environment

Within the simulated environment, the observations of every vehicle can be precisely obtained. To take uncertainty into account, Gaussian noise was added to the speed and distance measurement to simulate the performances of a perception under real world conditions. For the SV a noise $N(0,0.5)$ is added for the measurement and a noise of $N(0,1)$ for the OV vehicle.

Best configurations were found via an iterative process, with random configurations tested in different scenarios. Then, the search range was narrowed to best performing configurations and tested again with fewer configuration parameters free. Currently, the process of selecting configuration range with successful outcomes is done manually, however in the future, statistical model checking methods [26] or reinforcement learning could be used to improve this optimization.

4 Evaluation Criteria Applied to Road Intersection Crossing

In this section, the evaluation criteria for Road Intersection crossing is formulated. The evaluation of safety of road intersections layouts has been the subject of major studies. These highlight the complexity of this road segment, which is proportional to the number of entries (legs), the type of intersection traffic control type, the type of roads accessing the intersection and the operational area. Other factors include lighting, presence of pedestrian crossings, bus stops, and so on [27]. Safety is studied from the point of view of the intersection configuration, the traffic volume and the incidents that have occurred over large periods of time. The main issue here is that most evaluation studies focus on driver-centric vehicles, whilst our concern is on fully computer-controlled ones. Evaluations reflect society concerns such as safety, non-compliance to traffic rules or signals, usage of attention devices while driving, young driver behaviour, etc. Other centre on user needs, like the HMI in relation to the crossing of intersection needs when using additional sensors [28].

4.1 Key Performance Indicators, a Rationale

Currently, there are no widely accepted performance measures and transferable methodologies, thus it is very difficult to assess the effectiveness of decision-making frameworks or other capabilities related to AVs. Each actor has developed its own performance indices with the aim to evaluate certain autonomous driving features. These are mostly used internally and as a result only refer to individual operational design domains. Thus, currently it is difficult to have an objective conclusion to the capabilities of different functions like in the case for decision-making associated to the crossing of road intersections. In this work, a set of Key Performance Indicators (KPIs) to assist with the evaluation is formulated. It is the product of what it is done empirically on the test sites, the way this is assessed when operating in public roads.

To assess the performance of the decision-making framework, a set of performance criteria are provided. The purpose is to define indicators that observe specific variables that allow to decide for the success or failure of the approach. When has an intersection been negotiated successfully? Is it when there are no collisions? What about the comfort 'feel' of the on-board passengers? What about the perceived behaviour by the other road users? What about the compliance with the traffic rules? It is not only the capability of crossing intersections in nominal conditions, but to be able to negotiate despite disturbance that can be the product of the unexpected behaviour of other road users, occlusion or road conditions. Thus, it is important to define a set of criteria that provides an evaluation that reflects the needs of all the stakeholders e.g. end-users, entities sharing the same workspace, public bodies and manufacturers. In a similar manner as driving assistance systems for manually driven vehicles are evaluated according to longer term trajectory prediction, applicability on arbitrary intersections, real-time capability and dealing with human uncertainty [29].

In this work, the evaluation addresses only nominal conditions.

4.2 Definition of the Key Performance Indicators

KPIs can be defined as measures used to reveal how successful the functionality/operation of a system has been achieved. These can constitute an effective system of performance measurement, for this purpose they must be well defined and the process for their assessment standardised with the participation of all the stakeholders. The KPIs are applied into different expectations of the system capabilities.

The evaluation of the proposed decision-making system requires that vehicles within the test environment react to each other's actions. Consequently, such vehicles are also able to avoid collisions with the SV. This reflects what occurs in public roads, the behaviour of one vehicle will affect the behaviour of the other within the vicinity. In simulation, few collisions were observed even if some undesired behaviours occurred. It motivated us to propose some key performance indicators to evaluate a decision-making system in relation with the tested scenario. As explained in [30], it is a disputable design choice to implement the metrics used to judge a system based on its rewards function alone. The system would tend to optimise the metrics only. Therefore, the KPIs were not implemented as part of the reward function.

For a decision-making system, the evaluation can be made through: safety, navigation, trust and comfort. For each, a set of KPIs have been defined as applicable to the crossing of road intersections only:

Safety It is unsafe and forbidden by the highway code to stop within an intersection. Thus, a first KPI can be formulated as the time the SV spends stopped within the intersection. Therefore, tests where this KPI is not null are regarded as a failure of the SUT.

Navigation It highlights the performance of the system to adapt its behaviour to the current situation. First the travel time is observed. A slow vehicle might be safe but will reduce the traffic flow. In the context of our experimentation, a vehicle driving at 8 m/s travels the distance necessary to cross a typical intersection in 6.5 s. Considering that deceleration is required when at road intersections, a travel time lower than 15 s is considered a maximum when SV has priority. However, when this is not the case, a travel time lower than 20 s is a success because the SV vehicle is required to decelerate. In both scenarios the SV can come to a stop before the intersection, if required by the situation. The time stopped before entering the intersection is used as a second KPI for the navigation category. The best possible outcome for both scenarios is that the SV has adapted its speed to negotiate and did not come to a stop. However, if the SV stopped, it should not wait for a long time (in scenario 1 the waiting time should be enough to let the OV cross). Thresholds for this category could be learnt from data obtained while driving.

Trust Drivers tend to maintain a certain time gap between them and other obstacles. As a rule of thumb, a gap of 4 s is a minimum period for passengers to feel comfortable. With a smaller gap the trust of the driver in the system might be reduced and lead to dangerous system disengagements. At road intersections the gap between the two vehicles can be small during the approach. However, when the SV enters the intersection this gap is required to be sufficient or non-existent (the OV has already crossed the intersection). Therefore, this KPI is formulated as the time gap when the SV enters the intersection. It could be considered successful when the SV lets the other vehicle pass first or the SV enters the intersection with a gap higher than 4 s.

The KPIs are evaluated from a comfort and safety perspective. It should be noted that there shall be other KPIs, that could include weather & ground conditions, the ability to yield to an emergency vehicle, etc. What determines their pertinence is the ODD where the vehicles are to be operated.

5 Validation Results

Applying the defined KPIs for the analysis of the SUT performance enables a systematic evaluation of the results. It provides good insights into the performance of the decision-making framework.

The SUT is tested for three scenarios: In the first scenario, the SV must yield to a vehicle approaching from its left as it arrives to the intersection. In the second scenario, the priority is given to the road in which the SV is travelling, thus the OV must stop. In the third scenario, the priority remains with the SV with the OV having to yield if necessary. These scenarios are illustrated in Fig. 4.

The differences between scenario 2 and 3 relate to the OV velocities to perform the correct manoeuvre when interacting with the SV as it must yield.

In this configuration, multiple SUTs will be used. As a baseline, the driving model within the simulator is used. This model plans its actions to avoid collisions at any cost using a linear estimation of the trajectory of the OV based on its speed. The POMDP model will be applied with two different configurations. These are selected because they outperform other solutions but also because they perform differently in the different portrayed scenarios.

For each test, the vehicles' initial velocity is selected randomly. The SV vehicle starts 40 m before entrance to the intersection, whilst the OV start at a random

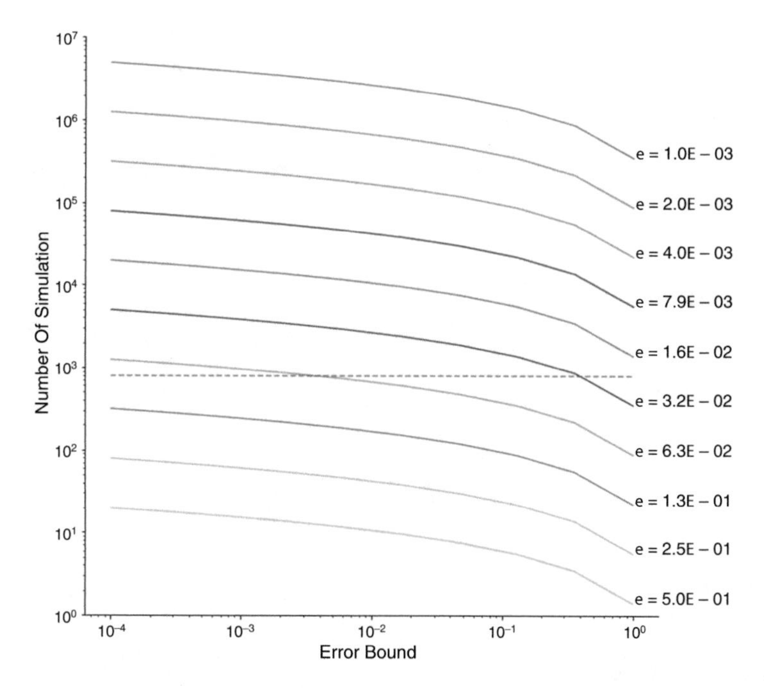

Fig. 13 Number of simulations required by the Chernoff bound to achieve a certain degree of error and to keep a certain bound around it. The dashed line represents the achievable value for 800 simulations

position before the intersection on lanes perpendicular to between 30 and 50 m. Each configuration of the system was tested 800 times. Application of the Chernoff bound help us to know how many simulations would be required to ensure that the error in the result would be below a certain margin [26]. The number of simulations N is given by:

$$N = \frac{\log(2) - \log(\delta)}{2\varepsilon^2},$$

with ε the error and δ the bound around error. Figure 13 shows the evolution of the number of simulations given these two parameters. With the 800 simulations an error 0.06 and be sure that the real value is within a bound of 0.05.

5.1 Results from System Under Test

Table 1 shows causes of failure found with the proposed KPIs. The first configuration is capable to cross the intersection for each test in each scenario, as for the baseline, no failures related to travel time. However, this is also the main cause of

Table 1 Percentage of successful test and causes for failure out of 800 experiments

Metric	Configuration	1	2	3
Success rate	1	75%	78%	86%
	2	82%	47%	56%
	Baseline	2%	75%	42%
Acceptable safe stop	1	0	4%	3%
	2	1%	8%	8%
	Baseline	91%	0	6%
Failure safe stop	1	0	0	0
	2	0	1%	1%
	Baseline	0	0	0
Failure travel time	1	0	0	0
	2	0	25%	17%
	Baseline	0	0	0
Failure jerk	1	3%	14%	8%
	2	4%	7%	8%
	Baseline	0	0	0
Failure gap	1	19%	0	0
	2	11%	1%	0
	Baseline	6%	25%	52%
Failure unsafe stop critic	1	3%	4%	3%
	2	2%	11%	10%
	Baseline	1%	0	0

failure of the second configuration for scenario 2 and 3. In scenario 1, it appears that the baseline is most likely to stop at the giveaway intersection instead of slowing and pass as the proposed system can do. However, it is balanced with the gap related failure caused by SV entering the intersection while the other vehicle is close or within. It is observed that the failure related to the gap is less important for configuration 2 than configuration 1. However, compared with the baseline in scenario 2 the POMDP can maintain a sufficient gap.

In some tests, the POMDP stopped the vehicle within the intersection (categorized as unsafe stop), even if illegal, it might have helped the SV to react to the OV unexpectedly crossing the intersection. Further investigations should be made on the scenario that caused these stops.

The results have shown that the decision-making framework when operating in Configuration 1 has the tendency to stop within the intersection. For Configuration 2 the focus should be on reducing gap related failures.

5.2 Discussion

SUT evaluation using KPIs provides a better understanding of weaknesses and strengths of the system in the proposed scenario. They can be used to determine whether a system is sufficiently mature to be tested on the test ground or must be improved.

We can use a classical analysis using only the number of accidents or accumulated rewards over the execution for the assessment. For the former, because the agents in the simulation can interact, they are going to avoid collisions. For the latter the accumulated reward is measured how well the solver performed during a test rather than how well it behaves in the scenario.

It can be argued that the KPIs should have been implemented in the reward function of the POMDP. This would have resulted in a system having a higher success rate though lacking some driving features that are complex to formalize, such as interactions. Furthermore, the design of the reward function is often done using a deep neural network where it would not be difficult to add domain knowledge. With these considerations, it is difficult to estimate if the list of KPI is exhaustive.

With the current testing framework, the distribution of the KPI value is unknow. Tests that are close to the KPI threshold need to be measured. To improve the analysis of the result, other methods used in the project could be applied. In particular, statistical model checking that is design to evaluate stochastic systems.

This method also addresses another issue of our current approach, that is to randomly sample the scenario space. It is sufficient for simple scenario involving few vehicles, but as complexity increases so does the number of tests. The ability to efficiently sample the scenario will be required to avoid an exponential grow of the number of simulations triggered by scenario that are too different to be validated together.

6 Perspectives

Testing of highly automated CPS is emerging a domain by itself. The evaluation of one of the simplest cases for crossing road intersections has demonstrated the complexity of implementing the SUT, of generating credible scenarios, and on the assessment of the systems' performance.

Testing using a nominal basis (ignoring the effects of board sensors, road surface conditions, occlusion of GNSS signals, errors in maps, the random behaviour of other road users, etc.) is still complex. The exploration space becomes very large if all conditions are to be considered. This results in a large number of required simulations. Another issue resides on the validation of the simulations itself. How can we validate the simulation models and scenarios to provide a high level of confidence on the simulation results?

The need for a framework that allows for the separation between the SUT and the testing system is very important. The evaluation is not as straightforward as was initially thought. There are no norms, and literature is centred on specific functions like positioning or perception. The KPIs defined were empirically defined from experience. Their exhaustiveness cannot be demonstrated. In depth studies are needed on the definition of the KPIs and discussions including all stakeholders are needed to define the most suitable indicators. Traditional methods that address single driving assistance functions might be insufficient.

Additionally, safety is a major concern. Real-world tests can be too hazardous and costly to perform. For the decision-making framework proposed with the approach taken, it has not only been possible to validate it but also to improve the algorithms.

As CPS introduce more and more functions based on machine learning methods, the issue of validation gains a higher level of complexity, it is not only necessary to validate but to understand causes for failure. Further, CPS should be evaluated as they operate to assess their capabilities online as a mean to warrant their operability.

CPS including safety-related functions need advanced validation methods. They need to be designed for validation with advanced techniques to make them acceptable by society in terms of safety, performance, etc. With foresight looking into the deployment of autonomous vehicles without involving driver supervision, despite substantial efforts across the automotive industry and technological companies, Level 4 SAE vehicles are still to be deployed as non-experimental units. This alone demonstrates the need for further work in the area of validation for the emergence autonomous systems and their acceptance by society.

References

1. Maurer, M., Gerdes, J.C., Lenz, B., Winner, H.: Autonomous driving, vol. 10, pp. 978–973. Springer, Berlin (2016)
2. Russo, F., Comi, A.: From the analysis of European accident data to safety assessment for planning: the role of good vehicles in urban area. Eur. Transp. Res. Rev. **9**(1), 9 (2017)
3. Dickmanns, E.D.: Developing the sense of vision for autonomous road vehicles at UniBwM. Computer. **50**(12), 24–31 (2017)
4. Thorpe, C., Hebert, M.H., Kanade, T., Shafer, S.A.: Vision and navigation for the Carnegie-Mellon Navlab. IEEE Trans. Pattern Anal. Mach. Intell. **10**(3), 362–373 (1988)
5. Pomerleau, D.A.: Alvinn: an autonomous land vehicle in a neural network. In: Advances in Neural Information Processing Systems, pp. 305–313 (1989)
6. Korosec, K.: Waymo's self-driving cars hit 10 million miles. In: Tech Crunch. https://techcrunch.com/2018/10/10/waymos-self-driving-cars-hit-10-million-miles/?guccounter=1&guce_referrer_us=aHR0cHM6Ly93d3cuZ29vZ2xlLmNvbS8&guce_referrer_cs=Hez2-a408-CA_FgHrLmVQg (2018). Accessed Mar 2019
7. Ibañez-Guzman, J., Laugier, C., Yoder, J.D., Thrun, S.: Autonomous driving: context and state-of-the-art. In: Handbook of Intelligent Vehicles, pp. 1271–1310. Springer, London (2012)
8. Dominguez, E., et al.: High Accuracy Positioning Engine with an Integrity Layer for Safety Autonomous Vehicles, Conference: ION GNSS, September 2018

9. Bresson, G., Alsayed, Z., Yu, L., Glaser, S.: Simultaneous localization and mapping: a survey of current trends in autonomous driving. IEEE Trans. Intell. Vehicles. **2**(3), 194–220 (2017)
10. Li, F., Bonnifait, P., Ibanez-Guzman, J., Zinoune, C.: Lane-level map-matching with integrity on high-definition maps. In: 2017 IEEE Intelligent Vehicles Symposium (IV), pp. 1176–1181. IEEE, Los Angeles, CA (2017)
11. Armand, A., Ibanez-Guzman, J., Zinoune, C.: Digital maps for driving assistance systems and autonomous driving. In: Automated Driving, pp. 201–244. Springer, Cham (2017)
12. Albus, J.S.: 4D/RCS: a reference model architecture for intelligent unmanned ground vehicles. In: Unmanned Ground Vehicle Technology IV, vol. 4715, pp. 303–311. International Society for Optics and Photonics (2002)
13. Paden, B., Čáp, M., Yong, S.Z., Yershov, D., Frazzoli, E.: A survey of motion planning and control techniques for self-driving urban vehicles. IEEE Trans. Intell. Vehicles. **1**(1), 33–55 (2016)
14. Schwarting, W., Alonso-Mora, J., Rus, D.: Planning and decision-making for autonomous vehicles. Ann. Rev. Control Robot. Auton. Syst. **1**, 187–210 (2018)
15. Lefèvre, S., Carvalho, A., Borrelli, F.: A learning-based framework for velocity control in autonomous driving. IEEE Trans. Autom. Sci. Eng. **13**(1), 32–42 (2016)
16. Gong, Z., Guzman, J.I., Scheding, S.J., Rye, D.C., Dissanayake, G., Durrant-Whyte, H.: A heuristic rule-based switching and adaptive PID controller for a large autonomous tracked vehicle: from development to implementation. In: Proceedings of the 2004 IEEE International Conference on Control Applications, 2004, vol. 2, pp. 1272–1277. IEEE, Taipei (2004)
17. Broggi, A., Debattisti, S., Grisleri, P., Panciroli, M.: The deeva autonomous vehicle platform. In: 2015 IEEE Intelligent Vehicles Symposium (IV), pp. 692–699. IEEE (2015)
18. NHTSA: Automated driving systems: a vision for safety. US Dept. of Transportation, DOT HS 812 442, September 2017
19. Feig, P., Schatz, J., Dörfler, C., Lienkamp, M.: Test protocol driver assistance systems parking and maneuvering. Technische Universität München, Garching b. München (2017)
20. Ibañez-Guzmán, J., Le Marchand, O., Chen, C.: Metric evaluation of automotive-type GPS receivers. In: Proc. FISITA, Munich, Germany (2008)
21. Kaelbling, L.P., Littman, M.L., Cassandra, A.R.: Planning and acting in partially observable stochastic domains. Artif. Intell. **101**(1), 99–134 (1998)
22. Thrun, S.: Monte carlo POMDPs. In: Solla, S., Leen, T., Müller, K.-R. (eds.) Advances in Neural Information Processing Systems 12, pp. 1064–1070. MIT, Cambridge, MA (2000)
23. Ross, S., Pineau, J., Paquet, S., Chaib-draa, B.: Online planning algorithms for pomdps. J. Artif. Int. Res. **32**, 663–704 (2008)
24. Liu, W., Kim, S.W., Pendleton, S., Ang, M.H.: Situation-aware decision making for autonomous driving on urban road using online POMDP. In: 2015 IEEE Intelligent Vehicles Symposium (IV), pp. 1126–1133. IEEE, Seoul (2015)
25. Barbier, M., Laugier, C., Simonin, O., Ibañez-Guzmán, J.: Probabilistic decision-making at road intersections: formulation and quantitative evaluation. In: 2018 15th International Conference on Control, Automation, Robotics and Vision (ICARCV), pp. 795–802. IEEE (2018)
26. Legay, A., Delahaye, B., Bensalem, S.: Statistical model checking: an overview. In: Runtime Verification, pp. 122–135. Springer, Berlin (2010)
27. Bonneson, J., Laustsen, K.: Intersection safety evaluation: InSAT guidebook. National Cooperative Highway Research Program. Transportation Research Board of the National Academies (2014)
28. Jagtman, H.M., Marchau, V.A.W.J., Heijer, T: Current knowledge on safety impacts of Collision Avoidance Systems (CAS). In: Critical Infrastructures–Fifth International Conference on Technology, Policy and Innovation, Lemma, Delft, The Netherlands, June 2001
29. Shirazi, M.S., Morris, B.T.: Looking at intersections: a survey of intersection monitoring, behavior and safety analysis of recent studies. IEEE Trans. Intell. Transp. Syst. **18**(1), 4–24 (2017)
30. Amodei, D., Olah, C., Steinhardt, J., Christiano, P.F., Schulman, J., Mané, D.: Concrete problems in AI safety. ArXiv e-prints (2016)

Validation of Automated Valet Parking

Hasan Esen, Maximilian Kneissl, Adam Molin, Sebastian vom Dorff, Bert Böddeker, Eike Möhlmann, Udo Brockmeyer, Tino Teige, Gustavo Garcia Padilla, and Sytze Kalisvaart

1 Introduction

Automated Valet Parking (AVP) as a functional extension of the parking assist is estimated to be one of the first commercially available automated driving functions at SAE level 4, see SAE's taxonomy and definitions for terms related to automated driving [1]. An early introduction is possible, as the driving operation is in a confined area and supervised through the parking house infrastructure.

As depicted in Fig. 1, the valet parking system task is to park the vehicles into assigned parking bays or hand them back to their owners at a dedicated pick-up location. When the driver approaches a parking lot with AVP functionality, he or she stops, leaves the car, and activates the AVP system. The Parking Area Management (PAM) locates the free parking slots and defines the path for the vehicle guiding it to the parking lot. The vehicle guidance proceeds under a supervisory control, in

H. Esen (✉) · M. Kneissl · A. Molin · S. vom Dorff · B. Böddeker
DENSO AUTOMOTIVE Deutschland GmbH, Eching, Germany
e-mail: h.esen@denso-auto.de; m.kneissl@denso-auto.de; a.molin@denso-auto.de; s.vomdorff@denso-auto.de; b.boeddeker@denso-auto.de

E. Möhlmann
OFFIS e.V., Oldenburg, Germany

U. Brockmeyer · T. Teige
BTC Embedded Systems AG, Oldenburg, Germany
e-mail: Udo.Brockmeyer@btc-es.de; Tino.Teige@btc-es.de

G. G. Padilla
HELLA Aglaia Mobile Vision GmbHs, Berlin, Germany

S. Kalisvaart
TNO Integrated Vehicle Safety, Helmond, Netherlands

A. Leitner et al. (eds.), *Validation and Verification of Automated Systems*,
https://doi.org/10.1007/978-3-030-14628-3_16

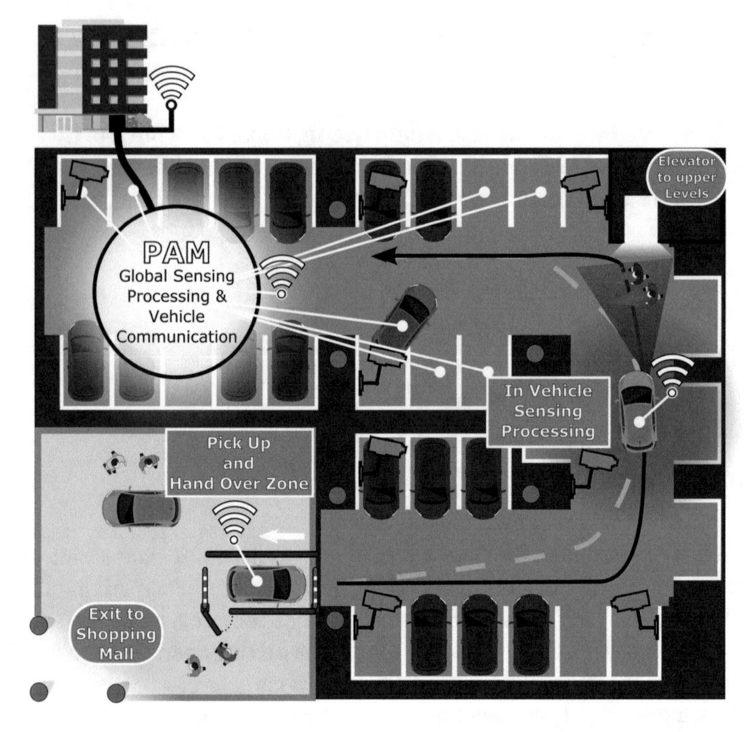

Fig. 1 Automated valet parking system

which the vehicle and the PAM continuously exchange real-time data and where mitigating operations can be commanded by requiring e.g. emergency braking.

The validation of the previously described AVP system is the subject of this chapter. The focus for validation is based on functional safety and safety of the intended functionality of AVP related to ISO 26262 [2] and ISO PAS SOTIF 21448, respectively. As an intermediate result, a safety analysis for AVP has been presented in [3]. To combat the enormous testing efforts for validating the safety of this driving function, we propose a test system architecture for AVP that combines computer simulation, automated test case generation, and semi-automated generation of test observers derived from a safety analysis. This architecture has been realized in an integrated test system demonstrator for virtual validation of AVP.

The remainder of this chapter is organized as follows. Section 2 gives an overview of the test architecture and the system under test used for validating the valet parking system. Section 3 describes the demonstrator for virtual testing developed in the ENABLE-S3 project. Finally, Sect. 4 gives a conclusion.

2 Test System Architecture

A test system architecture, which is an instantiation of the ENABLE-S3 generic test architecture, composed of validation & verification (V&V) management, test management, and test platform layers is developed to validate the safety of the AVP system. The complete test architecture envisioned in the use case is illustrated in Fig. 2. More information about each layer's functionality is given in respective Sects. 2.1–2.3. The safety concept is detailed in Sect. 2.4.

2.1 Test Platform

The key functional blocks of the test platform are summarized as follows.

Environment is the park area or its virtual model with static/dynamic objects.

Park area management (PAM) represents the infrastructure that allocates the parking space to vehicles and coordinates the automated parking procedure.

Autonomous vehicle control system represents the complete perception-cognition-action loop needed for automated parking within a car including pedestrian/object detection, localization, mapping, trajectory planning and tracking.

Communication represents the data exchange between PAM and the vehicles.

Vehicle dynamics represents the dynamical model to simulate the vehicle's behavior, where a simple model is sufficient due to the low speed profile; alternatively it is represented by the real test vehicle.

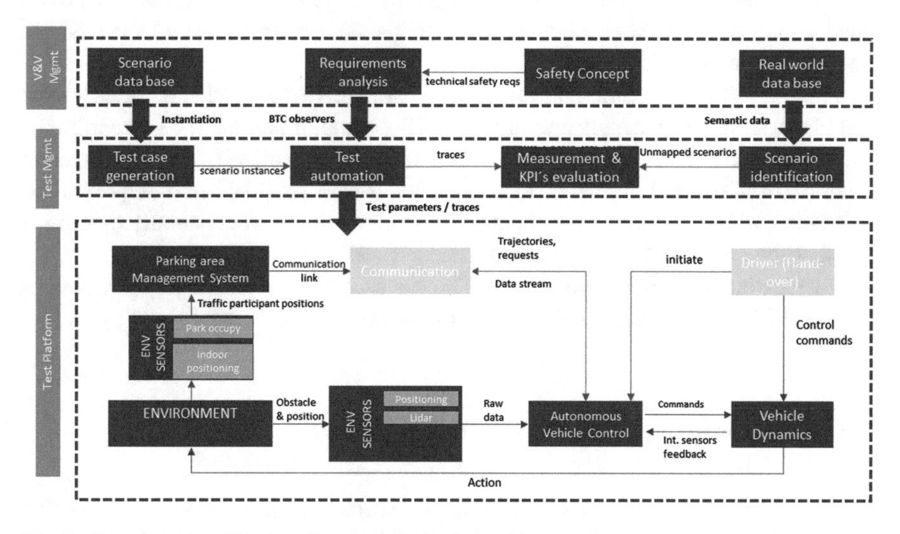

Fig. 2 Generic test architecture for automated valet parking

Environment sensors include both on-board sensors at the vehicle and infrastructure sensors available at the PAM. The sensors are either modeled in the virtual environment or represented by physical sensors.

Human driver initiates automated parking by handing over the vehicle to the PAM system and by taking over the vehicle after it has been returned to the drop-off/pick-up location.

A key component is the distributed System Under Test (SUT), which is part of the test platform layer. It is the aggregate system composed of multiple autonomous vehicles in the parking lot and the PAM system. As the main focus of ENABLE-S3 is on V&V of automated systems using virtual testing, models of test platform components are developed by using the simulator Virtual Test Drive (VTD) from VIRES and a ROS-based development framework for rapid prototyping of the SUT.

2.2 Test Management

The virtual testing applies scenario-based testing using abstract scenarios described by a set of base tiles. By stitching the tiles, multiple concrete sceneries (the static part of a scenario) are implicitly defined in the scenario database.

Test Case Generation performs the actual stitching of the tiles and, hence, the basis for the scenario of every test case. On this generated static parking environment, automated and non-automated vehicles are placed randomly. While the task of the automated vehicles (SUT) is to navigate to a free parking lot or the pick-up location, the non-automated vehicles may occupy some of the parking bays or are declared with a specific driving behavior. Together with expected behavior (such as not colliding with other traffic participants) this forms a single test case to be executed in the test platform.

Test Automation component iteratively executes generated test cases in the test platform by preparing/triggering the VTD simulator and the SUT accordingly. Moreover, the test automation retrieves the (semi-)automatically generated testing criteria from the requirements database and translates them into requirements observers, which monitor the test case during execution, and eventually determine whether the test run has been accomplished successfully.

Measurement & KPI Evaluation: The test results are reported to the Measurement & KPI evaluation block for further evaluation.

2.3 Validation & Verification Management

This layer has three main responsibilities: (i) Define a scenario database, (ii) define the testing criteria (SUT requirements) for validating the safety of the AVP, which are derived in the safety concept (Sect. 2.4), and (iii) generate real-world scenarios.

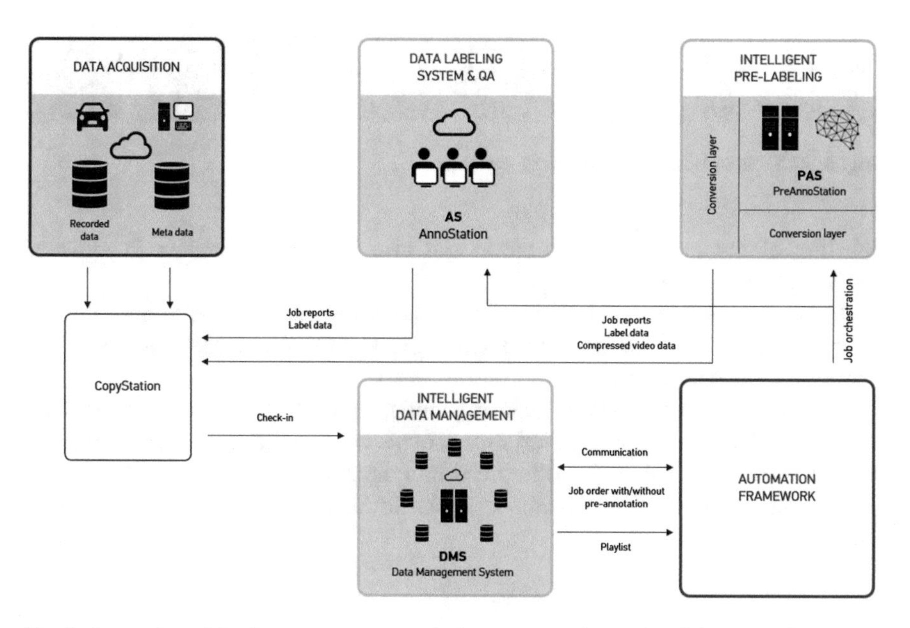

Fig. 3 Interaction of the data management tool, the pre-annotation tool and the annotation tool

Scenario database defines the collection of abstract scenarios containing static parking environments and dynamic content defining the behaviour of other traffic participants. We provide the construction of synthetic parking environments using a tile-based approach that is capable of generating a large number of scenarios.

Based on the results of the safety analysis, the **requirements database** collects the partly informal safety requirements into a list of formalized testing requirements. These will be the input for the requirements observers generated in the test automation to evaluate the safety of each individual test execution.

Real-world scenario generation: To guarantee a sufficiently large coverage of the synthetically defined scenarios above with respect to the real world, observations thereof can be used to estimate the degree to which the scenario database can be used as a proxy for real-world valet parking.

The tool chain for ground truth generation from real data from HELLA Aglaia is shown in Fig. 3. Because this tool chain operates with real data, it can be used to provide support for validating the representativeness of abstract scenarios used in the simulation. To facilitate the exchange of real-world and virtual scenarios, OpenSCENARIO/OpenDRIVE are used as a common scenario description format.

Pre-annotation tool: The main goal of this tool is the partial automation of the annotation process, and thus reduce the effort from performing the complete annotation manually by a human operator to only review and correction.

Annotation tool: This tool shall provide efficient GUIs with enhanced usability to the human operator for performing the review/correction tasks and additional

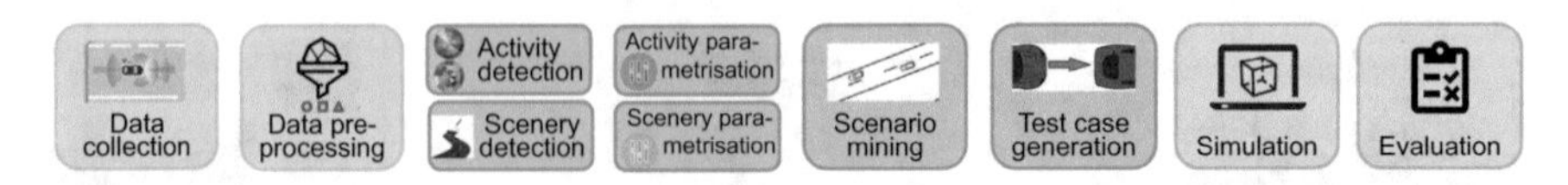

Fig. 4 TNO StreetWise scenario mining pipeline

labeling of not pre-annotated objects and also to provide support for efficient workflow and intelligent preprocessing.

Data management tool: It provides capabilities for systematic and efficient storing and for efficient scenario based retrieval of real recorded data based on corresponding metadata.

Using the TNO StreetWise scenario mining pipeline (Fig. 4) [4], large scale automated detection of single road user activities and multiple road user scenarios can be performed. This will further feed into the real-world scenario database for estimating coverage of the synthetic scenarios with regards to real world valet parking.

2.4 Safety Concept

The introduction of a new technical system, such as AVP, requires an adapted safety concept. In contrast to current systems, the automated valet parking has no human driver in the control loop and therefore disrupts existing hierarchies used in safety design.

Even though SAE level 4 systems are considered as a revolutionary leap ahead in technology, the principles of safe design demand for cautious evolution of well-understood techniques and methodologies. Therefore, the challenge lies in bridging the gap of the radical advancement in technology and the conservative nature of safety design principles.

Current automotive safety standards, such as ISO 26262 and the upcoming ISO PAS 21448, implicitly assume a driver to be available in all situations. Since the automated valet parking does not hold up to this prerequisite, those standards cannot be simply applied without adjustments. We assume that a human operator of the parking house will not intervene in individual vehicle trajectories. However, to satisfy the imperative of relying on proven and well-tested methods, existing approaches shall be taken over as much as possible.

The ISO 26262 provides a systematic approach to assess safety risks (hazards) and elaborate safety requirements to reduce these risks to acceptable levels.

In the hazard analysis and risk assessment (HARA) potential malfunctions are analysed and assessed according to Severity S, Exposure E, and Controllability C. Severity represents expected injuries if an accident happens, Exposure indicates how likely the scenario occurs, and Controllability describes how controllable the situation for the accident participants is. Especially this parameter has to consider the missing driver of the vehicle, so that most cases have to be classified as "uncontrollable". Dependent on the classification of Severity, Exposure, and Controllability, an abstract classification of the safety risk is defined by Automotive Safety Integrity Levels (ASIL) A to D with ASIL D representing the highest and ASIL A the lowest risk. If the scenario is not safety relevant, the classification Quality Management (QM) is used. For each hazardous event a safety goal is derived which inherits the hazard's ASIL. Finally, safety goals are refined into lower-level safety requirements. These can be allocated to architectural components. The validation of each component can be performed according to those safety requirements and a safety concept, as well as a safety-architecture can be developed. This process has been conducted for the AVP system with the outcome of a set of top level safety goals and the derived safety requirements. These have been applied to the decomposed architecture between vehicle and PAM.

To guarantee the absence of unreasonable risks, a safe trajectory shall be provided by the control system at any time. This implies that a minimal risk state shall be reachable from any conducted manoeuvre. Safety in this context demands that the trajectory avoids hazardous events due to intended behaviour and maintains the necessary safety margins to compensate potential system errors caused by malfunctioning components or software modules.

No.	Description	ASIL
SG 1	The valet parking function shall not be active outside of a PAM managed parking area.	D
SG 2	The system shall prevent collisions between vehicles and persons.	C
SG 3	Removed	–
SG 4	The system shall not start moving during embarkment and disembarkment.	C
SG 5	The system shall prevent collision with other vehicles.	B
SG 6	In case of a collision or firer the system shall notify a human supervisor.	B
SG 7	The integrity of the communication between the PAM and the Vehicle shall be ensured.	B
SG 8	The system shall ensure that the vehicle stays within the (statically defined) drivable area of the parking area during automated operation.	B
SG 9	Removed	–
SG 10	The valet parking function shall be disabled when people are inside the vehicle.	A
SG 11	The system shall prevent collision of the automated vehicles with objects.	A

3 Integrated Virtual Demonstrator and Test

Even though the platform supports hardware-in-the-loop tests, we start with virtual testing only. The objective of the demonstrator is to show the functionality of the complete integrated test system intended to validate the safety of the valet parking system in a virtual environment. The current demonstrator integrates the test platform and test management layers. Therefore, the sections below are dedicated to those layers. Components of the V&V layer are developed and run in a stand-alone manner. The integration of the V&V layer into the remaining part is on-going work.

3.1 Test Platform Demonstrator Components

The virtual simulation in the test platform is mainly based on Vires VTD for both the environment and the vehicle dynamics. An interface has been developed that allows for data exchange between the virtual simulation and the SUT which is based on the Robot Operating System (ROS). This interface (VTD-ROS Bridge) has been adopted to cater for multiple automated vehicles monitored and controlled by the valet parking system. It is a central element, as it transforms data from the simulation environment Vires VTD into a ROS data format that can be read by the SUT, and vice versa. Furthermore, a physics-based ideal lidar model has been established to serve as interface between SUT and VTD for localization and free-space detection. Figure 5 provides a simplified illustration of the connection between the simulation environment and the SUT components. The SUT implementation is discussed in the following.

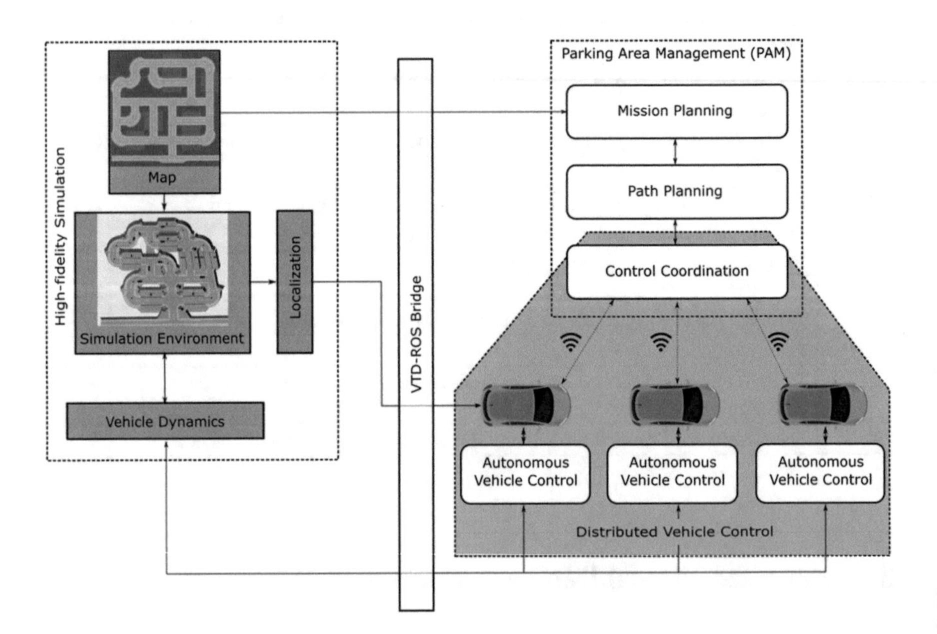

Fig. 5 Virtual simulation environment (left) and system under test (right)

3.1.1 System Under Test (SUT)

In order to coordinate multiple vehicles in the parking area in an efficient way with high throughput, a distributed optimal decision-making and control system is developed.

The PAM system, which has a global scenario overview, assigns each vehicle an appropriate parking lot in the Mission Planning unit and calculates a required path from the current to the destination position in the Path Planning unit. These units compute their results based on information provided by OpenDRIVE map data. The Control Coordination determines the dependencies between vehicles and generates coupling messages shared with the local Autonomous Vehicle Control units. The model predictive control (MPC) based autonomous vehicles are responsible for determining local feasible trajectories given the coupling message received from the PAM and share in turn its position predictions. MPC takes the vehicle constraints and safety factors into account. The system computes the trajectories for each vehicle in a distributed manner by considering the decisions of neighboring vehicles. Thus, collisions at critical areas (e.g. intersections) can be avoided and smooth velocity profiles are generated. Conceptual and technical details of this structure can be found in [5]. The local Autonomous Vehicle Control has been developed in MATLAB/Simulink and is translated into a ROS node by using the MATLAB Coder Toolbox.

The final vehicle maneuvering into the parking bay is done by a separate, dedicated maneuvering component, which is based on hybrid A_* path planning and implemented in the local Autonomous Vehicle Control unit.

3.2 Test Management

The test management is based on the OFFIS StreetTools tool suite, which was developed for automated testing AVP. The central part is OFFIS StreetRun, which is responsible for automated generation and execution of test cases. It manages and configures individual plugins that together provide the functionality for test case generation, execution, evaluation, and result reporting. The main components used in the demonstrator and their functionality are explained below. Figure 6 shows StreetRun during the execution of test cases and reporting of the results.

3.2.1 OFFIS StreetArt

StreetArt is a tool for the automated generation of synthetic parking sites. In the first place, these parking sites are tile-based, and thus, very simple but yet serve as the basis for the automated generation of synthetic scenarios. OpenDRIVE was selected as the exchange format for road networks. Figure 7 shows the tiles and a randomly generated scenery.

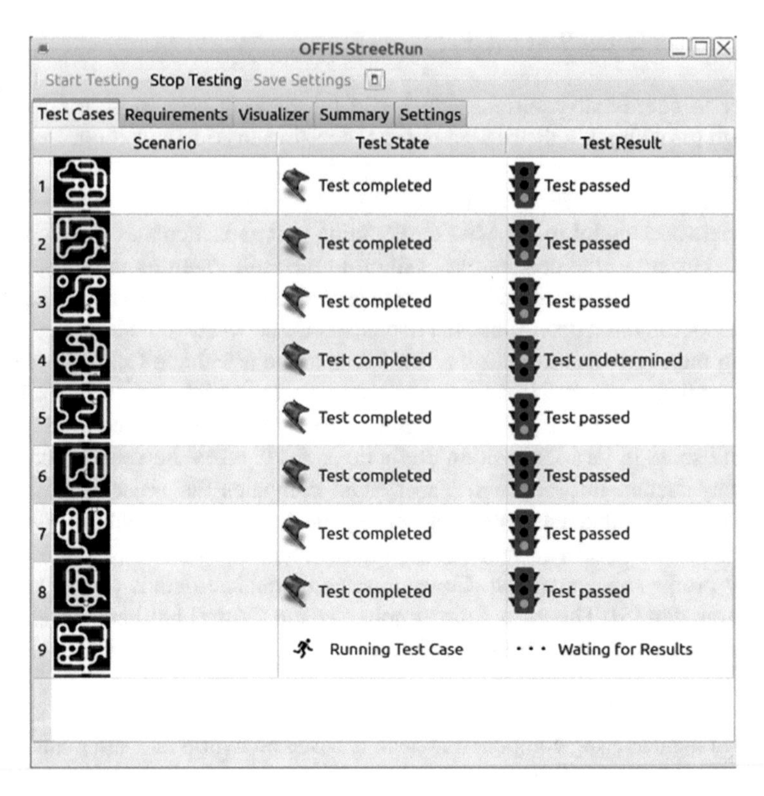

Fig. 6 StreetRun: List of test cases and exemplary results in the test management. On the left, the auto-generated parking environment is displayed. The center column shows the test automation phase. On the right, the test result according to the requirements observer is shown

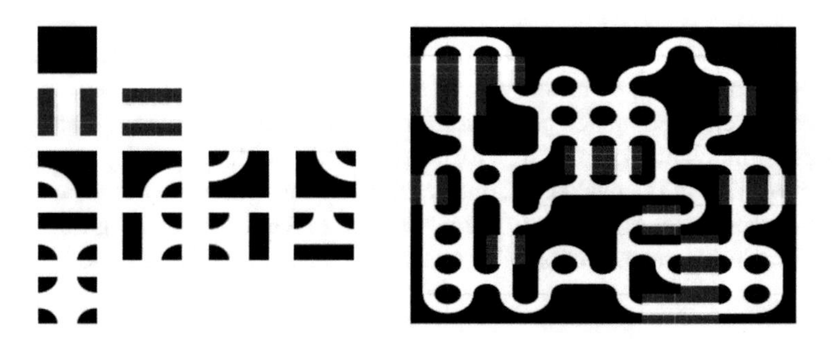

Fig. 7 StreetArt: A set of bases tiles (left), a randomly generated parking area (right)

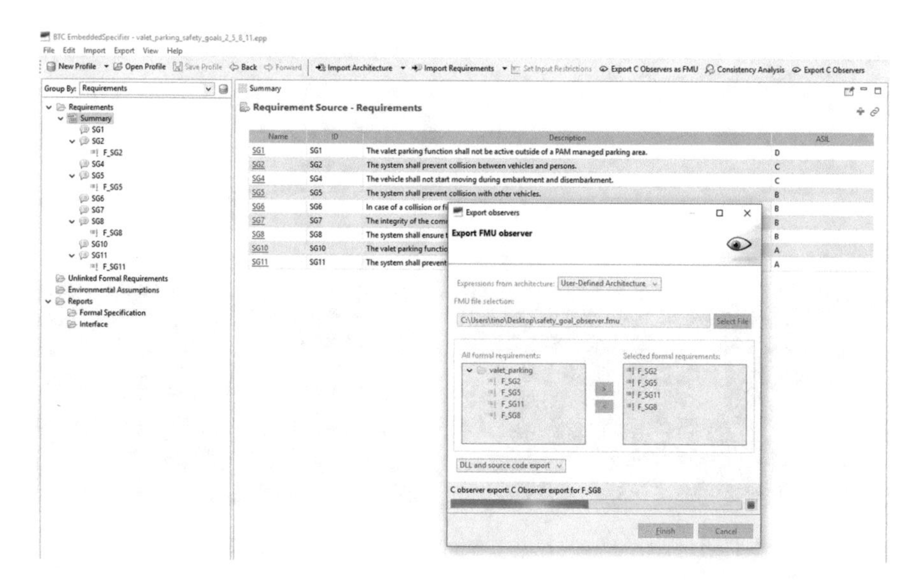

Fig. 8 BTC EmbeddedPlatform: Formalizing requirements and FMU observer generation

3.2.2 BTC FMU Observer Generation from Safety Goals

In a first step, the textual Safety Requirements derived from the Safety Goals are imported into BTC EmbeddedPlatform,[1] where the informal requirements are formalized by the help of the graphical specification language Simplified Universal Pattern [6]. After the formalization step, the safety requirements are automatically exported as FMU observers as shown in Fig. 8. This FMU is then ready for integration into the VTD simulation framework for online evaluation of the safety goals during simulation of AVP scenarios. For the integration in the test system, a StreetTools plugin and an FMU interface for VTD has been developed.

3.2.3 Map Population

Unimore Map Populator is a plugin integrated into the OFFIS StreetArt that populates the maps generated by the latter. It fills the maps with the objects and the actions related to the parking lots: parked cars, moving cars looking for an empty lot or the ones that are exiting their lots and re-enter the traffic again.

The Unimore Map Populator uses the parking lots as defined in the map and randomly places vehicles on the map. The vehicles are associated with different

[1]BTC EmbeddedPlatform is a platform for specification, testing, and verification of requirements for Simulink® and TargetLink® models and production code provided by BTC Embedded Systems AG, cf. https://www.btc-es.de/en/

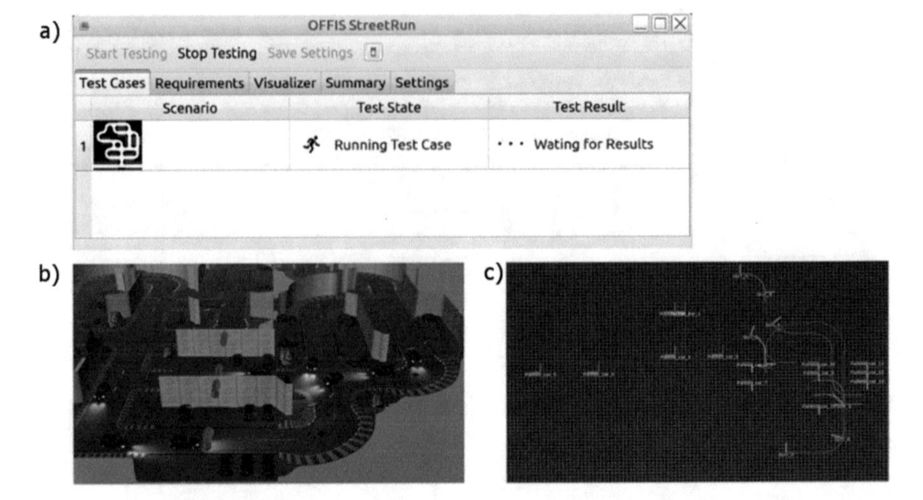

Fig. 9 Integrated test automation and test platform for AVP testing the multi-vehicle coordination in auto-generated parking environments. (**a**) frontend of test automation OFFIS StreetRun, (**b**) 3D Visualization in simulator VIRES VTD), (**c**) visualization of Valet Parking SUT using ROS RViZ, coloured lines indicate paths to assigned parking lot

behaviors, such as cars entering/exiting their lots (e.g., triggering the exiting car to leave its lot when the tested car is in vicinity) or cars moving around and may make sudden stops to simulate the behavior of an undecided driver.

The Unimore Map Populator permits a user defined ratio of parked cars, ratio of exiting cars, range from the automate vehicles and delay after which a parked car exits, number of moving cars, and number of suddenly stopping cars.

3.3 Integration of Test Platform and Test Management

The test management blocks consisting of the scenario generation (StreetArt), test automation (StreetRun), and the test platform have been combined with the SUT for testing the valet parking system in an automated fashion. The SUT using an infrastructure-supported vehicle coordination mechanism has been integrated into the test system. BTC FMU observers have been included into the test system that use formalized safety requirements that result from the completed safety concept. This enables the execution of automatic test series that return a safety evaluation of the valet parking system (Fig. 9).

4 Conclusion and Future Work

A fully integrated test system for validating AVP has been proposed that integrates (i) auto-generation of synthetic scenarios, (ii) a semi-automated mechanism to generate testing requirement observers resulting from the safety analysis, and (iii) test automation to execute tests on a multi-vehicle test platform. The proposed architecture has been used to test parking scenarios with multiple automated vehicles coordinated by a parking area manager in synthetic parking environments.

The approach and tools have been proven to be capable of testing safety properties of AVP. Moreover, the high degree of automation in the virtual test system and the diversity of the scenario generator enable the virtual validation of AVP for a large number of scenarios.

The proposed test architecture together with its integrated realization for virtual testing of AVP manifests the basis for further investigations.

To enhance flexibility and extensibility of the test platform, it is envisioned to extend the integrated test system with standardized interfaces based on FMI/FMU and the Open Simulation Interface [7].

Advanced "intelligent" test strategies need to be developed to reduce the number of testing trials. These iterative sampling strategies select test cases based on the results of prior executions to drive the System-under-Test to critical scenarios.

Finally, enabling the transfer of virtual testing results into statements validating the safety of the AVP system in operation, it is crucial to develop a quality indicator that determines the "representativeness" of synthetic worlds and the computer simulation related to the real-world AVP system. The representativeness of the virtual model and test cases is the final building block to judge the safety of the intended functionality, based mainly on virtual validation.

References

1. SAE International: Taxonomy and Definitions for Terms Related to Driving Automation Systems for On-road Motor Vehicles (2016)
2. ISO: ISO 26262: Road vehicles – Functional Safety. International Organization for Standardization, Geneva. International Standard (2011)
3. Schönemann, V., Winner, H., Glock, T., Otten, S., Sax, E., Böddeker, B., Verhaeg, G., Tronci, F., Padilla, G.G.: Scenario-based functional safety for automated driving on the example of valet parking. In: Future of Information and Communication Conference (FICC) 2018. SAI, Singapore (2018)
4. Elrofai, H., Paardekooper, J.-P., de Gelder, E., Kalisvaart, S., Op den Camp, O.: StreetWise Scenario-Based Safety Validation of Connected and Automated Driving. White paper. http://publications.tno.nl/publication/34626550/AyT8Zc/TNO-2018-streetwise.pdf (2018)
5. Kneissl, M., Molin, A., Esen, H., Hirche, S.: A feasible MPC-based negotiation algorithm for automated intersection crossing. European Control Conference (ECC) (2018)

6. Teige, T., Bienmüller, T., Holberg, H.J.: Universal pattern: formalization, testing, coverage, verification, and test case generation for safety-critical requirements. In: MBMV, pp. 6–9 (2016)
7. Hanke, T., Hirsenkorn, N., van-Driesten, C., Garcia-Ramos, P., Schiementz, M., Schneider, S., Biebl, E.: Open Simulation Interface: a generic interface for the environment perception of automated driving functions in virtual scenarios. Research Report (2017)

Validation of Railway Control Systems

Tomas Fischer, Klaus Reichl, Peter Tummeltshammer, Thai Son Hoang, and Michael Butler

1 Introduction

Modern railway infrastructure is typically layered and complex. *Operational control centers* enable the railway network operator to control the railroad traffic over a large area (e.g. for a country). *Route control* is done by classical interlocking systems. Interlocking controls *field elements* such as signals, points and axle counters. Finally, through the introduction of ETCS Level 2 the *train control* is handled through radio block centers (*RBCs*), which communicate with the train on board units (*OBUs*) (Fig. 1).

A route control system is a safety-critical cyber-physical system, where common principles are well established and adopted on a broadly generic infrastructure but with an abundance of feature variations across national boundaries. In order to be able to get a configurable, yet certified, product it is therefore essential to adopt an efficient product development process that allows a verified core product to be adapted to specific solutions. At the same time the European Train Control System ETCS [1] aims at standardization, local sub-systems have to be interfaced and integrated within the overall railway control. We propose a model-based approach that will support such development processes. This approach makes large use of the methodologies presented in chapter "Systematic Verification and Testing", where the tools used in the associated process are described.

T. Fischer · K. Reichl · P. Tummeltshammer (✉)
Thales Austria GmbH, Vienna, Austria
e-mail: tomas.fischer@thalesgroup.com; klaus.reichl@thalesgroup.com; peter.tummeltshammer@thalesgroup.com

T. S. Hoang · M. Butler
ECS, University of Southampton, Southampton, UK
e-mail: t.s.hoang@soton.ac.uk; mjb@soton.ac.uk

A. Leitner et al. (eds.), *Validation and Verification of Automated Systems*,
https://doi.org/10.1007/978-3-030-14628-3_17

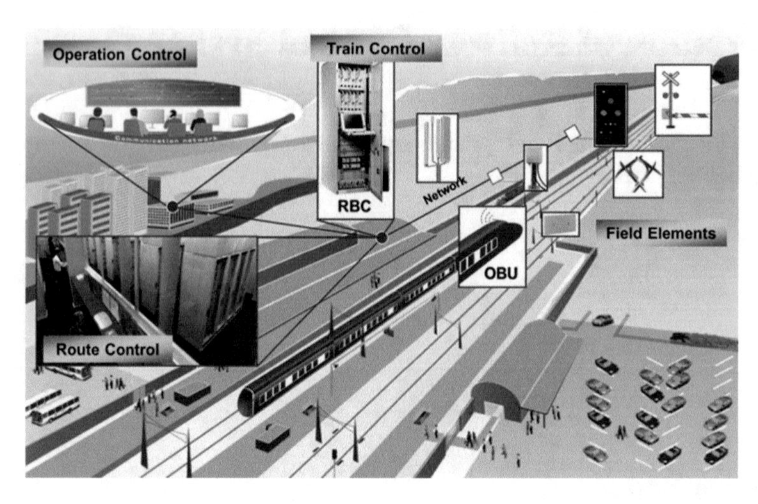

Fig. 1 Modern railway infrastructure

The main technical challenges which need to be solved to this end are: The automation of the translation step between a formalized specification understood by domain experts (signaling engineer) and a mathematical model which allows for proofs of functional and safety properties of the solution. Furthermore, the solution has to be split into a generic part (domain model), a customer-specific part (railway operator's procedures and rules) and concrete instances of track topologies (stations and open lines). Finally, verification results need to be aggregated back to the specification language domain experts (signal engineer, verifier, validator) can interpret.

2 Specific Requirements, Processes and Tools of Railway

Today's railway safety process is guided by the CENELEC [2, 3] standards. CENELEC/TC 9X is responsible for the development of European Standards for Electro Technical Applications related to the Rail Transport Industry of the European Union.

CENELEC states on formal methods to

- apply formal methods to requirements and high-level designs where most of the details are abstracted away
- apply formal methods to only the most critical components
- analyze models of software and hardware where variables are made discrete and ranges drastically reduced
- analyze system models in a hierarchical manner that enables "divide and conquer"
- automate as much of the verification as possible

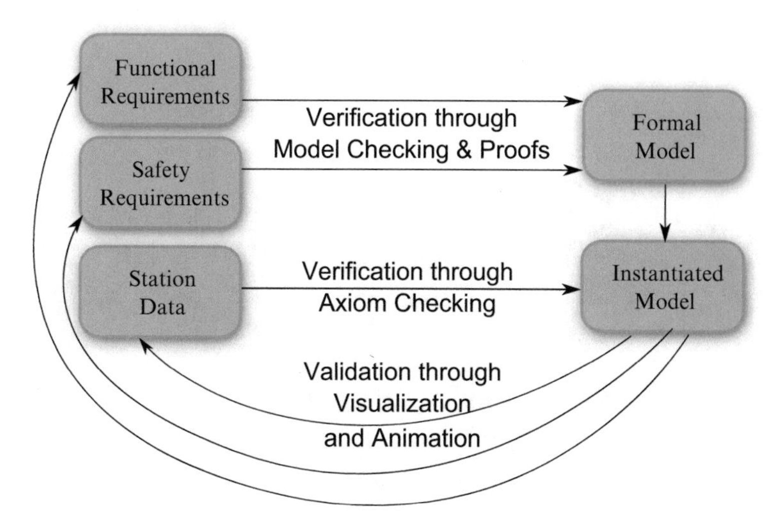

Fig. 2 Verification and validation through formal modeling

Therefore, the goal is to test novel approaches for verification and validation of large and complex systems, such as railway command and control systems through the introduction of formal methods in the process.

Model driven engineering, and especially formal modelling, offer a very promising and efficient method of showing the correctness of a complex system. The evolution of such systems can be more cost efficient or even only enabled by using a formal approach. While in today's industrial applications formal methods are mostly used for verification (i.e. for showing that the system model fulfils properties such as completeness and consistency) these methods are useful for validation as well (i.e. correspondence of the model with the customer needs) (Fig. 2).

The **validation** process ensures that the model is compliant to the customer requirements. It means that the functional properties express the required behavior, the safety properties express constraints derived from the hazard analysis and the data model is neither too general nor too restrictive. This serves as a proof of correct specification and correct implementation of requirements for the resulting software-intensive sub systems.

The **verification** process shall ensure that the model is unambiguous and complete. Formal models allow automatic reasoning, e.g. model finding (search for counterexamples), model checking, and model proof.

Automated test case generation helps to assure that the final implementation is compliant to the model, thus assists in both implementation verification as well as validation. Note that care must be taken to ensure that the generated test cases deliver expressive results in order to carry the trust gained in the model through formal methods to the implementation. This can be done using coverage analysis on execution traces of the model or the corresponding implementation driven by the automatically generated test cases (Fig. 3).

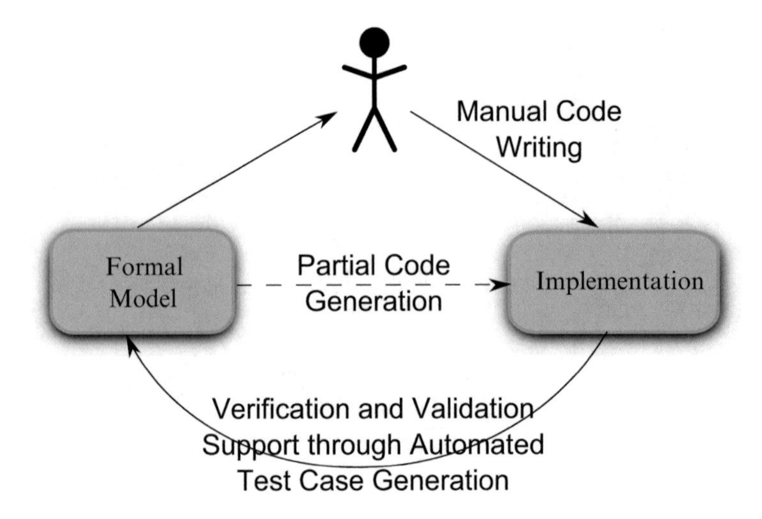

Fig. 3 Manual implementation and automated test case generation

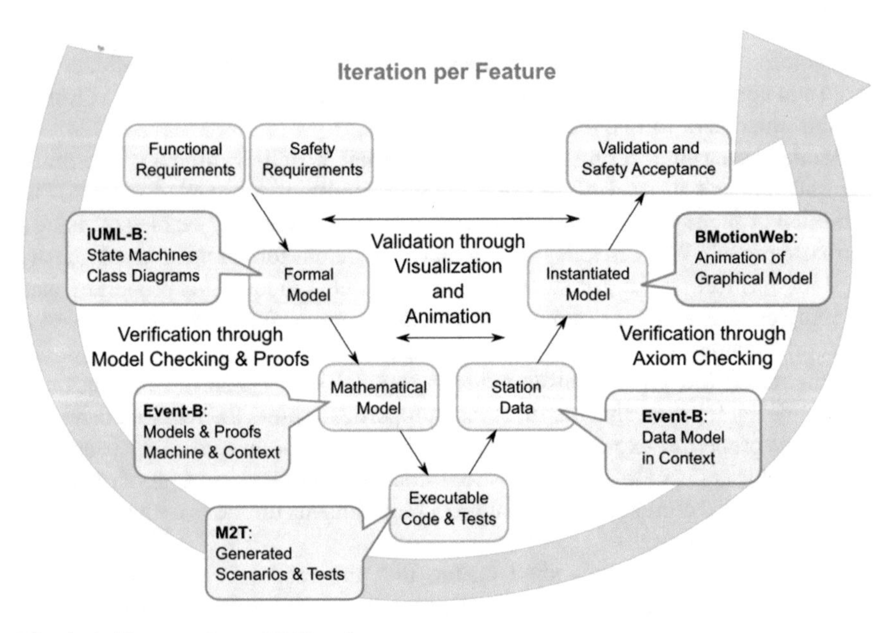

Fig. 4 Artifact mapping to V&V cycle

The above figure (Fig. 4) shows how tool capabilities map to the artifacts in the V-cycle. Introducing a formal model and a mathematical model allows for a better split between the domain expert and the formal modelers.

The formal specification uses DSL (Domain Specific Language) principles and is derived from the customer specification including the traceability between the DSL Model elements and the (informally written) customer specification.

It is optimized for the domain experts and fits domain elements as found in the customer's specification. Parts are automatically translated into the Mathematical Model and the Executable Code. As of today, the DSL is limited to signaling problems but can be generalized to fit other railway specific challenges (virtual fixed block, moving block).

The mathematical model allows for formal verification and station data validation. This is achieved by mathematical reasoning and axiomatic definition of the data rules and through (partially) automatic proves. Here structuring techniques like (De) Composition and generic inclusion apply.

Automated test case generation and model validation allow the final validation of the implementation with respect to conformance to the proven correct model. Acceptance test scenarios can be brought into the model and thus be executed on the model and on the implementation accordingly.

The approach we have taken is an instantiation of the V&V pattern "systematic Verification and Testing" described in more detail in chapter "Systematic Verification and Testing". We apply this in two different example applications for evaluation.

3 Use Case: Railground

3.1 Description

According to Pachl [4], a railway system consists of the following parts: First, infrastructure including trackworks, signaling and stations. Second rolling stock with rail vehicles, and third operating rules for safe and efficient operation. In this use case we will take a closer look at the first (signaling aspect) and the third (comprising the "software") parts of a railway command and control system.

The basic elements which constitute a railway network are tracks, points and crossings (T, P and C in Fig. 5). Axle counters at the start and end of a connected set of elements, form a network of blocks, which usually start and end with a signal (S in Fig. 5). The red line in Fig. 5 denotes a block being occupied by a train. These blocks are classically used for safe train operation, since

- A train entering and leaving a block will always be detected by the axle counters (vacancy detection) and
- The trains on the network can be controlled through signals.

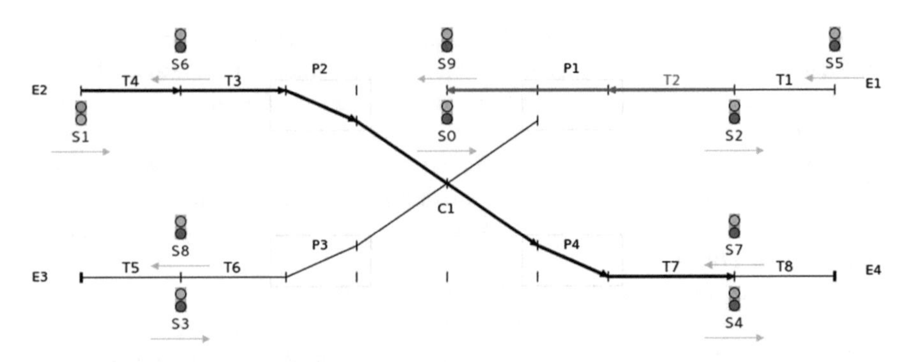

Fig. 5 Model animation of simple railway network example

To allow a train to pass through a network of blocks, a route (bold black line in Fig. 5) can be set, which ensures that the train can safely use this path and

- all points are set correctly as well as
- no other path conflicts with the one set.

3.2 Application and Evaluation of the Approach

In order to evaluate the above presented model, during the Transport Research Arena 2018 (TRA'18) a model railway track was constructed according to the model specification in Fig. 6.

The model was written using the graphical formal modeling methodology of iUML-B [Reference to chapter "Systematic Verification and Testing" Sects. 3.2 and 3.3] and animated using ProB and BMotion Studio [Reference to chapter "Systematic Verification and Testing" Sect. 3.6]. This model railway serves as an enabler to execute generated test cases [Reference to chapter "Systematic Verification and Testing" Sect. 3.5 (MoMuT)] in the form of scenarios on a real testbed track, thereby closing the gap between the model-based animation and a real implementation of the signaling solution.

4 Use Case: Virtual Fixed Block

4.1 Description

The European Rail Traffic Management System (ERTMS) is a system of standards for management and interoperation of signaling for railways by the European Union. The aim of ERTMS is to replace the different national train control and command systems in Europe with a seamless European railway system. The advantages of

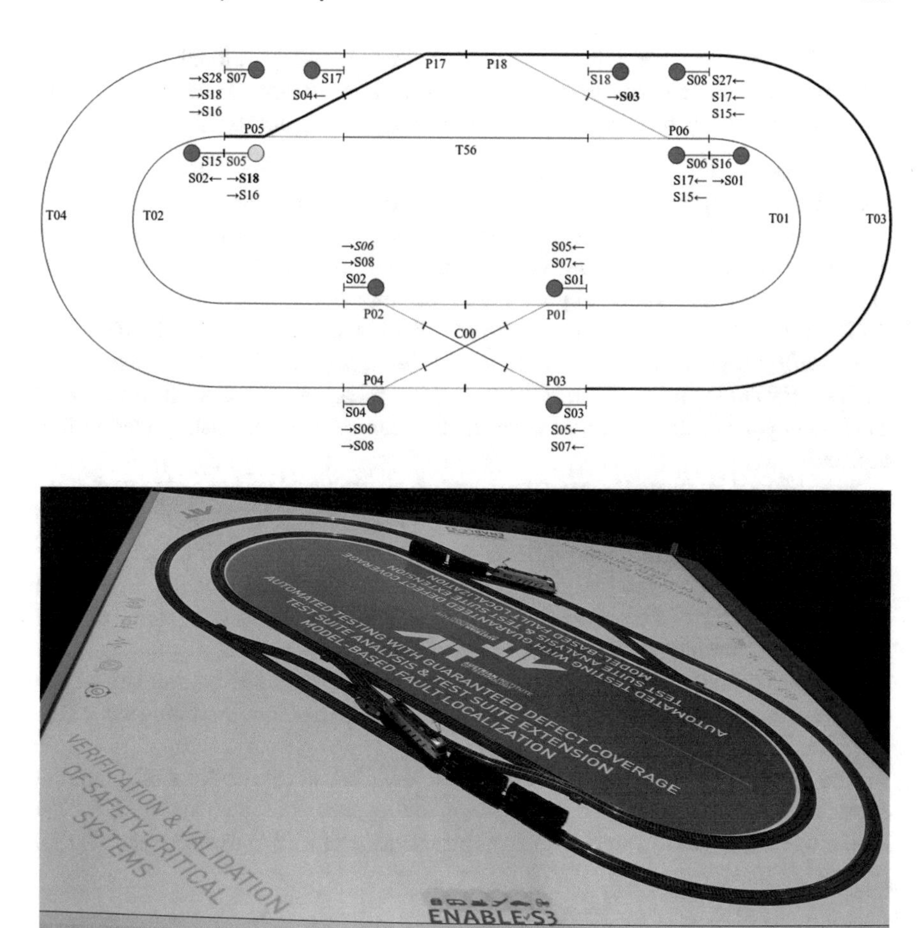

Fig. 6 Model animation and demonstrator presented by AIT at TRA 2018

ERTMS include increased capacity, higher reliability rates, improved safety, and open supply market.

There are three signaling levels for ERTMS:

Level 1 Communication between trains and trackside equipment by means of transponders called Euro-balises. Trackside equipment is needed for detecting train location and train integrity (train completeness) and lineside signals are required.

Level 2 Communication between trains and trackside equipment is provided by the Global System for Mobile Communications Railway (GSM-R). Track-side equipment is needed for determining train location and integrity while lineside signals are optional.

Level 3 The train determines its location using fixed positional transponders and supervises its integrity using the on-board Train Integrity Monitoring System (TIMS). This means that trackside detection equipment is not required.

As of today, systems up to ERTMS Level 2 are in operation. There are different options depending on levels of maturity in terms of definition and development, leading to several ERTMS Level 3 types. This use case focuses on Level 3 Hybrid which uses fixed virtual blocks, which is the most mature and is developed using existing technology solution augmented for optimization.

The Level 3 Hybrid has so far been specified in a first version in EEIG [5]. It copes with different train configurations (TIMS-equipped, ERTMS without TIMS, and non-ERTMS) and uses a limited amount of trackside detection. In the case of TIMS-equipped trains, the capacity of the line can be increased using fixed virtual blocks.

To achieve this, trackside detection sections are divided into several Virtual Sub-Sections (VSS). Depending on the train's equipment, the "occupied" and "free" status of the VSS is computed differently based on the train position information and the trackside detection information. Due to the discrepancy of the timing and spatial information of the trackside detection two additional (internal) status of VSS are specified: "ambiguous" and "unknown" (Fig. 7).

Despite of limited number of states for each VSS, the huge complexity of the state transition together with timing constraints require rigorous modeling technique to assure the correctness of the specification.

The goal of this use case is to get a detailed understanding of potential shortcomings of the examined specification and to stress out the benefits of formal modeling techniques in comparison to expert reviews as well as to discover potential pitfalls.

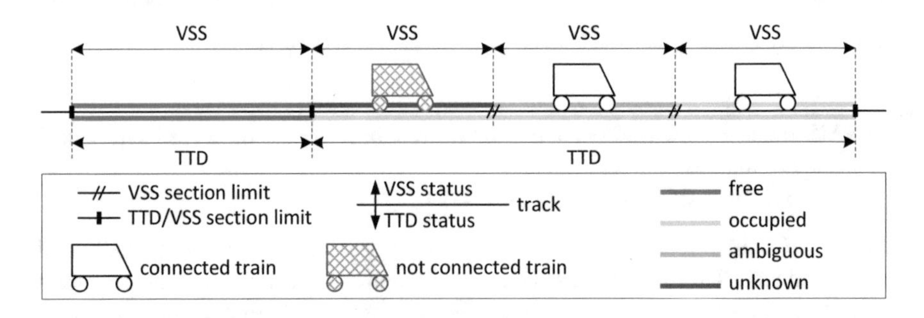

Fig. 7 Hybrid ERTMS L3 Section conventions from EEIG [5]

4.2 Application and Evaluation of the Approach

Basis for this work is [5], an informal specification done by railway experts together with acceptance scenarios explaining the various situations.

Acceptance scenarios and examples [Reference to chapter "Systematic Verification and Testing" Sect. 3.1 (scenario modeling)] are implemented according to the informal specification in order to validate the formal model and the implementation later on.

The test suite to be executed on the implementation comprises the generated acceptance scenarios (user acceptance tests for validation) and generated test cases out of the formal model. Tools used here are Cucumber for BDD and MoMuT for test case generation [Reference to chapter "Systematic Verification and Testing" Sects. 3.4 (Cucumber) and 3.5 (MoMuT)]. As the project ENABLE-S3 is still ongoing, the work on this topic is not finished as well.

In the context of ERTMS the model was already shown to external railway domain experts (ERTMS users group) and had a positive impact on the evolution of the definition of the informal specification.

5 Summary

In this chapter we described how the principles of formal modeling, formal proofs, BDD and automated test case generation can be mapped to two different railway related use cases.

It was shown, that the application of the methodologies is not only recommended by railway certification given by the safety standards, but they offer the possibility to ease the verification and validation process by introducing automation steps and using methods to create the results which are understandable by domain experts.

References

1. C. Directive: 96/48/ec of 23 July 1996 on the interoperability of the trans-European high-speed rail system. Off. J. L. **235**(17), 09 (1996)
2. European Standard EN 50126: Railway Applications – The specification and demonstration of Reliability, Availability, Maintainability and Safety (RAMS). CENELEC September 1999
3. European Standard EN 50128: Railway applications – Communication, signalling and processing systems – Software for railway control and protection systems. CENELEC June 2011
4. Pachl, J.: Railway operation and control (2002)
5. EEIG ERTMS Users Group, Brussels, Belgium: Hybrid ERTMS/ETCS Level 3: Principles. Ref. 16E042 Version 1A (July 2017)

Reconfigurable Video Processor for Space

L. Armesto Caride, A. Rodríguez, A. Pérez Garcia, S. Sáez, J. Valls, Y. Barrios, A. J. Sanchez Clemente, D. González Arjona, Á. J.-P. Herrera, and F. Veljković

1 Overview and Motivation

One of the main problems of autonomy in space applications is that once a mission is in orbit it is very difficult, sometimes even impossible, to replace a processing module on board. The vast majority of space digital systems are based on Application-Specific Integrated Circuits (ASICs) or antifuse-based FPGAs. These two technologies imply non-recurring engineering (NRE) costs and lack of flexibility requested in the New Space currently being defined. On the other hand, Static Random-Access Memory (SRAM)-based FPGAs as programmable devices with shorter design cycle and reduced NRE offer themselves as highly flexible and convenient platforms. Modern FPGAs also offer a possibility to be reprogrammed which makes them perfectly suitable for remote long-term space missions.

L. Armesto Caride (✉) · F. Veljković
TASE–Thales Alenia Space España S.A, Tres Cantos, Madrid, Spain
e-mail: laura.armestocaride@thalesaleniaspace.com; filip.veljkovic@thalesaleniaspace.com

A. Rodríguez · A. Pérez Garcia
UPM—Universidad Politécnica Madrid, Madrid, Spain
e-mail: alfonso.rodriguezm@upm.es; arturo.perez@upm.es

S. Sáez · J. Valls
ITI—Instituto Tecnológico de Informática, Valencia, Spain
e-mail: ssaez@iti.es; jvalls@iti.es

Y. Barrios · A. J. Sanchez Clemente
ULPGC—Universidad de Las Palmas de Gran Canaria, Las Palmas, Spain
e-mail: ybarrios@iuma.ulpgc.es; ajsanchez@iuma.ulpgc.es

D. González Arjona · Á. J.-P. Herrera
GMV Aerospace and Defence S.A.U, Madrid, Spain
e-mail: dgarjona@gmv.com; ajpherrera@gmv.com

A. Leitner et al. (eds.), *Validation and Verification of Automated Systems*,
https://doi.org/10.1007/978-3-030-14628-3_18

Increased flexibility of FPGA devices opens many possibilities for space missions. One of those possibilities is applied to autonomous navigation in space, where different vision-based algorithms, as HW accelerator modules, can be interchanged depending on the phase of the mission and in order to adapt to the specific environmental conditions and the navigation to perform in order to reach target satellite, moon, planet or asteroid.

However, in space applications, there are unique environmental challenges that need to be accounted for. Despite high performance, flexibility and low design costs, volatile nature of SRAM-based FPGAs makes them highly susceptible to radiation effects. Depending on the type of effect different types of failures may occur in the system. Radiation can produce both destructive effects and non-destructive event such as Single Event Upsets (SEU) and SE Transients (SET).

The Reconfigurable Video Processor for Space will take the benefit of ENABLE-S3 developments for the integration and validation of an autonomous navigation system. The processor features the in-flight reconfiguration with a two-fold objective: mitigate HW upsets caused by the radiation and protect the payload by re-using the same HW platform for different vision-based algorithms depending on the phase of the mission.

Through this Use Case of ENABLE-S3, space community should take the benefits of a preliminary method for in-flight reconfiguration in modern multiprocessors system-on-chip (MPSoCs) and a potential reduction in terms of efforts invested in always exhausting radiation simulations and emulations. Different image/video processing and navigation algorithms will be implemented and used as hardware accelerators in a reconfigurable architecture capable of scheduling the tasks and configuring only those modules necessary at certain point of operation. An updated version of the current system is intended to feature in future payload data processing equipment for video processing and navigation sensors based on cameras. This type of video processors is commonly in charge of conditioning, processing and compressing the images acquired by an Earth Observation Satellite before their transmission to ground or by any kind of spacecraft for navigation purposes.

Embedded systems combine HW/SW dedicated data processing units which are integrated into a larger electronic system or are specifically designed for stand-alone operation. Embedded systems are usually constrained for real-time performances. Not only ad-hoc implementation of algorithms into HW modules is necessary but also a specific HW architecture design. Hence, architecture of the system and algorithms implementation are constraining or driving each other's task. This differs from general purpose processor systems as they are designed explicitly to meet efficiently the real-time constraints of a specific development. It is translated into HW-SW-System inter-dependences.

The main objectives of ENABLE-S3 are adapted to the space domain and addressed in this UC:

- Reduction of expensive testing time in nuclear facilities (Objective 1),
- Optimization of the test setup through lab testing and prior to the real radiation testing (Objective 2)
- Strengthen the critical parts of the design in early steps of development by real and unlimited fault injection in the laboratory (Objective 3)
- Optimization of the component qualification efforts as a consequence of the first three objectives (Objective 4), reducing the number of visits and the corresponding testing time in the beam.

To make sure that the Objectives are accomplished, several KPIs, which are defined and explained in this document, are measured including the testing costs, elapsed time and associated costs (Table 1):

The KPIs involving Vision-based Navigation (VBN) and Hyperspectral Image Compression Algorithms are the ones impacting the efforts in the design and validation at early stages including the analysis, development and the qualification process. The main benefits involve the adaptation and validation acceleration; on-the-fly reconfiguration of different VBN systems; robustness to single error events in space electronics; a support for formal verification and analysis with a reduction of V&V efforts; shorter development life cycle. These reductions in design and validation are translated into a cost reduction which can lead to up to 50% of savings considering simplified radiation test campaigns and HW setup which could be partially covered by simulation.

Table 1 KPIs in enable S3

KPI description	
Testing effort (person hours)	(dKPI1) Average duration of testing activities
	(dKPI2) Average duration of the testing environment setup
Time elapsed (days per product)	(tKPI1) Calendar duration of testing activities
	(tKPI2) Calendar duration of the testing environment setup
Costs (personnel needed or costs of test equipment)	(cKPI1) Decrease in testing personal cost
	(cKPI2) Decrease in testing equipment cost
	(cKPI3) Decrease in operational costs for testing
Quality (stage of development cycle)	(iKPI1) % of (costly to fix) defects found at later stages of development process
	(iKPI2) % of defects resolved at the early stages of development

2 Detailed Description of the Automated System and Platform

The architecture proposed in the Use Case comprises the Reconfigurable Video Processor (RVP) as the System Under Test (SUT), as well as the Test System (TS) infrastructure, as shown in Fig. 1. The SUT has been implemented in a Zynq UltraScale+device that contains an FPGA-fabric tightly coupled with a Processing System featured with application-oriented and real-time specific processors. The mission specific tasks in the SUT are accomplished by a set of hardware accelerators running in the FPGA side exploiting the benefits of the ARTICo3 infrastructure.

2.1 SUT: Reconfigurable Video Processor Architecture

The ARTICo3 architecture [1] is a hardware-based processing architecture for high-performance embedded reconfigurable computing. Using Dynamic and Partial Reconfiguration (DPR) in SRAM-based FPGAs as its technological foundation, it supports and enables run-time adaptable implementations of data-parallel algorithms. Moreover, this user-driven application adaptively generates a solution

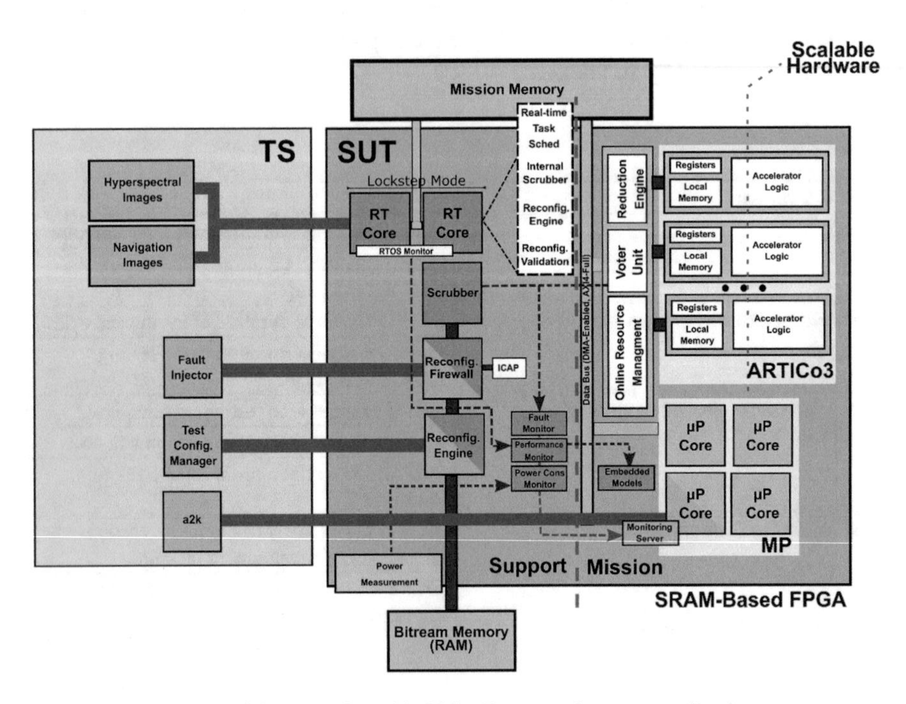

Fig. 1 Block diagram of the Reconfigurable Video Processor for space applications

space defined by a dynamic tradeoff between computing performance, energy consumption and fault tolerance.

DPR-powered module replication in the reconfigurable partitions (or slots) is combined with an optimized and Direct Memory Access (DMA)-capable data path, whose internal structure can be dynamically modified to support different processing profiles. Hence, *performance-oriented* or *fault-tolerant* processing can be selected on demand. Assuming that multiple copies of a given hardware accelerator have been loaded in the FPGA fabric using DPR, the former would use different input data for each copy to exploit Single instruction, multiple data (SIMD)-like execution, whereas the latter would deliver the same input data to each copy and vote the obtained results to mask faults. In addition, an embedded monitoring infrastructure allows users to analyze system performance and error rates.

Traditionally, both the design and management of dynamically reconfigurable hardware systems have been tasks that exhibit high levels of complexity. To reduce development effort and make the ARTICo^3 architecture accessible to embedded system engineers with little or no previous experience on hardware design, two additional components complement the processing architecture and constitute a fully-fledged framework: a toolchain to automate system generation from source code to application binaries, and a runtime library to hide parallelism deployment and DPR from programmers.

ARTICo^3-based processing follows a processor-coprocessor approach, where an application runs on a host microprocessor and data-parallel computations are offloaded to the reconfigurable computing engine. The ARTICo^3 toolchain takes an already partitioned hardware/software application as input and generates both the executable to be run on the host and the configuration files to program the FPGA. While the host code needs to be specified using C/C++ descriptions, hardware accelerators can be designed using low-level HDL (useful for legacy or highly optimized designs) or high-level C/C++ descriptions, since High-Level Synthesis (HLS) is also supported within the toolchain.

Two different types of parallelism can be exploited using ARTICo^3: on the one hand, the execution model provides transparent data-level parallelism by using several replicas of the same hardware accelerator working in a *performance-oriented* profile; on the other hand, DPR enables task-level parallelism by using different hardware accelerators. The ARTICo^3 runtime library hides all low-level details of FPGA reconfiguration, as well as any parallelism management other than the initial partitioning of the application. A simple yet powerful C API acts as an interface between user code running on the host microprocessor and the hardware accelerators. As part of this API, functions to manage shared memory buffers, the internal monitoring infrastructure, or additional register-based configuration in a transparent way are also provided.

2.1.1 Fault Mitigation Techniques in the SUT

The ARTICo^3 framework can be used to develop highly configurable systems for a wide range of Cyber-Physical System (CPS) applications. However, some of its features might not be enough in the space scenario, targeted in this Use Case. In a safety-critical context, although ARTICo^3 itself can provide fault tolerance in the reconfigurable partition, additional mechanisms need to be implemented to ensure fault tolerance at system level.

To enhance the fault-tolerance capabilities of the SUT, DPR is exploited at different levels of granularity for correcting errors in the configuration memory of the FPGA. Transient faults are fixed by rewriting the affected parts of this memory. Moreover, if a part of the FPGA is permanently damaged, accelerator relocation from one reconfigurable slot to another can be triggered to mitigate the problem. Hence, DPR is used with a twofold objective: to adapt the system dynamically (functional reconfiguration in ARTICo^3) and to fix faults. As a result, both the configuration interfaces and their arbitration are essential elements of the system.

Techniques for correcting transient faults in memories are generally called scrubbers. In the RVP they have been implemented in different fabrics: real-time processors (ARM Cortex R5), platform management unit (PMU), and dedicated hardware cores inside the FPGA. The real-time processors are used in lockstep mode, which renders both hardware and temporal redundancy, and the PMU is a fault-tolerant triple-redundant processor. These processors, thanks to their particular implementations, are the most reliable elements against radiation inside the system, so they have been used to allocate the tasks that will ensure the survival of the system in the deep space. On the other hand, the scrubber implemented in the FPGA fabric is based on information redundancy codes, offering high performance in terms of fault detection and correction times but being more error prone since it is also part of the implemented circuitry. This variety of scrubbers, implemented with different approaches and in different fabrics, increase the design diversity, ensuring that they will not fail under the same circumstances, thus decreasing common-mode faults.

One of the main sources of design complexity comes from the fact that the execution of these support tasks must be multiplexed with the execution of the mission tasks developed by the RVP. The problem of scheduling both types of tasks is not trivial and must be solved by fulfilling real-time constraints and in a deterministic manner. The implementation of this scheduling is complex due to the high amount of resources and algorithms. Hence, the RTEMS real-time OS is used in the SUT to coordinate the accesses to shared resources and schedule the execution of different algorithms. The decision to select RTEMS was an easy choice due to the following factors: it is a certifiable OS that has been already used in many space missions, and it generates significantly larger added value to industrial partners for extending previous expertise in the field to reconfigurable platforms.

2.2 *Algorithms in Space Scenario*

The other key aspect of the application is the set of algorithms running in the SUT. These algorithms must be tailored depending on the target application. In addition, considering that these algorithms are intended to be accelerated by hardware, and the SUT is thought to work in a space environment, where the processing resources are limited, the algorithms must be optimized in terms of computational complexity, memory footprint and power consumption.

The specific algorithms usable by the RVP architecture should support HW/SW partitioning, so that certain parts (or tasks) can be offloaded to and executed in hardware accelerators. To take further advantage of the hardware-based execution of tasks, the algorithms should comply with at least one model of computation in which parallelism is defined at different levels. In this regard, task-level parallelism, seen as different hardware accelerators performing computations at the same time, and data-level parallelism, seen as several copies of the same hardware accelerator performing the same operation on different sets of data (in an SIMD-like fashion), should be given a higher priority.

2.2.1 Hyperspectral Image Compression Algorithms

Explanation about why compression techniques are mandatory is missing: hyperspectral sensors produce large amounts of data, which may exceed the transmission bandwidths. In the image compression state-of-the-art, two categories can be clearly identified: lossless and lossy compression techniques. On one hand, lossless compression preserves all the information presented in the original image; consequently, the achieved compression ratios are limited to 4:1 or less, clearly insufficient for long term and large distance space missions with high-resolution hyperspectral sensors on-board. On the other hand, lossy compression techniques yields higher compression ratios introducing losses in the data, this is, the recovered image is not identical to the original one [2].

In this work we present an implementation of a lossy extension of the CCSDS (Consultative Committee for Space Data Systems) 123.0-B-1 lossless standard, specifically thought for compressing multispectral and hyperspectral images in space applications. This algorithm provides a trade-off between its compression efficiency and the design complexity. The output data constitute a variable-length encoded bitstream from which the original image can be fully recovered [3]. It uses a scheme based on prediction and entropy coding of the resultant prediction residuals, i.e. the differences between each input sample and its corresponding predicted value.

The prediction stage sequentially processes all the input samples in a single step, using a spatial and spectral neighborhood of previously processed samples to predict the value of the current sample. Therefore, if $s_{z,y,x}$ is a sample located in the spatial coordinates (y, x) and band z, the predicted sample $\hat{s}_{z,y,x}$ is calculated using previously pre-processed neighboring samples of $s_{z,y,x}$ in the current and in the P

previous bands. The number of previous bands P used for prediction is a tunable parameter between 0 and 15. The predictor first computes a *local sum* $\sigma_{z,y,x}$ of the neighboring samples of $s_{z,y,x}$ in the current band. Two possible configurations for the local sum computation can be selected: neighbor-oriented or column-oriented. Then, the local sums are used to compute the *local differences* $d_{z,y,x}$. The central local differences together with the directional local differences conform the $U_{z,y,x}$ vector. Finally, the prediction is calculated by computing the inner product, $\hat{d}$, between the local differences vector $U_{z,y,x}$ and the weight vector $W_{z,y,x}$. This weight vector quantifies the contribution of each value of the local differences vector in predicting the corresponding sample. The *prediction residual* is calculated as $\Delta_{z,y,x} = s_{z,y,x} - \hat{s}_{z,y,x}$ and then it is mapped to unsigned integer values, $\delta_{z,y,x}$. These mapped prediction residuals are passed to the entropy coder to generate the compressed bitstream together with the header, that includes all the necessary information to decompress the bitstream at the ground stations [3]. A sample-adaptive encoder is used, where the mapped prediction residuals $\delta_{z,y,x}$ are encoded using a Golomb power-of-two variable-length binary codeword. The codes are adaptively selected based on statistics that are updated after each sample is encoded [4].

Several modifications are introduced on the CCSDS 123.0-B-1 lossless standard in order to extend it to a near-lossless to lossy scheme, including quality and bit rate control [5]. A key part of the development is the quantifier that works within the prediction stage with a level of flexibility defined by the user and is able to adapt to the input image samples. The bit rate control is based on [6], but applying some modifications to try to obtain a hardware-friendly description, simplifying the algorithm complexity and reducing its latency. We propose to apply a unique quantification step to each spectral line. The calculation of these quantification steps is done considering that the variance of the prediction residuals between two adjacent lines are highly correlated. The median of the residuals is used as an estimator to characterize the distribution of residuals. A median is computed for each band, and after all the medians have been obtained, the quantization step for the next line is computed. However, with the goal of optimizing the hardware implementation, the median of partial medians is computed instead per band.

2.2.2 Computer Vision-Based Navigation Algorithms in Space Scenario

Guidance, Navigation and Control (GNC) engine in the spacecraft will allow autonomous navigation to planets or asteroids surface, to orbits around those bodies or servicing operations between satellites already in orbit. The navigation system will be used to estimate the position of the spacecraft with respect to the target asteroid, moon, planet or debris. Accurate positioning is required to enable pinpoint landing ability or grapping of debris. To do that following vision-based navigation approach, the system will rely among other sensors on a navigation camera. The core of the navigation system consists of the vision-based relative and absolute navigation. The solution provided by the navigation camera is supported by the measurements of altimeter to compensate for poor observability of the depth in monocular camera. The system consists of two main parts: Navigation filter and Image Processing. The image processing is a very demanding task in terms of computational load and performance; therefore, it is developed as a HW accelerator in the FPGA.

The Vision-Based Navigation (VBN) system is designed as an Image Processing and Navigation (IP&N) Architecture following a HW/SW Co-design methodology. VBN is implemented partly in HW and partly in SW. Image processing techniques are very computationally intensive and therefore, HW acceleration in FPGA acquires plenty of importance for this kind of developments. The IP&N is deployed to a System-on-Chip reconfigurable HW which forms part of the SUT. ARTICo3 reconfigurable architecture is used to configure different Vision-Based Navigation algorithms at different points of the operation. In addition, thanks to ARTICo3 the robustness to radiation induced failure events is significantly increased. The time consumed in the reconfiguration is a main constraint due to the space mission characteristics as the systems are built as hard real time, so the Image Processor shall deliver the processes data within a defined period. The HW platform is based on a system-on-chip to accommodate the HW/SW partitioning and therefore internal communication is used between the SW processor cores and the HW IP cores. The HW IP cores will be provided with images from a navigation camera. For the ENABLE-S3 validation, we use pre-generated set of image representatives of a valid descent and landing trajectory and satellite rendezvous maneuvers. Image data, as well as other spacecraft sensor data, will be exchanged by Ethernet link to the target system. Image and sensor data is used in the FPGA IP cores as well as in the navigation filter. The GNC system, running on a processor is also interfacing the image processing. Nevertheless, the amount of data is reduced compared to the camera interface.

Three different algorithms are selected for ENABLE-S3 to evaluate the use of a common reconfigurable HW avionics.

Absolute Navigation Algorithm

The absolute navigation technology uses image processing techniques in order to find relevant surface features, called landmarks, in the image collected by the Navigation Camera. The landmarks are then mapped and compared with a database present on-board and previously generated. The mapping of multiple landmarks allows the Navigation Filter to derive accurate position estimation with respect to the target surface. The image processing provides edges that are later on post-processed to extract landmarks of the surface. The selected algorithm is the widely known Canny edge detector.

Relative Navigation Algorithm

The Image Processing part of the algorithm is responsible for extraction of the landmarks/features points in the image and tracking them. Following the displacement of up to 100 distinct points observed in the images, the navigation filter is able to accurately calculate position of the spacecraft. Image processing algorithm uses detectors based on the selection of good features which find corners using Harris Corner or Minimum Eigen Value corner algorithms [4].

Stereo Vision Algorithm

To obtain 3D information regarding the target body, stereo vision gathering information from two separated cameras is used. The system utilizes extraction of lines which are matched in the two stereo views; these matches are then utilized to estimate pose. The selected algorithm to be HW accelerated is proposed by Harris and Stephens already.

2.3 *Test System Tools*

2.3.1 Radiation Sensor: Fault Injection Rate Tool (FIRT) and Fault Injection Engine (FIE)

The traditional platforms intended for the development of space digital systems face the following main problems: Increased non-recurrent engenieering costs (ASICs), low performance (Anti-fuse based FPGAs) and significant lack or flexibility (in both platforms) which becomes an important asset in modern space applications.

SRAM-based FPGAs are being introduced in the market with a good pace as they feature cost-effectiveness, high performance computing and reprogrammability which significantly increases the flexibility of the whole system. The configuration memory content of an FPGA determines the final functionality of the device. It is stored in a static random access memory (SRAM). This volatile characteristic makes the SRAM-based FPGAs very sensitive to radiation effects. When a particle hits an SRAM-based device, the content of one or several configuration-memory cells can change. When it happens, the implemented functionality may change as the content of the memory determines the behavior of the chip (See Fig. 2).

The *FIE* is a mix of the hardware on the actual device in which the faults will be injected and a software which is capable of controlling this hardware in order to perform the fault injection as desired by the user.

Commercial parts, as a vast majority of SRAM-based FPGAs cannot provide a reliable hardware for space environment since they haven't been designed following

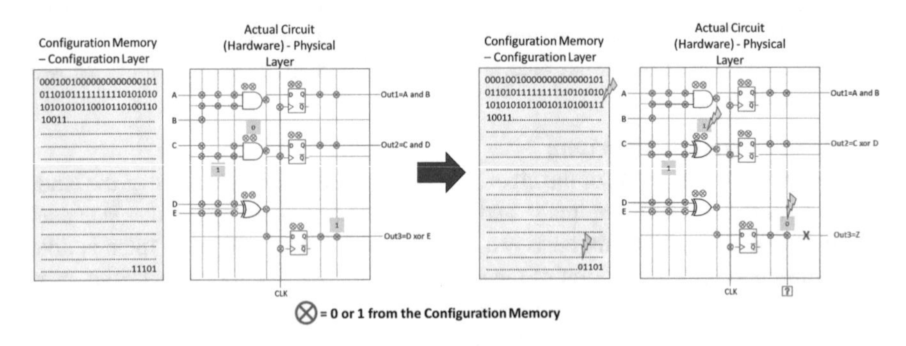

Fig. 2 An example of a change in the configuration memory and its effect on the actual hardware

secure radiation hardened and fault tolerant design processes. Verification and validation are essential steps in the development process of any autonomous system and as such represent key targets of ENABLE-S3. In this work, the system will be put to several types of tests to assess the correctness of its behavior. This evaluation will be performed through several phases where the first one is the sensitivity analysis of the configuration memory of the device. This evaluation will be performed as a part of radiation damage simulation campaign at the Radiation Effects facility. A large number of particles will be applied to the chip in order to emulate the radiation environment during a sufficiently long period of time in space and be able to extract the statistics necessary to calculate the probability of an error in different orbits.

The *FIRT* should lead a fault to be injected trough the FIE, generating the error files according to the fault injection rate and distribution computed by the tool. It uses the following parameters as its inputs for the calculations such as the FPGA to be used, its number of cells, LETth, Cross section, width, surface, the environment conditions requested by the mission (selected orbit) and others.

2.3.2 a2k Tool Suite

The a2k tool suite is used to assist in the V&V process. This web-based toolchain provides features to help in the design process of high integrity systems with real time constraints. It allows a flexible modelling and analysis of the system under development because of different stages offered by the tool.

The engineer defines the hardware platform of the target system including the programmable devices part of the ARTICo3 architecture, the model of the application software that is to be executed on the platform, jointly with its real-time constraints, and finally the deployment information, as a nexus between both platform and application model. Following the constraints defined by the engineer, the tool applies the task partitioning into threads, the priority assignment for each thread and, lastly, the processor allocation of the thread's activities to the different processors and processing devices.

The engineer is then able to select from different analyses offered by the application, allowing him to analyze different aspects of the model and check the feasibility of the configurations under study.

To improve the V&V process, the a2k tool suite has been extended in this work to include new features such as: an event-based discrete simulator that eases the analysis of the real-time behavior of complex systems and a monitoring tool that collects the tracing information of the system for a subsequent analysis.

a2k Model-Based Simulator Tool

Schedulability analysis is one of the most important evaluations of a system, especially in hard real-time systems, in which missing a deadline can lead to catastrophic consequences. A complementary approach to performing off-line analysis, when the system is too complex to use analytical methods, is the evaluation of the real-time system through simulation of its real-time behavior. Simulation offers advantages that makes it an interesting tool: (1) there is no probe effect (no disturbance) on the system due to instrumentation, which can be problematic in real-time systems, and (2) one can study the system behavior under the same execution conditions and just varying the characteristics in which the engineer is interested.

The previously mentioned points are some of the reasons that led to the development of a discrete-time event-based simulation tool to extend the a2k tool suite. A smooth integration with the tool is achieved by employing the models defined by the a2k tool as the simulation components. A deterministic simulation is obtained by utilizing the worst-case execution time (WCET) and the best case execution time (BCET) of the activities. The engineer might opt, nonetheless, to make a non-deterministic simulation by employing a random execution time of the activities between those two threshold times.

The simulation returns a series of results that have been computed during the simulation process and that can be shown on the a2k interface. For example, a Gantt diagram with the representation of when an activity has begun or ended can be displayed. Also, statistics like response time, jitters and deadline misses can be summarized in a table that is also shown in the tool. All this information not only allows identifying unfeasible system configurations, but also helps the engineer to better understand the run-time behavior of the concurrent system.

a2k Real-Time Monitoring Tool

To check the fulfilment of the constraints of a real-time system, the a2k tool has been extended to provide a run-time monitoring capability that allows the collection of several statistics and the later display of them in a unified interface.

First, temporal behavior is monitored with the purpose of finding system events not taken into account in the model, such as the effect of the interruptions or the operating system overheads, and of obtaining a better approximation of the system times defined by real execution time and real response time of the activities. It allows the engineer to discard unfeasible temporal configurations of the application and to find out the candidates to be optimized, while the real software of the system (on-board and payload applications) is being executed.

Run-time monitoring also includes the capability of collecting application specific statistics, such as error counters, communication baud rates or the use of shared resources, which help having a better understanding of the System Under Test execution.

The above is achieved thanks to the introduction of small-overhead tracepoints in the code, for instance, in the real-time scheduler context switch or at the beginning and the end of an activity execution. These tracepoints can be post-processed and workload temporal characteristics computed.

Run-time monitoring provides metrics about system faults, functional results and temporal behavior in order to check the system feasibility and to help the V&V of the system in an iterative design process.

3 Verification and Validation Approach and Results

The demonstrators developed in this work evaluate the results expected for Enable-S3 reducing the cost-intensive validation and verification efforts by virtual and semi-virtual testing.

The reconfigurable video processor demonstrator is intended to show the feasibility of the following elements for Earth observation (EO) applications:

- Autonomous failure detection and reconfiguration features
- Autonomous video adaptability for space variable environment
- In flight reconfigurable Autonomous Video Based (AVB) navigation
- Compliance to real time and dependability space constraints
- Radiation hardness through self-healing and self-reconfiguration techniques

A demonstrator has been implemented integrating some of the main characteristics of the SUT and the Test System. It consists of a Parametric Hyperspectral Image Compressor implemented on a fault tolerant FPGA-based system, validated with fault injection and based on ARTICo3 HW acceleration in the Reconfigurable Video Processor. It therefore shows the integration of three components of the use case: CCSDS-123.1 Compressor (ULPGC- Universidad de Las Palmas de Gran Canaria), ARTICo3 architecture (UPM- Universidad Politécnica Madrid) and fault injection (TASE- Thales Alenia Space in Spain)—Fig. 3.

The hyperspectral image compressor included in the demonstrator is based on the CCSDS-123.1 standard, implemented on a fault tolerant system based on ARTICo3 reconfigurable architecture. The compression rate can be modified at run-time to show adaptable operation of the compressor which uses real hyperspectral images obtained from a camera contained in a server. Fault emulation is implemented by the fault injection mechanism described above to show the system reaction capabilities when dealing with those unexpected events.

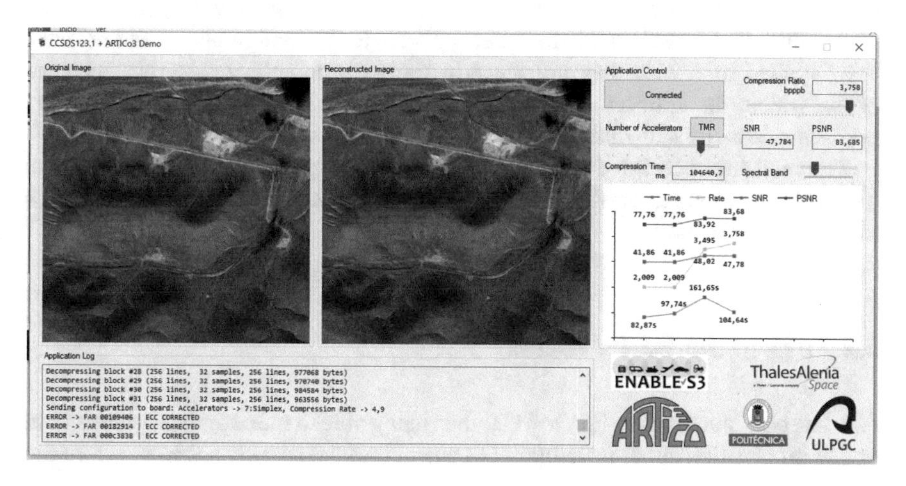

Fig. 3 ENABLE-S3 Year 2 demonstration

The demonstration includes fault mitigation techniques to show fault masking capabilities, fault detection and faster repair times.

The development of the demonstrator has been done according to the architecture developed during the first year of ENABLE-S3 [7].

3.1 Fault Injection in the Platform and Reconfiguration Techniques Verification

The demonstrator includes tests involving fault injection in the configuration memory through universal asynchronous receiver-transmitter (UART). Fault injection is automated and performed both randomly and in a deterministic manner to be able to affect all non-dynamic bits of the configuration memory. In order to add additional value to the setup, and affect only those bits of the configuration memory used by the implemented design, the information on the location of these bits will be extracted from the .EBD and .EBC files which are created during the implementation process in Xilinx software tool called Vivado. Consequently, it is expected to reduce the test execution effort by up to 60%.

To assess fault protection of combined fault mitigation techniques with HW reconfiguration techniques, a fault injection process was developed which combines, on one side, an addition of faults that mimic a realistic fault injection rate and, on the other side, all existing RFM techniques (reconfiguration and fault mitigation). Under this context, performance and fault monitoring infrastructure could help to ensure the correct behavior and timing of the system. Therefore, performance and fault monitors should be also made externally accessible in order to facilitate validation, as well as for the resource manager in order to take the appropriate measures (reconfigure, change scrubbing rate, etc.).

The performance monitoring infrastructure could be able to evaluate the correct behavior of the system as well as all the tasks compliant with the real time constraints. The usual validation mechanisms for reconfigurable systems have been studied to ensure they are the best tools to measure the correct behavior of the system. In addition, new validation techniques are developed based on the performance metrics obtained during the task execution performed by the SUT.

The fault injection rates will be measured and compared with the performance lost by the system to ensure that the graceful degradation mechanism is good enough to protect the system.

3.2 a2k in the V&V Process

A2k has been integrated with the SUT, thus being able to monitor the system and to obtain information about its temporal behavior and the statistics obtained from the

ARTICo3 architecture. The tool provides the real period of the execution flows, as well as the activation jitters or measures of the frequency of certain system events.

The ability to modify the functional configuration of the application on-the-fly, while extracting results eases the V&V process of comparing the functional behavior in a real execution with the expected one. This capability can be combined with other external tools such as the fault injector for the programmable devices or the space navigation simulator.

The integration of the a2k tool in this demonstrator contributes to the improvement of the processes that analyze and validate the compliance of real time and dependability space constraints of the applications. This is achieved by either reducing the number of tests required in a first phase of analysis which discard the unfeasible test cases or by reducing the duration of the tests employing the model-based simulation, faster than a real time execution, which can help identifying unfeasible configurations during early stages of development.

3.3 Image and Video Processing Verification

Different results have been obtained in terms of speed up depending on the number of hardware accelerators implemented on the ARTICo^3 architecture. In Fig. 4 the execution time in seconds is shown, depending on the number of accelerators used to compress an image.

The use of multiple accelerators is intended to split the hyperspectral images into portions on the ARM cores (software domain) and distribute them among the different accelerators. Taking into account that the algorithm, in its purely software version running on an ARM Cortex-A9, takes around 555 s to compress an image

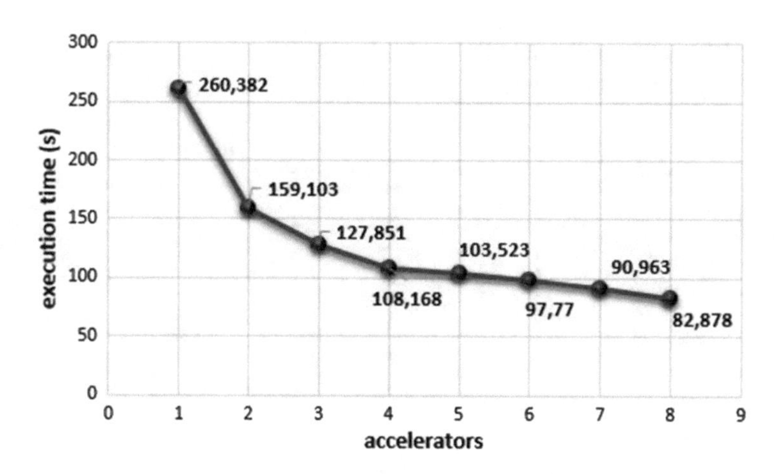

Fig. 4 Execution time depending on the number of used accelerators

with a size of $Nx = 512, Ny = 512, Nz = 256$, we obtain a maximum of speed up of $\times 7$ when eight accelerators are used.

Furthermore, in order to verify the efficiency of the implementation in terms of compression ratio, it has been tested with different hyperspectral images coming from different types of sensors, such as AVIRIS (aircraft), CRISM (spacecraft) and a subset of others captured directly by a Headwall Hyperspec VNIR E-Series hyperspectral camera, available in our hyperspectral image acquisition facilities. In Fig. 5, we show the compression ratio achieved for each image in terms of *bits per pixel* (*bpp*) against the quality of the reconstructed image after the decompression step. This quality is measured with the Signal-to-Noise Ratio (SNR), obtaining higher values (a maximum of 90 dB for the CRISM image) when the images have been pre-processed or calibrated after the decompression. In all the cases, the design reaches the desired compression ratio specified by the user with accuracy. In Fig. 5, beyond the limits defined for each image, the compressor works in a lossless mode.

3.4 Space Vision-Based Navigation Verification

Validation of the proposed system is accomplished within different testing frameworks: Model-in-the-loop, SW-in-the-loop, Processor-in-the-loop and HW-in-the loop. Results show the capability of GNC to meet precision landing requirements and stable behavior even in harsh environment allowing robustness to different illumination conditions. Real-Time performance can only be achieved by designing a proper HW architecture for the Vision-Based Navigation System and the corresponding reconfiguration platform. A HW demonstrator of the system prototype will be implemented using flight-representative HW including equivalent space-grade version of the architecture boards. The difference from space-grade version and its equivalent version lies in its radiation protection to guarantee single event effects (SEE) latch-up Immunity and SEU mitigation as well as guaranteed operation over full military temperature range.

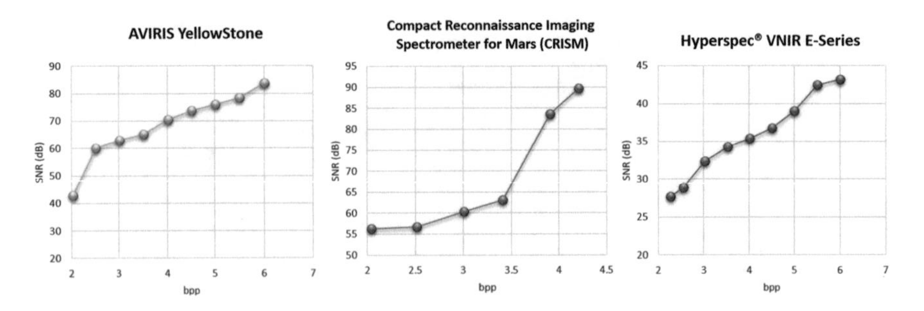

Fig. 5 Compression ratio vs. SNR for different kind of images

One of the most important tasks for the validation campaign is the generation of input data to use. A set of images will be generated in order to compose a representative spacecraft trajectory approaching a target asteroid or debris in order to feed the vision-based navigation system.

The generated dataset and the different validation methods in the loop will allow the verification of the GNC results produced using the IP results as inputs as well as the sensor data to determine the error rate. This implementation and validation methodology will cause a cost reduction due to an early error detection process.

The Test System (TS) is composed of several modules in charge of the spacecraft sensor data generation needed for both the navigation filter and for managing the control actuators (Reaction Wheels, reaction control system- RCS) that receive the actuators' commands from the SUT. For this purpose, a specific simulator (GMV-DL-Simulator) has been built to implement the Test System functionalities allowing the operator to control the input data of the Test system as well as validating the results of the test. The GMV-DL-Simulator closes the loop in such a way that the output data from the SUT determine the subsequent inputs of the SUT. The GMV-DL-Simulator also provides the capability of introducing pixel errors in the input image to the HW IP so that the operator is able validate the SUT behavior under hardness simulated conditions (Fig. 6).

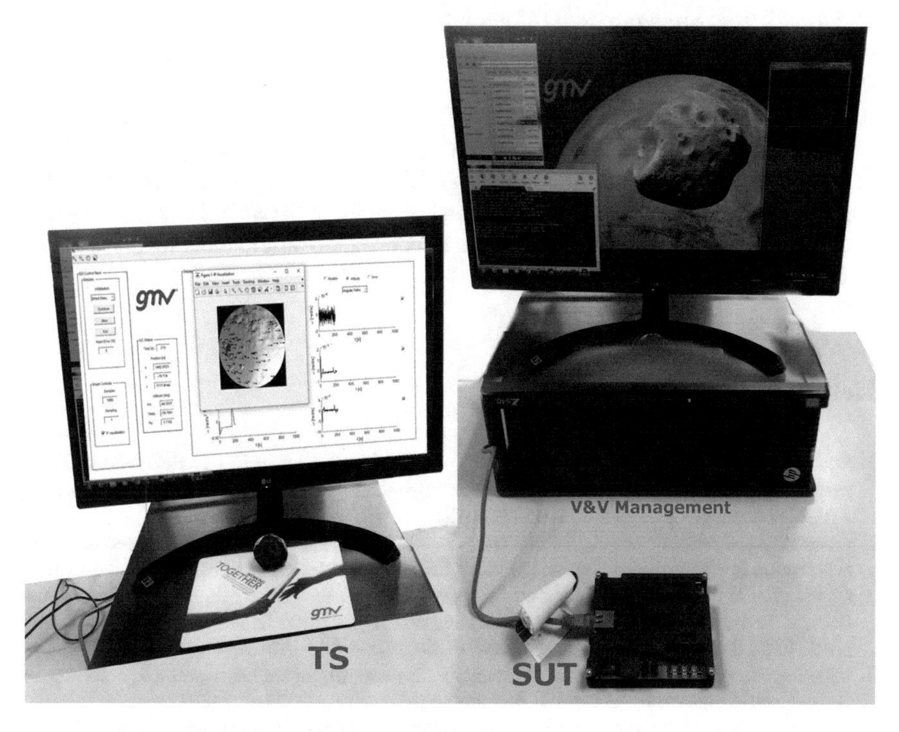

Fig. 6 Test system (GMV-DL-Simulator) and SUT

4 Conclusion

According to the obtained results, we can conclude that the proposed solution provides an efficient solution in terms of memory footprint and logic resources utilization for onboard applications, without compromising the compression quality. In addition, we provide an approximation to the development of robust designs against radiation over COTS FPGAs, devices of special interest for nowadays and future space missions.

A lossy extension of the CCSDS 123.0-B-1 Lossless Multispectral and Hyperspectral Image Compression algorithm is implemented on a COTS FPGA-based SoC, running over a reconfigurable, scalable and fault-tolerant architecture named ARTICo3. This solution has been implemented on a Xilinx Zynq UltraScale+ FPGA-based MPSoC.

The possibility of injecting real faults in the FPGA according to the expected failure obtained by analyzing the radiation environment allows to evaluate the behavior of the designs installed in the SUT in the different stages of development: avoiding visits to the BEAM, developing test set-ups adapted to the failures expected, reducing testing time, and, in general lines, providing robustness against radiation to the design and improvement the main KPIs requested in the ENABLE S3 Project.

The a2k Tool Suite provides different functionalities with respect to the requirements of the UC8. It provides modelling, analysis and simulation tools that allow the evaluation and understanding of the temporal behaviors of the final system. Additionally, it permits unfeasible scenarios or configurations to be discarded, thus decreasing the number of tests required which in turn brings savings in both testing time and personnel cost. The monitoring tool offered by the framework, helps the engineer in checking and observing the real behavior of the SUT. Lastly, it serves as a test management platform, that is able to coordinate all the tools involved in the test.

References

1. Rodríguez, A., Valverde, J., Portilla, J., Otero, A., Riesgo, T., de la Torre, E.: FPGA-based high-performance embedded Systems for Adaptive Edge Computing in cyber-physical systems: the ARTICo3 framework. Sensors. **18**, 1877 (2018)
2. Transon, J., d'Andrimont, R., Maugnard, A., Defourny, P.: Survey of hyperspectral earth observation applications from space in the Sentinel-2 context. Remote Sens. **10**(2), 1–32 (2018)
3. Christophe, E.: Hyperspectral data compression tradeoff. In: Prasad, S., et al. (eds.) Optical Remote Sensing. Advances in Signal Processing and Exploitation Techniques. Springer, Berlin, pp. 9–29 [Online]. http://link.springer.com/10.1007/978-3-642-14212-3 (2011)
4. Valsesia, D., Magli, E.: Fast and lightweight rate control for onboard predictive coding of Hyperspectral images. IEEE Geosci. Remote Sens. Lett. **14**(3), 394–398 (2017)
5. The Consultative Committee for Space Data Systems, Lossless Multispectral and Hyperspectral Image Compression, CCSDS 123.0-B-1 Recommended Standard, no. May. 2012

6. Valsesia, D., Magli, E.: A hardware-friendly architecture for onboard rate-controlled predictive coding of hyperspectral and multispectral images. In: International Conference on Image Processing, pp. 5142–5146. IEEE, Paris, France (2014)
7. ENABLE-S3 Consortium.: Generic Test Architecture. https://www.enable-s3.eu/media/publications/ (2017)

Maritime Co-simulation Framework: Challenges and Results

Arnold Akkermann, Bjørn Åge Hjøllo, and Michael Siegel

1 Introduction

This technique, which relies on the exchange of data between separate executables, has been referred to as co-simulation or external coupling [1]. Various co-simulation realizations have different implications regarding stability, convergence, accuracy, efficiency and ease of implementation.

The maritime co-simulation is a spatially separated co-simulation between a MiL/SiL[1] based maritime runtime environment (OFFIS[2]), a physical satellite emulator (GUT[3]) and a HiL[4] own ship simulation (AVL SFR[5]).

The main task of our co-simulation is to test the navigational safety of a navigation planning system. In addition, the possibilities and limitations of this system with regard to remote control of a ship from shore will be investigated. The

[1]MiL: Model in the Loop, SiL: Software in the Loop.

[2]Institute for Information Technology (Oldenburg, Germany).

[3]Gdansk University of Technology (Gdansk, Poland).

[4]Hardware in the Loop.

[5]AVL Software and Functions GmbH.

A. Akkermann (✉) · M. Siegel
OFFIS e.V., Institut für Informatik, Oldenburg, Germany
e-mail: arnold.akkermann@offis.de; Michael.Siegel@offis.de

B. Å. Hjøllo
NAVTOR AS, Egersund, Norway
e-mail: bjorn.hjollo@navtor.com

A. Leitner et al. (eds.), *Validation and Verification of Automated Systems*,
https://doi.org/10.1007/978-3-030-14628-3_19

associated ship (hereafter referred to as own ship) must correspond in dynamics and behavior to a real ship.

In this context, OFFIS and AVL SFR have developed a maritime co-simulation which simulates the own ship including its physical behavior (AVL SFR), its sensory behavior (OFFIS) and its intelligent behavior (OFFIS). In addition, the maritime co-simulation is used to provide the surrounding traffic that could be observed from the perspective of the own ship.

Skjong et al. [2] and Sadjina et al. [3]: project Virtual Prototyping of Maritime Systems and Operations (ViProMa) developed a framework for overall maritime system design by use of distributed co-simulation technology based on FMI. They developed an open-source platform (Coral) with master/slave network communication. As the network communication is not standardized via the FMI, a proprietary interface with a slave provider is used within their framework.

Also the European Joules project[6] analyzed the applicability of FMI in the maritime domain.

Chu et al. [4] use FMUs for developing a maritime crane model. For distributed simulation they put a Java wrapper around the FMUs to export functionality via RMI.[7]

Methods for the development of platforms for real-time simulation have been extensively studied for many years. However, research in this area has recently accelerated following advances in terms of speed and the ease of development associated with new hardware platforms. Methodologies for real-time simulation have included the use of hardware, such as digital signal processors, general-purpose processors and even reconfigurable computational solutions employing FPGAs[8] [5–7].

However, there are more approaches to combine FMI with HLA for distributed co-simulation. There are two approaches, use HLA RTI as master, which is not defined in FMI or Awais et al. [8] propose two algorithms for using HLA in conjunction with FMI. One based on fixed time steps and one based on discrete event-based simulation.

Overviews of commonly used inter-process communications protocols and co-simulation frameworks can be found in Yahiaoui and Trcka [9].

A challenge is to integrate the data exchange with the internal data structures, time integration algorithms and program flow of the individual simulators.

[6]http://www.joules-project.eu/Joules/index.xhtml

[7]Remote Method Invocation.

[8]Field Programmable Gate Array.

2 Demonstrator Setup for Distributed Co-simulation

2.1 General Requirements for Co-simulation in Maritime Domain

The development of vessels is a complex task as a vessel comprise a lot of subsystems from a wide range of engineering domains. Further, the designer must consider extreme requirements, e.g. significant power consumption, and has to find a tradeoff between costs, risk factors and environmental impacts. In addition to the difficult development, the maritime domain typically provides one-of-a-kind solutions. While the automotive or aerospace industry can invest a lot of resources in one prototype, since it lays the ground work for mass production, in the maritime industry a vessel is tailored and rarely mass produced [2].

Moreover, testing costs a lot and is time-consuming. Traffic and environmental situations required in this context are often not producible and even less reproducible. A traditional monolithic simulation approach is inflexible, costly and inefficient. A modular approach, in which models of the relevant subsystems are interconnected and simulated together, is needed to reduce costs and development time. The use of co-simulation enables multi-domain simulations, which makes it possible to test a vessel design, including all its subsystems and equipment, using different modeling and simulation software suited for specific systems [2].

Full-system simulation nevertheless remains challenging. Typical maritime systems are complex and difficult to model and simulate as there are interactions between several different domains having different time scales (e.g. slow dynamics of large mechanical ships and fast response of electronic bridge components). Further, models span a wide range of complexities and accuracies. Thus, understanding how different subsystems interact and how they influence the overall system behavior is a demanding task [3].

These facts, and our geographically distributed simulation, form the fundament for our general requirements for a co-simulation framework:

- interoperability of different simulation models, simulation tools as well as hardware components
- reusability of simulation components
- support of soft real-time operation
- possibility to use already existing software (also commercial of the shelf software)
- integration of geographically distributed components
- integration of time independent submodules (e.g. for coordinate transformations)
- possibility to use simulation components as black boxes (to protect IP).

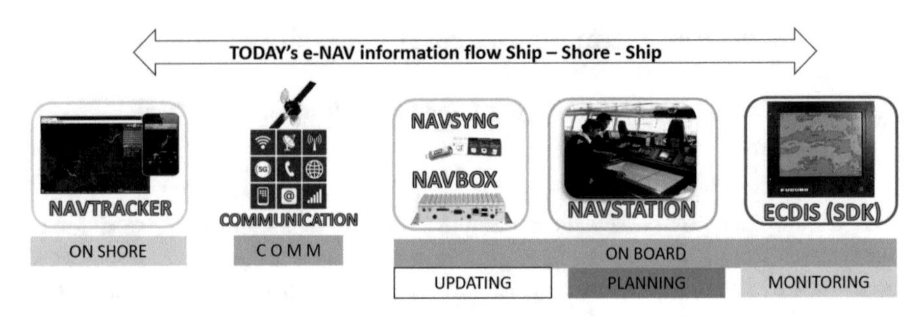

Fig. 1 Coarse Architecture NAVTOR SBB. Source: NAVTOR

2.2 *Description of the System Under Test*

The System under Test (SuT) is the Shore Based Bridge (SBB) from NAVTOR[9] (Fig. 1).

The SBB is designed for:

- "One click" user interface for navigator to exchange routes between ECDIS (Electronic Chart Display and Information System) and Planner (NavStation)
- Highly automatic passage planning (detailed navigational plan made before leaving port)
- High quality routing due to full resolution of weather/wave/sea data (NavStation)
- Complete 2D wave spectra (not a few wave trains only) (NavStation)
- Benefit from probability-based weather forecasts (NavStation).

The SBB ECDIS integrates a variety of real-time information, making it an automated decision aid capable of continuously determining a vessel's position in relation to land, charted objects, navigation aids and unseen hazards. An ECDIS includes electronic navigational charts (ENC) and integrates position information from the Global Positioning System (GPS) and other navigational sensors, such as radar, fathometer and automatic identification systems (AIS[10]). It may also display additional navigation-related information, such as sailing directions. Further, the SBB includes a NavStation which includes the key features of passage planning. As a subscriber of the NAVTOR ENC Service, the NavTracker allows to track own vessel's position, costs of e.g. ENC and review update reports. With NavBox integrated in the bridge networks, the vessel will automatically receive the latest

[9]https://www.navtor.com

[10]Like almost every technology, AIS is subject to specific restriction and limitations, too. Because of the dual character of AIS data (disengageable, dependent on the human initiated processes) and the dependency on other onboard devices (for example the GPS receiver) there is still a margin for errors in both the static as well as the dynamic data. Insofar, a possibility cannot be ruled out, that AIS data is wrong or not meaningful during important maneuvers of a vessel. Source: ANNUAL OF NAVIGATION 19/2012/part 1. DOI: https://doi.org/10.2478/v10367-012-0001-0.

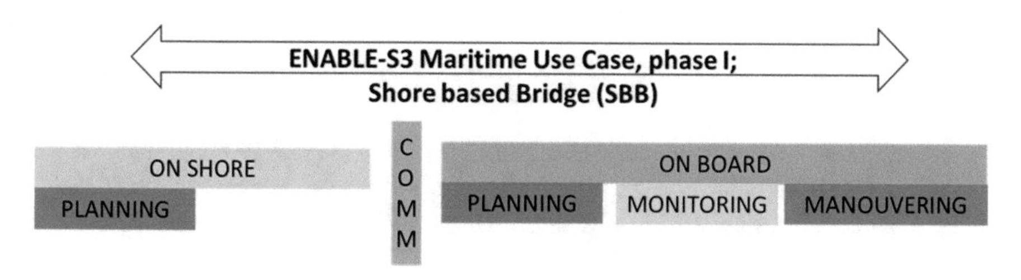

Fig. 2 Deployment SBB Phase I

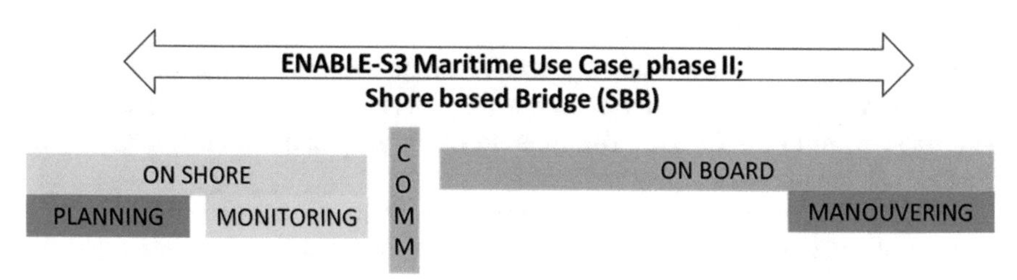

Fig. 3 Deployment SBB Phase II

updates of charts and navigational publications. This saves time and always ensures full compliance to maritime regulations. Furthermore, the routes and passage plans for the own ship, together with sensor data available on board, are transmitted ashore via the NavBox. The integration of SBB in the co-simulation ashore and on the own vessel is shown in Fig. 5.

Today the ships voyage is planned on board with the NavStation and afterwards this route is transferred to the (ECDIS). This system is a navigation information system which, with adequate back-up arrangements, can be accepted as complying with the updated chart required by regulation V/19 & V/27 of the 1974 SOLAS (Safety of Life at Sea) convention, by displaying selected information from a system electronic navigational chart (SENC) with positional information from navigation sensors to assist the mariner in route planning and route monitoring, and by displaying additional navigation-related information if required. In ECDIS a safety check for the planned route is accomplished. The voyage of the ship, both track and planned route, can be observed by a web-based NavTracker via communication link (see Fig. 1).

Within the European project ENABLE$_*$S3 the SBB is developed further in the following stages:

- Duplicate planning function to Shore (Phase I, see Fig. 2)
- Move permanently the Planning and Monitoring function to shore (Phase II, see Fig. 3)
- Steering a ship autopilot from ashore (Phase III, see Fig. 4).

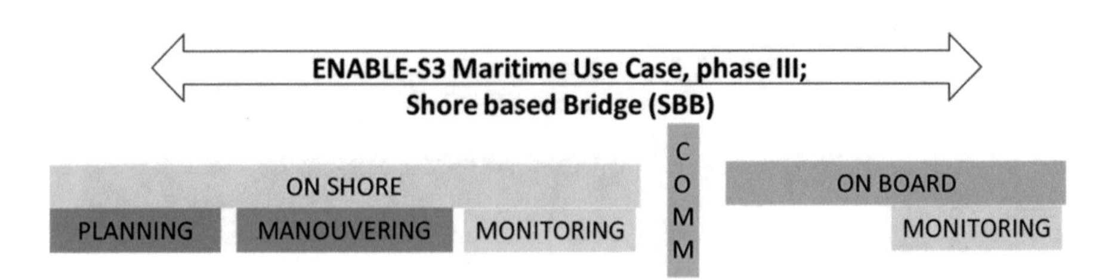

Fig. 4 Deployment SBB Phase III

In phase I, the complete planning process for the required route is carried out by SBB ashore and finally transferred to the ship. Here the planning is checked again and transferred to the ECDIS. The steering of the ship and the monitoring of the route is carried out exclusively on the ship.

In Phase II, both planning and monitoring of the route will be carried out with support of SBB ashore. For this purpose, all the ship's sensor information must be available ashore. The ship's steering remains on the ship.

In Phase III, the planning, steering and monitoring of the vessel's route and the surrounding traffic will be carried out ashore. The crew of the ship will be substantially de-stressed and can potentially be reduced in numbers. For double navigational safety, the monitoring function additionally remains on board the ship.

To test the SBB the following six test cases have been defined:

- Test case 1: Passage planning and deployment (reduced crew)
- Test case 2: Passage monitoring (sensor & traffic data in near real time)
- Test case 3: Reaction to monitoring information (sensor data & alarms may lead to change of route)
- Test case 4: Acknowledgement on board (controlled transfer of autopilot control from shore or own ship)
- Test case 5: Remote vessel guidance (remote control of autopilot/virtual handles)
- Test case 6: "Fail-safe" and "fail-operational "(fail-secure to avoid a disaster).

In this paper we focus on test cases 1, 2 and 5.

2.3 Maritime Co-Simulation Framework

In the maritime sector, in contrast to automotive engineering, only very small series are developed and built. The development costs amortize therefore only very slowly and there is permanently a high cost pressure. Due to this pressure, the development of these systems must be parallel and distributed, i.e. distributed among different teams and/or external suppliers, each working in their own domain with their own specific tools. Each participant, who is often far away, develops a partial solution into a system that has to be integrated with all other partial solutions. If the integration is carried out later in the development process, it will be less effective

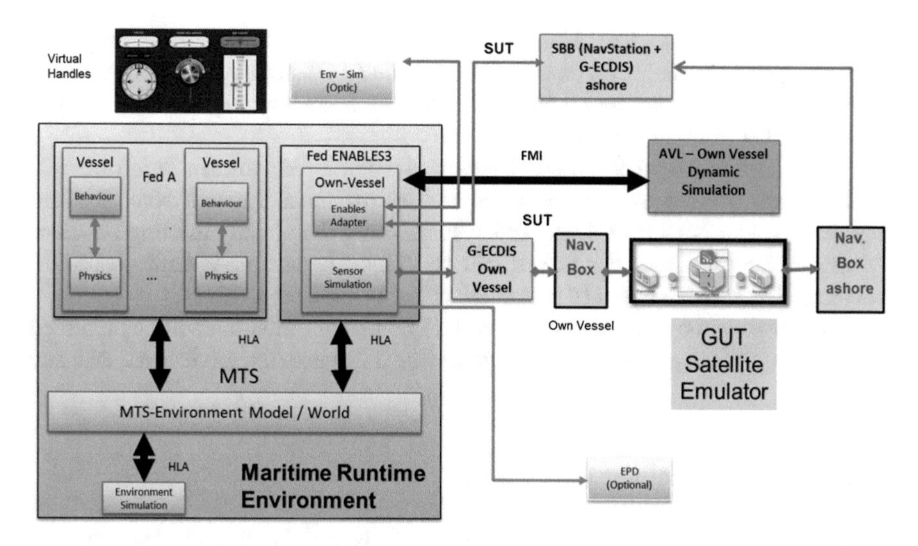

Fig. 5 Architecture co-simulation with integrated SUT

[10]. Here we see a broad field of activity for the use of a distributed co-simulation to reduce costs on the one hand and to increase the quality of the products on the other. This means that not only finished products such as autopilots, GPS or AIS can be integrated into the co-simulation, but also their models. This does not enable only the early detection of the undesirable developments and errors, it also allows existing requirements to be validated and verified. In order to fully exploit the theoretical possibilities of a distributed maritime co-simulator (Fig. 5) and to test it in practice, it was distributed over two locations and a flexible, open, standards-based architecture was chosen to establish it.

Further for effective testing of the SBB, it must be integrated into the co-simulation so that it is possible at any time to have a:

1. Time synchronization between the Maritime Runtime Environment and the SBB
2. Comparison of the current data of the Maritime Runtime Environment with the target data/results of the SBB
3. Fast time simulation of longer ship approaches within special scenarios
4. Accurate evaluation/comparison of individual NMEA[11] sentences.

Sensor measurements from the own ship are provided either to the on board ECDIS system using standard maritime communication channels, or to the shore based ECDIS using similar standard communication channels. Those measurements

[11]National Marine Electronics Association. The NMEA 0183 Standard, a communication standard defined by the NMEA organization (www.nmea.org), defines a communication protocol that enables navigation instruments and devices to exchange data with each other. The NMEA 0183 Interface Standard defines electrical signal requirements, data transmission protocol and time, and specific sentence formats for a 4800-baud serial data bus.

provided directly to the SBB Planning Station on shore are the same measurements that are globally available for example through the AIS distribution system (Satellite based AIS, see also Marine Traffic), but with a NavBox you may see the complete coverage of surrounding vessels.

The ECDIS on board of the simulated own ship processes the sensor information in the same manner as it is done on real ships. In addition, it shall forward the messages to NAVTOR's backend system, using the already existing NavBox. Messages between two NavBoxes will be transmitted using satellite based IP services. The satellite based IP communication is simulated by GUT.

After passing the satellite simulation the available information will be presented at the SBB, allowing a nautical officer to decide if he need to intervene and take any action.

2.3.1 Maritime Runtime Environment

The Maritime Runtime Environment (Fig. 5) is the fundamental part of our distributed co-simulation. It is realized as a Model in the Loop (MiL)/Software in the Loop (SiL) simulation based on HLA (High Level Architecture), compatible to the eMIR-platform.[12] The data model does contain the current state of simulation technics.[13] It contains a world editor which is an Eclipse-based editor that allows set up a static scene according to a predefined scenario. This model contains the fundamental components of all used resources, actors and environmental factors. Further, the part Maritime Traffic Simulation (MTS) is a flexible simulator for implementing, executing and observing the behavior of multiple vessels in a realistic context. Routes for vessel traffic can be modelled or imported via standards defined by NMEA. Different infrastructures are available for the individual blocks of the Maritime Runtime Environment and also for the entire co-simulation in order to optimally fulfil the respective requirements for the simulation. Simulation of vessel movements can be coupled with sea charts and environmental simulations. The environment simulation provides a simulation of maritime environmental factors such as wind, wave, current, tide, etc. Further, the MTS contains an unlimited

[12]Open initiative of German maritime industry for improving safety and efficiency in maritime transportation systems.

[13]HLA standards: IEEE 1516-2000: High Level Architecture (Framework and Rules), IEEE 1516.1-2000: High Level Architecture (Federate Interface Specification), IEEE 1516.1-2000: Errata (16. Oct. 2003), IEEE 1516.2-2000: High Level Architecture (Object Model Template (OMT) Specification, IEEE 1516.3-2003: Recommended Practice for HLA Federation Development and Execution Process (FEDEP). Our model is based on the S-100 standard. S-100 is the document that explains how the IHO will use and extend the ISO 19100 series of geographic standards for hydrographic, maritime and related issues. S-100 extends the scope of the existing S-57 Hydrographic Transfer standard. Unlike S-57, S-100 is inherently more flexible and makes provision for such things as the use of imagery and gridded data types, enhanced metadata and multiple encoding formats. It also provides a more flexible and dynamic maintenance regime via a dedicated on-line registry.

set of vessels. Each of those vessels can be characterized with different features such as shape describing the hull of the vessel, current vessel velocity, position and orientation in global coordinate system. Other features include for example the vessel dynamics that is manipulated through the powertrain simulation on HiL.

The sensor simulation (Fig. 5) is a real-time generator for AIS/ARPA/GPS data for simulating data of real sensors on board. The simulation includes on the one hand sensors that give information about the vessel state, such as the actual rudder position, rotational speed, orientation, position, etc. On the other hand, there are sensors for environmental perception, e.g. the AIS which is commonly used in the maritime sector. Both types of sensors deliver their data to the SBB in the form of an NMEA data set.

The ECDIS of the simulated own vessel is fed with AIS information by the Maritime Runtime Environment. The radar and sensor data of the own ship and the surrounding vessel traffic is in the form of NMEA 0183 sentences, so no real sensors are required. Additionally, it forwards the NMEA messages to NAVTOR's backend system at shore for monitoring this data on the G-ECDIS (Generic ECDIS). The G-ECDIS processes the sensor information onboard as real ships. Messages between two NavBoxes (one ashore and one on board) will be transmitted using satellite based IP services.

The "Lobo Marinho" (Ro-Ro Cargo Ship) is specified as own vessel. This ship with its maneuvering behavior is simulated by the HiL simulation. The surrounding ship traffic, the sensor data of the own ship and the entire environment of the respective scenario are simulated by the Maritime Runtime Environment. The influences on the communication like run times and disturbances between the SBB (shore station) and the own ship, are taken into account by a satellite emulator (GUT).

2.3.2 HiL Simulator at AVL SFR

The following figure (Fig. 6) illustrates the HiL Simulation of the own ship simulation.

The HiL simulator consists of the ship and engine model in loop with the Electronic Control Unit (ECU) for engine control. It is real-time capable and thus a CAN interface (Controller Area Network) was implemented for the signal transfer between HiL and Model.CONNECT. Model.CONNECT provides then the communication with other components, running by OFFIS.

The concept of "master" and "slave" in the FMI is very different from that of HLA's. When a simulation package imports an FMU, it becomes its "master" and the loaded component becomes its "slave". FMI for co-simulation provides a mean to utilize models using an API, in a form where slave acts as a black box to master. It can react to inputs and gives outputs at discrete time steps. While using an FMU conforming to FMI for co-simulation, one does not need to know which integration method is actually applied to solve the model. Parameters to this black box can be set in initialization phase and cannot be changed afterwards, while the inputs can

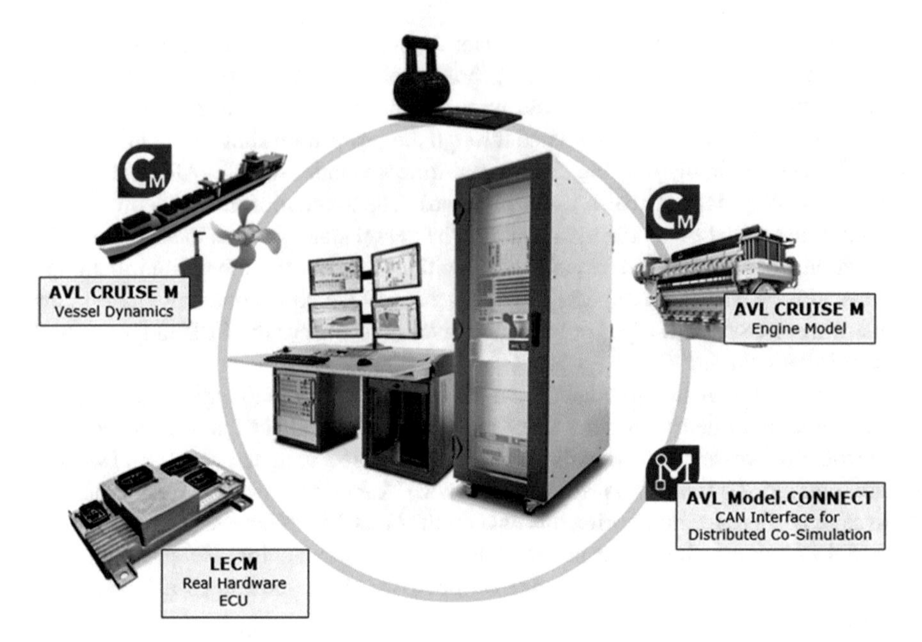

Fig. 6 HiL simulator AVL

be changed between discrete time steps. In our case the HiL is set up so that the models on the HiL start the simulation as soon as the HiL with the zero set points is switched on.

2.3.3 Interface Between the Maritime Runtime Environment and HiL

A dedicated VPN network has been established between AVL SFR and OFFIS for the distributed co-simulation platform. The VPN is encrypted to ensure data security at both sites.

To setup the simulation of the test scenarios, Model.CONNECT is used as co-simulation environment. Model.CONNECT controls the simulation and provides mechanisms for an intelligent coupling and advanced simulation mechanism by using of the ICOS[14] blocks. The following figure (Fig. 7) shows the first implementation (Var. 1) of the distributed co-simulation architecture for this part in detail.

Via a VPN connection data will be exchanged by a proprietary UDP connection. The Maritime Runtime Environment only needs to create an interface to Model.CONNECT ①. This interface is realized as FMI which is supported by Model.CONNECT by default. The connection to the ICOS-Remote-Server

[14]Independent Co-Simulation.

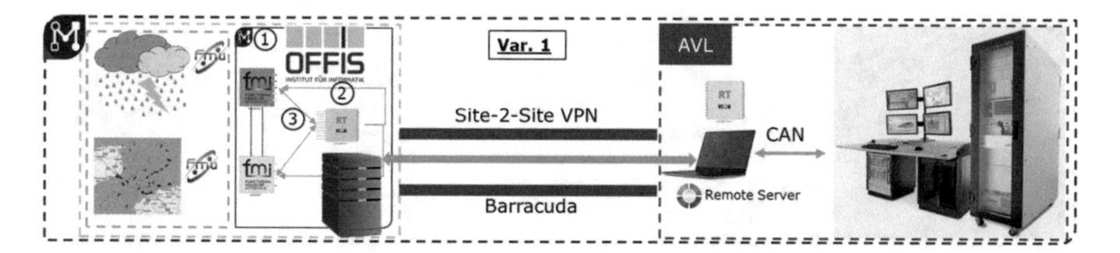

Fig. 7 First implementation (Var. 1) distributed co-simulation architecture

(AVL-side) must be realized via RT[15]-ICOS-wrapper ②. In Model.CONNECT FMUs/other Models can be connected to RT-ICOS-wrapper ③. On AVL side the access to the HiL is realized via ICOS Remote Server thus there is the possibility of connection ECU, injectors, flaps etc. The Cruise M models are running on the HiL in real-time. However, the real-time capability of the entire architecture depends on the network infrastructure.

During the implementation of the Var. 1 architecture the question arose why a hard real-time capable HiL should be integrated over a secure VPN (Barracuda Server) with a soft real-time capable maritime runtime environment. With regard to the implementation of the planned test scenarios, no advantages of this implementation could be identified. This was especially true as the internal powertrain behavior should not be tested within the scenarios.

In general, it is not possible to be faster than the network infrastructure. One of the tasks of our research work is to investigate which time delays hinder or render impossible the execution of the panned test scenarios. Usually the lags were about 20 ms. In addition, the processes running in milliseconds within the powertrain simulation of the HiL are not required for the scenario execution, but the surge, sway and yaw velocity.[16] These speeds are essentially determined by set points rudder position, engine speed and propeller position of the Maritime Runtime Environment. Both the update rates for the settings of the Maritime Runtime Environment (setpoint data) and the feedback from the own vessel simulation are in the range of seconds. Thus, the HiL itself does not have to be integrated, but a software component which represents the real dynamics and behavior of the "Lobo Marinho". Consequently, a second communication architecture (Var. 2) was developed, which is illustrated below (Fig. 8).

---

[15]Real time.

[16]Three translations of a ship's center of gravity in the direction of the x-, y-, and z-axes:

- surge in the longitudinal x-direction, positive forward
- sway in the lateral y-direction, positive to the port side
- heave in the vertical z-direction, positive upward

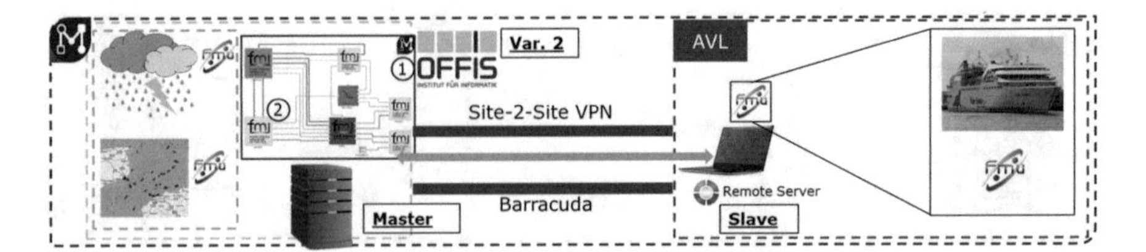

Fig. 8 Second implementation (Var. 2) distributed co-simulation comArchitecture

On AVL-side the Cruise M models in form of a FMU are running on a simulation computer. The access to simulation computer is realized via ICOS-Remote-Server.

In the Maritime Runtime Environment Model.CONNECT is running ①. The Connection to the ICOS-Remote-Server is implemented via FMU-wrapper ②.

After loading the FMU the Maritime Runtime Environment as master becomes the complete owner of the FMU and there is no question of "entity transfer". Similarly a "shared resource" does not exist in the FMI. An FMU can only read and write some values, it does not have an entity similar to a process. Naturally, there cannot be a concept of ownership management, when slave itself is owned by master.

Figure 9 illustrates the main components of the data exchange between the Maritime Runtime Environment and the AVL simulation computer to simulate the behavior of the "Lobo Marinho". The following data are provided by the Maritime Runtime Environment as set point data

- weather data
- rudder position
- engine speed
- propeller position.

As feedback to these control signals, the AVL simulation computer sends

- Surge (forward and backward movement of the center of gravity of a ship)
- Sway (movement along the transverse axis)
- Yaw (movement around the vertical axis)

speed (actual data).

The current speed of the own ship or its current position and orientation (pose) within the world are simulated in the Maritime Runtime Environment.

2.3.4 Satellite Emulation

The simulation of a proportionate real maritime world includes the simulation of the satellite connection between the ship and the shore station. This simulation is performed by a satellite emulator, which simulates the occurring delays in the communication as well as the possible disturbances within this connection. This emulator is located in Gdansk.

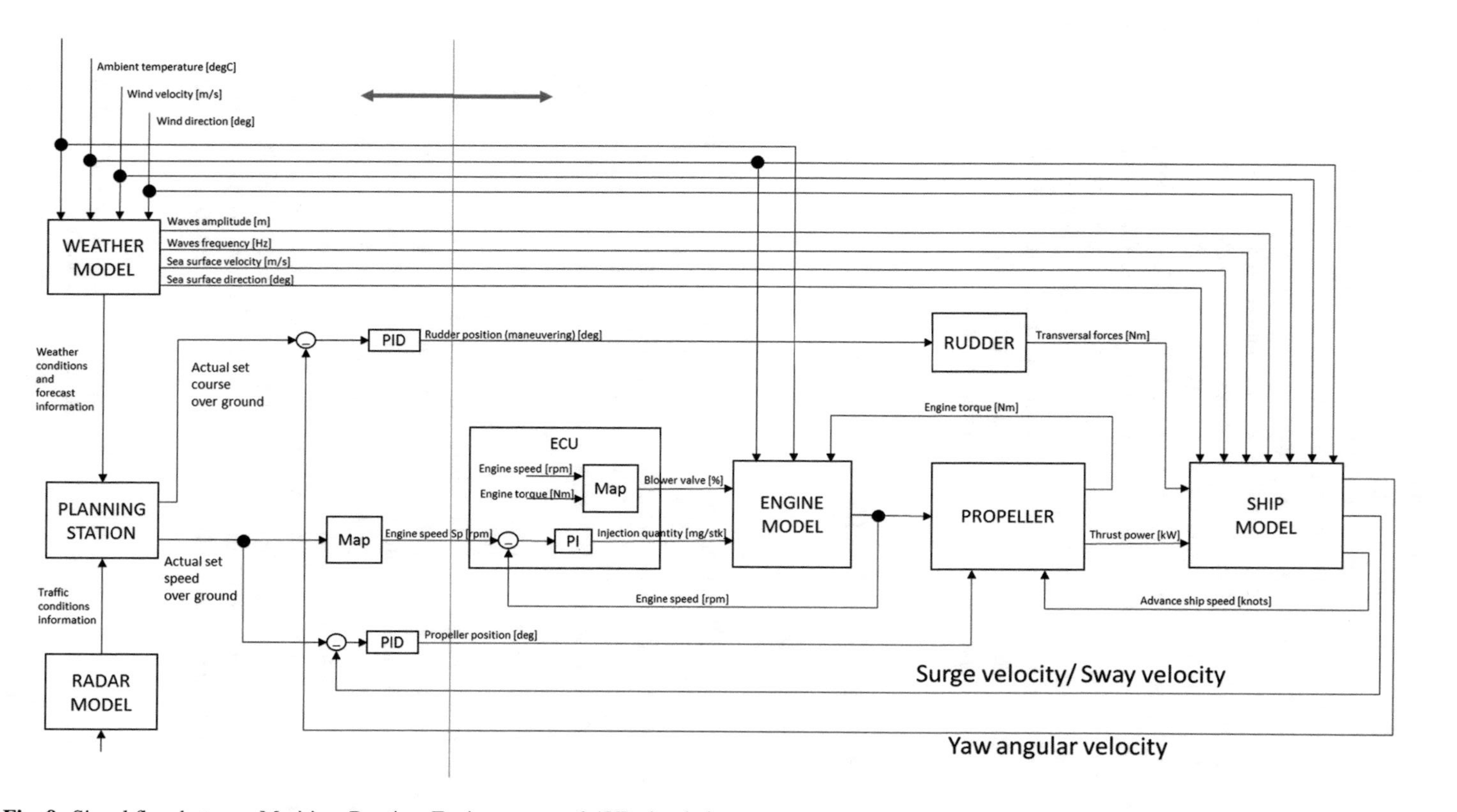

Fig. 9 Signal flow between Maritime Runtime Environment and AVL simulation computer

The simulation will be done in the PhyWise Tool. Two different configurations will be available: software simulation and hardware simulation with devices prototypes. Additionally each of the configurations will allow simulation of communication with additional jamming signal source.

In Fig. 10, the PhyWise tool is placed between two components of a given ACPS system. In the case of maritime co-simulation, this tool is located in the communication channel between the two NavBoxes (ship and shore) (Fig. 5). The two NavBoxes display their data via the PhyWiSe-Tool, which has the main task of checking the security at the physical level. The PhyWise tool contains a special communication simulation component that is responsible for reflecting the characteristics of the communication channel as a model, enabling both MiL and HiL testing. The jammer simulation is responsible for generating deliberate interference for the needs of evaluation of secure communication and SBB resistance to such a threat. The interference detection receiver is a component that can detect, detect and locate both intentional and unintentional interference.

3 Experiences and Results

3.1 System Under Test

We have tested the SBB with over 60 scenarios using our co-simulation setup. Based on the results, we can say that the route planning of the NavStation is carried out fully navigationally safe (test case 1). The tools installed on the NavStation, such as Admiralty Total Tide[17] (ATT) or NavArea,[18] have proved to be extremely helpful and efficient. As an example, a time saving of more than 66% compared to conventional planning of a route from Barcelona to Las Palmas (Gran Canary) can be mentioned. In our opinion a network based route optimization tool is missing in the weather forecast. A corresponding concept for this has already been developed.

According to Winner et al. [11] and Winner and Wachenfeld [12], following basic statistics, between 100 million and 5 billion km of test driving are required in order to ensure that software for autonomous driving or in our case for voyage planning and remote vessel guidance is at least as save as a human driver. The navigational/mathematical analysis and reconstruction of the scenarios enabled to determine both the causes of the accidents (collisions, groundings, etc.) and their avoidance actions. Test case 5 (remote vessel guidance) shows that these accidents can be avoided by means of this function. The NavStation, as part of SBB, was continuously further developed and optimized based on results of scenario-based

[17]ADMIRALTY TotalTide (ATT) provides bridge crews with fast, accurate tidal height and tidal stream predictions for more than 7000 ports and 3000 tidal streams worldwide.

[18]NAVAREAs are the geographic areas in which various governments are responsible for navigation and weather warnings.

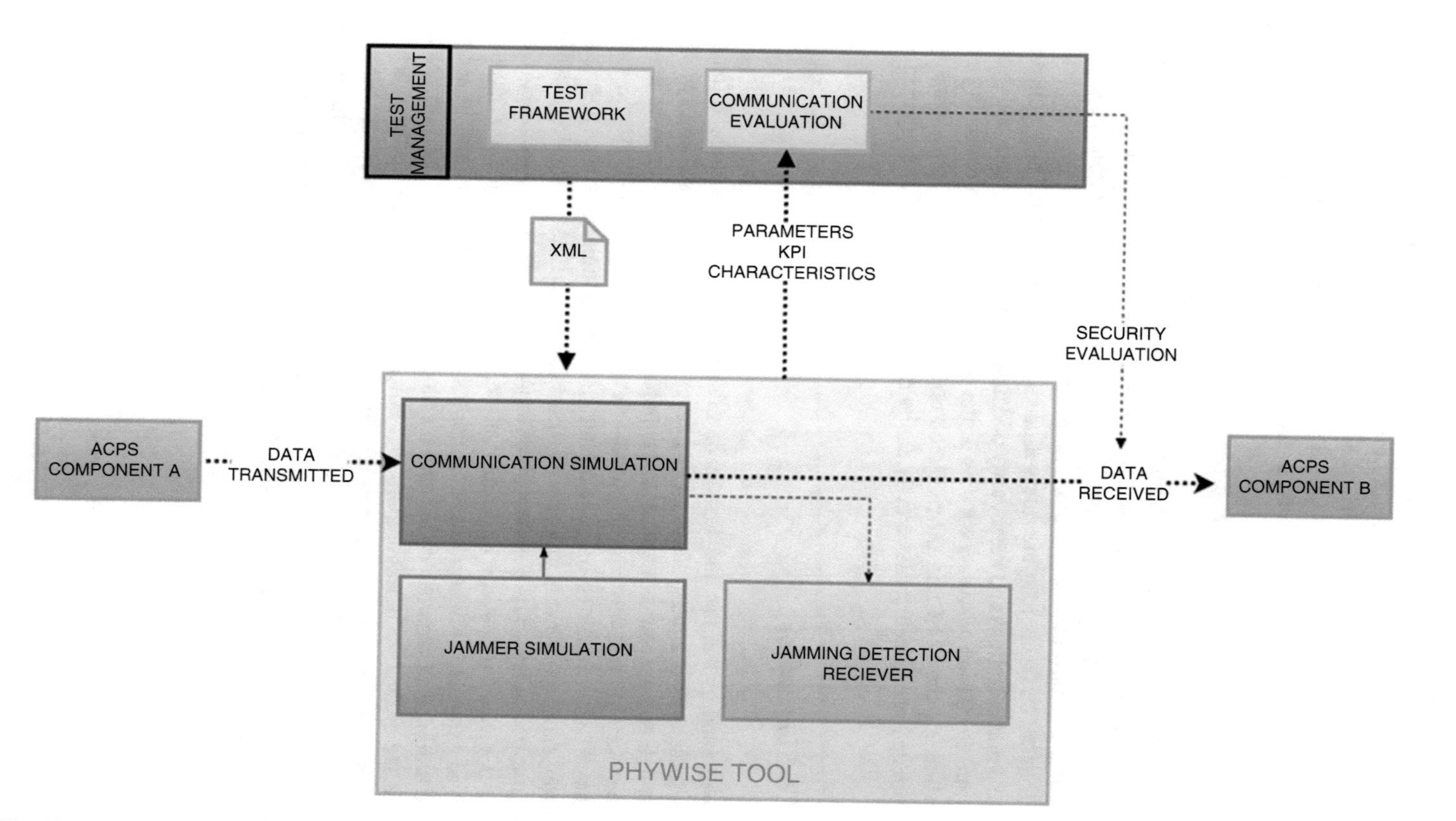

Fig. 10 Basic architecture PhyWise Tool. Source: GUT

testing. However, the situation analysis on the ECDIS, which is based on the monitoring data of the own ship, must be carried out by the respective operator. The SBB does not contain a tool for situation analysis. In addition, there is no extensive support with regard to the COLREGs[19] (Convention on the International Regulations for Preventing Collisions at Sea) in implementing the remote vessel guidance.

The lack of a genuine handshake procedure (request/acknowledge) in the data traffic between the shore station and the ship has proved to be a serious deficit at SBB. This makes it impossible, for example, to transfer autopilot control from shore back to ship or vice versa (test case 5). This leads to confusion as to where the autopilot is controlled from. The necessary coordination was made possible during the test phase by conventional voice communication.

A further deficit lies in the pure safety aspects in the use of SBB. These concerns both access authorizations and the security of data exchange between ship and shore.

The pure evaluation of the sensor data of the own ship and the surrounding ship traffic ashore by the G-ECDIS proved to be error-free. Correct NMEA sentences were evaluated completely and correctly, faulty ones were detected and rejected.

3.2 Co-Simulation

The co-simulation enables the testing of all scenarios. In particular, different systems such as assistance systems or ship types with different propulsion systems can be tested at numerous environmental conditions and conclusions can be drawn about the planning and monitoring process (test case 2) of the SuT. At the same time, the effects of delays and disturbances in communication on ship route monitoring and remote vessel guidance can be recorded and analyzed using satellite simulation. For delays >10 s neither monitoring nor the remote vessel guidance function is given. Further, the integration of vehicle dynamic simulation makes it possible to explicitly test ship engines, rudders and propulsion systems under all possible sailing and weather conditions on the planned voyage. The purpose of performance evaluation is to compare fuel efficiency or propeller performance at a given point in time, i.e. to compare the ship state at several conditions. Since a ship is exposed to external factors such as wind, waves, shallow water, changes in seawater temperature, etc., it is unlikely that the ship will ever be in exactly the same situation more than once. An outstanding advantage of the co-simulation is

[19]The International Rules shipping traffic rules were formalized in the Convention on the International Regulations for Preventing Collisions at Sea, 1972, and became effective on July 15, 1977. The Rules (commonly called 72 COLREGS) are part of the Convention, and vessels flying the flags of states ratifying the treaty are bound to the Rules.

the exact reproducibility, the starting and stopping of the scenarios at any time and the available analysis possibilities. This is rarely, if ever, not the case with real-world tests.

The different results occur between the mathematical-navigational calculation of the scenarios, the results within the maritime runtime environment and those in the SBB. In many cases, deviations occur in distances, bearings, courses and speeds above ground. To the best of our knowledge, these deviations can be attributed to the use of various projection models. While normal nautical charts and ECDIS use Mercator projection, the Maritime Runtime Environment uses the mathematical models of Vincenty.[20] The deviations that occur may not be relevant on the high seas with correspondingly low traffic volumes, but they are relevant in coastal waters or in narrow fairways. Remote vessel guidance is not permitted in these sea areas for this reason alone. It is an important question whether in the ECDIS time still Mercator projection is essential for marine navigation. Further research is required to move from a Mercator projection to calculations based on a spherical earth model. Vincenty's solution for the distance between points on an ellipsoidal earth model is accurate to within 0.5 mm distance and 0.000015″ bearing, on the ellipsoid being used.

Further safety-critical deviations in the results occur in the Closest Point of Approach (CPA) and Time to the Closest Point of Approach (TCPA) calculation. Comparing encounter parameters CPA and TCPA obtained from ARPA and decision-support system it can be stated that:

1. CPA presented by ARPA is always larger than the manual CPA calculation
2. TCPA values approximate each other.

Differences in CPA values are due to the fact that the length of the vessel is not included in the calculation. For example, information on the size of the vessel and the antenna position can take from the AIS system. In ARPA as well as in the formulas presented by Lenart [13], the ships are treated as points; therefore the falsification of the results, which can lead to a wrong assessment of the situation by the navigator, this is particularly true for the encounters of large ships. The largest CPA difference within the scenarios carried out was 0.21 nm at close range, which represents an error of more than 50% at a CPA value of 0.4 nm.

CPA results by the existing methods which only consider Speed over Ground (SOG) and Course over Ground (COG) are imperfect. A CPA calculation method taking account of SOG, COG, Change of Speed (COS) and Rate of Turn (ROT) is required.

[20]Vincenty's formulae are two related iterative methods used in geodesy to calculate the distance between two points on the surface of a spheroid, developed by Thaddeus Vincenty in 1975. They are based on the assumption that the figure of the Earth is an oblate spheroid, and hence are more accurate than methods such as great-circle distance which assume a spherical Earth (http://en.wikipedia.org/wiki/Vincenty's_formulae).

AIS can provide information on all four essential factors, as well as the position and other important information. AIS data is therefore suitable to be used to predict the positions and calculate CPA.

The results of CPA calculation in the Maritime Runtime Environment confirm this tendency. On smaller vessels the smallest passing distance on the level of 0.5 nm is often considered as safe. Within this short distance, the deviations described play a significant role. TCPA values are close to each other, as the vessel's size does not affect the moment of contact, only its value.

4 Conclusion

To test functionality in all possible situations, the automotive domain assumes that a high automation assistance system must drive 240 million kilometers, when the distance between two accidents with personal damage is 12 million kilometers. Such a number on test runs is hard to reach, even with fast-time simulation. One approach to reduce the number of necessary runs is to generalize situations and explicitly focus on critical, rare events. We have followed this approach in principle, but without ignoring "normal behavior" in maritime transport. The SBB was tested with the use of 60 scenarios that included both the simulation of:

- Close-range situations
- Collisions
- Groundings
- Hazards due to technical failures
- Normal traffic situations.

SBB's planning and monitoring function (evaluation of sensor data) could be carried out realistically and effectively by means of the co-simulation (test cases 1 and 2). Within the Remote Vessel Guidance function (test case 5), further research effort is required to develop and test target-oriented additional functions. The combination of scenario based testing and requirements based testing provides a more robust acceptance potential that gives the user and manufacturer confidence that the system meets their requirements. There is a risk that a legacy system which has not been subject of a validation program, will not meet regulatory expectations, e.g. IMO compliance. It is therefore necessary to check the conformity of existing systems. Basically, the owner of a legacy system should ensure that an appropriate validation package is available. Here we see another approach and advantage for scenario-based testing. In the further development of the NavStation, the conformity with the currently existing regulations was comprehensibly proven with the support of scenarios.

Co-simulation represents an enormous increase in effectiveness for the maritime domain. The first prototype is usually only available after 50–60% of the development time. With a development time of 5 years, the first 3 years would be without any experience of how the interaction of the new components works at all. This is exactly where co-simulation techniques come in: They bring together individual simulation models. Once the prototype of a component (e.g. the autopilot, the integrated bridge or the ECU) is finally available, the individual virtual model is removed from the overall simulation and replaced by the prototype on the test bench. This data then comes from the real component, but is fed into the same system and correlated with the other values still simulated. The model thus becomes more and more real step by step.

References

1. Trcka, M., Hensen, J.L.M., Wetter, M.: Co-simulation of innovative integrated HVAC systems in buildings. J. Build. Perform. Simul. **2**(3), 209–230 (2009)
2. Skjong, S., Rindarøy, M., et al.: Virtual prototyping of maritime systems and operations: applications of distributed co-simulations. J. Mar. Sci. Technol. 1–19 (2017)
3. Sadjina, S., Kyllingstad, L., et al.: Distributed co-simulation of maritime systems and operations. J. Offshore Mech. Arct. Eng. **141**, 011302 (2018)
4. Chu, Y., Hatledal, L.I., Sanfilippo, F., Schaathun, H.G., et al.: Virtual prototyping system for maritime crane design and operation based on functional mock-up interface. In: OCEANS 2015, pp. 1–4. IEEE, Genova (2015)
5. Boukerche, A., Lu, K.: A novel approach to real-time RTI based distributed simulation system. In: Proceedings of the 38th Annual Simulation Symposium, pp. 267–274. IEEE, San Diego, CA, USA, 4–6 April 2005
6. Monga, M., Karkee, M., et al.: Real-time simulation of dynamic vehicle models using a high-performance reconfigurable platform. In: Proceedings of the 2012 International Conference on Computational Science (2012)
7. You, T., Zhu, Y., et al.: Applied research of delaminated real-time network framework based on RTX in simulation. In: Proceedings of the 2nd International Conference on Information and Computing Science, pp. 389–392. IEEE, Manchester, UK (2009)
8. Awais, M.U., Palensky, P, et al.: Distributed hybrid simulation using the hla and the functional mock-up interface. In: 39th Annual Conference on Industrial Electronics Society. IEEE, Vienna, Austria, 10–13 November 2013
9. Trcka, M.: Co-Simulation for Performance Prediction of Innovative Integrated Mechanical Energy Systems in Buildings. Eindhoven University of Technology, Eindhoven (2008)
10. Tomiyama, T., D'Amelio, V., et al.: Complexity of multi-disciplinary design. CIRP Ann. Manuf. Technol. **56**, 185–188 (2007)
11. Winner, H., Wolf, G., et al.: Freigabefalle des autonomen Fahrens/the approval trap of autonomous driving. VDI-Berichte. **2106**, 17–29 (2010)
12. Winner, H., Wachenfeld, W.: Die Freigabe des autonomen Fahrens. In: Maurer, M., Gerdes, J., Lenz, B., Winner, H. (eds.) Autonomes Fahren, pp. 439–464. Heidelberg, Berlin (2015)
13. Lenart, A.S.: Manoeuvring to required approach parameters—CPA distance and time. Annu. Navigat. **1**, 99 (1999)

Validation of Automated Farming

M. Rooker, J. F. López, P. Horstrand, M. Pusenius, T. Leppälampi, R. Lattarulo, J. Pérez, Z. Slavik, S. Sáez, L. Andreu, A. Ruiz, D. Pereira, and L. Zhao

1 Overview and Motivation (TTC) (2 Pages)

Various aspects of smart farming technologies such as detection of the crop's needs and problems (e.g. fertilizer, water application and crop spraying according to the needs of the individual plants, rather than treating large areas in the same manner) have already been implemented. Like in other domains, there are several levels of automation depending on the number of automated tasks performed (supervision, classification, motion, etc.). Some early examples of automation do already exist: the straight line keeping products made by John Deere and CLAAS [1, 2], Beeline Navigator, AgGPS Autopilot, Autofarm, amongst others [3].

M. Rooker (✉)
TTTech Computertechnik AG, TTControl GmbH, Vienna, Austria
e-mail: martijn.rooker@tttech.com

J. F. López · P. Horstrand
Universidad de Las Palmas de Gran Canaria, Las Palmas, Spain

M. Pusenius · T. Leppälampi
Creanex Oy, Tampere, Finland

R. Lattarulo · J. Pérez
Tecnalia, Derio, Spain

Z. Slavik
FZI, Karlsruhe, Germany

S. Sáez · L. Andreu · A. Ruiz
ITI, València, Spain

D. Pereira
ISEP, Porto, Portugal

L. Zhao
DTU, Lyngby, Denmark

A. Leitner et al. (eds.), *Validation and Verification of Automated Systems*,
https://doi.org/10.1007/978-3-030-14628-3_20

What is currently missing is autonomous operation in order to optimize resources, increase the level of efficiency and reduce costs significantly. One main goal is to get the driver out of the cabin by developing driverless farming vehicles (i.e. tractors or harvesters). This could revolutionize every aspect of planting, watering and harvesting, not to mention optimizing the selective use of pesticides, reducing costs and increasing yields while maintaining high levels of respect with the environment.

Although farming vehicles (e.g. tractors and harvesters) are considered as off-highway vehicles, the direct application of technologies from passenger vehicles (e.g. ADAS) is not trivial and in most situation not even possible. Conditions and constraints from e.g. highway or city traffic are completely different than the ones from the agricultural domain. Highway traffic has clearly defined conditions and constraints, like speed limit, directions, road boundaries (markings on the road), etc. Additionally, environment conditions are completely different in farming. While harvesting, many dust, or crop particles are flying around, potentially causing a lot of interference in the sensor values, for which compensation has to be taken into account. Consequently, it is still under evaluation if sensors coming from the automotive branch are really applicable to the agricultural domain, or maybe other (e.g. hyperspectral sensing) or completely new sensors are necessary to guarantee safe and secure behavior of the farming vehicles.

Nevertheless, some methods and tools used in automated driving, such as techniques used for controlling the vehicle actuators, motion planning algorithms or software architectures, can be reused for farming. In [4], we present an automated driving architecture consisting of six blocks (sensor data acquisition, perception or vehicle and environment modeling, communication among different participants and infrastructure, decision or trajectory planning, control for tracking the trajectories and actuation or low level actuator control) applied to automated farming.

Within the ENABLE-S3 farming use case, the focus is on verification and validation of technologies for autonomous harvesting. Hereby, different concepts are considered, varying from Unmanned Aerial Vehicles (UAV, or drone) equipped with sensing technology (i.e. hyperspectral sensing) for crop detection and sensing, autonomous driving capabilities, communication, vehicle sensing technology, environmental and vehicle simulation, etc. The goal is to define and develop new V&V methodologies targeting farming technologies, based on existing ones coming from different markets (e.g. automotive) and a scenario-based approach to test and validate automated CPS at reasonable costs.

One of the major problems with testing of farming technologies for harvesting is that real-life testing and validation possibilities are very expensive and validation opportunities are very limited in time. To validate a realistic autonomous harvesting scenario with a harvester, the farming environment needs to be available. This means that the crops (e.g. corn or soy) have to be present at a harvesting field and ready for cropping. Weather conditions must be acceptable to actually perform the cropping (in raining conditions, harvesting doesn't often take place), so the time frame that the actual harvesting can take place is rather short, thereby significantly reducing testing time. Another constraint here is that once the field has been cropped, the harvesting environment is not available anymore to repeat the test under similar conditions. In the worst case, one has to wait another year for having the crops available again. Additionally, validating cropping technologies on real-life harvest can have a severe impact on available resources (i.e. the actual harvest), which can have a major ecological influence, next to the financial impact of having to compensate for the loss of harvest to a farmer.

Integrating simulation and virtual farming environments into the validation process will enable developers to increase testing possibilities, create continuous testing, repeatability of tests, significantly reduce costs and increase the verification and validation process. Nevertheless, the results of these tests are only as good as the data that is being provided to the tools. Therefore, one objective of the use case is to gather and include as much real-life data (e.g. sensor, vehicle, environmental) as possible in the simulation.

The work performed in the ENABLE-S3 farming use case aims to provide a first validation and test framework, consisting of an integrated co-simulation tool chain supporting environmental, sensor, vehicle and drone simulation. Additionally, a web-based toolchain for model-based analysis and simulation of farming systems is developed to provide flexible simulation of various components in the farming environment. Finally, vehicle internal deterministic communication is a core concept for safe and secure autonomous driving, also in off-highway scenarios. A runtime verification and automation testing framework for in-vehicle communication is developed to validate the timing of communication.

2 Detailed Description of the Automated Farming System and Platform

The farming use case introduces an architecture comprising of various Systems under Test (SUT) and a collection of Test Systems (TSY). Within the test cases, there will be simulated vehicles and in a second stage real vehicle (i.e. Renault Twizy electric vehicles simulating the harvester and tractor). The two vehicles will interact with each other on a real test track, simulating the situation that the tank of the harvester is full and needs to be unloaded. Both vehicles will autonomously decide where to drive and how to interact with each other while performing the parallel driving action. The harvester vehicle will perform the driving pattern for harvesting the field in an optimized manner. Additionally, a drone will be available that is intended to collect crop data in advance and during the harvesting process providing real-time information regarding obstacles (i.e. dynamic objects like humans and animals). The SUTs are depicted in Fig. 1. Additionally, Fig. 2 gives a schematic overview of the appliance of the SUTs within the farming use case. It shows a farming field, where the drone is performing a pre-flight over the field (left figure) for identifying the important areas. In the right side of Fig. 2, the harvester (yellow) is harvesting and the tractor (red) drives alongside the harvester

Fig. 1 SUTs in real-life

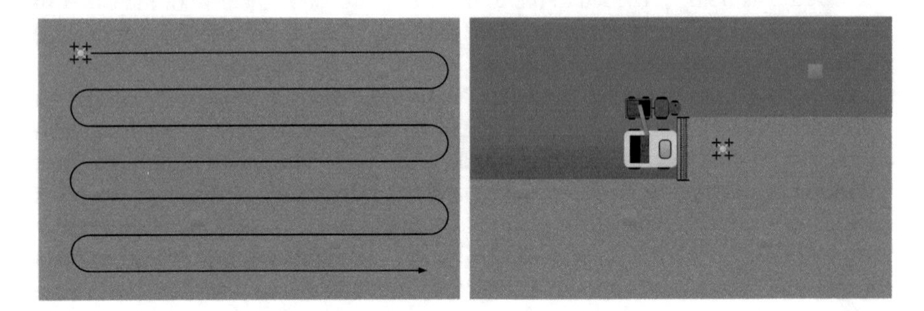

Fig. 2 SUT in application (sketched)

for unloading. The drone is flying in front of the harvester detecting obstacles. Section 2.1 will give a more detailed and technical overview of the SUTs applied in the use case.

The main focus of the ENABLE-S3 project is on the improved testing, verification and validation of the defined use cases. Test systems (TSY) are developed that will alleviate the users and reduce the amount of testing required, including costs. Additionally, it will enable users to perform continuously testing, thereby not being dependent on external variables, like weather conditions and hardware availability. A large part of the chapter is focused on the description of the developed systems that will perform tests and verification of various aspects of the farming demonstrators. Section 2.2 gives a detailed description about a subset of the developed TSYs within the farming use case.

2.1 Systems Under Test (SUT)

In the following subsections, an overview of the Systems under Test (SUT) within the farming use case will be presented, together with their applicability within the farming domain.

2.1.1 Hyperspectral Flying Platform

Originally viewed as a military tool, the Unmanned Aerial Vehicles (UAVs) market has shown an incredible growth in the last years in different domains, competing with traditional acquisition platforms such as satellites and aircrafts, and in the near future with high altitude platforms or pseudo-satellites (HAPS) [5]. For the particular case of precision agriculture, it is undeniable that the use of UAVs provides a huge benefit, due to their advantages in terms of cost, manageability and logistics, not to mention the facility to incorporate different sensors as payload.

In parallel to these developments, the creation of low-cost/low-weight multi-/hyperspectral sensors (originally employed in Earth Observation satellites) has accelerated the entry of UAVs in the precision agriculture domain. Hyperspectral imaging (HSI) is a powerful technique combining imaging and spectroscopy to survey a scene, extracting detailed information related to its chemical composition based on their reflected light in different narrow wavelengths or bands invisible to the human eye [6, 7]. In this way, a 3D data cube is created where x and y represent spatial dimensions while the third one is related to the spectral dimension, λ. Each pixel has a spectral signature associated to it which is related to each component in nature, similar to a fingerprint, and hence it is possible to identify features in an element which are not evident in the visible spectrum (see Fig. 3).

For the particular case of agriculture applications, the spectral signature of a portion of the crop gives information related to its maturity level and health as well

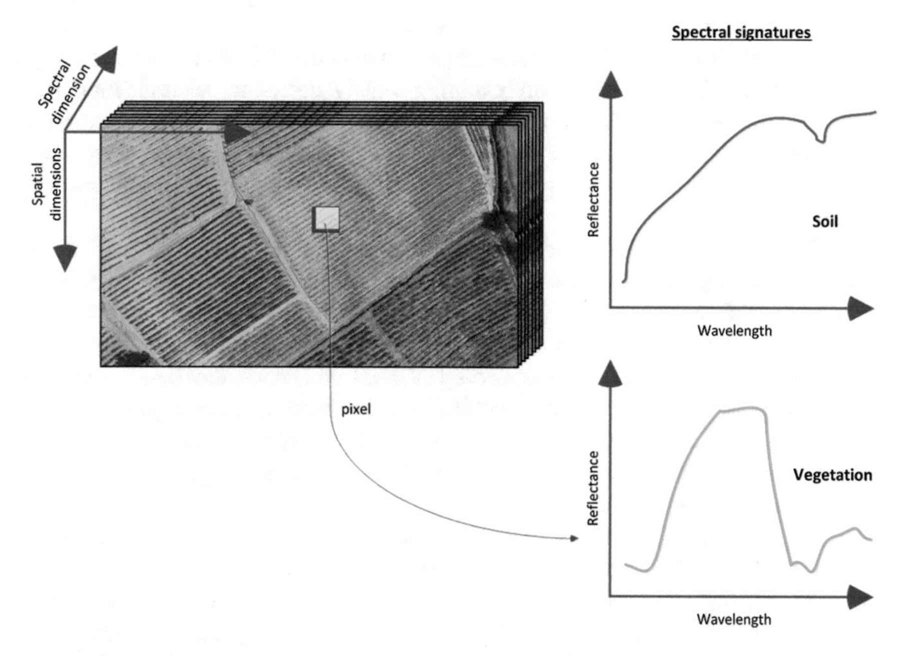

Fig. 3 Hyperspectral imaging concept. Every pixel represents a component of a mix of components

Fig. 4 Hyperspectral flying platform

as the humidity of the soil in the surrounding or the presence of plagues in their initial stages [8–10]. This fact, combined with the use of GPS and GNSS, creates a powerful tool able to help farmers in the management and control of the crop, both, qualitatively and quantitatively.

Within the ENABLE-S3 project, the use of hyperspectral technology is a key factor in order to monitor harvesting fields and give information to an autonomous harvester related to its status. However, as we are dealing with hundreds of bands (as opposed to multispectral sectors, which are in the order of tens), heavy computational hardware implementations are mandatory, especially when dealing with algorithms focused on compression, anomaly detection, identification and many more. The technical solution employed within ENABLE-S3 is based on a combination of a Dji drone (the Matrice 600 model) and a pushbroom hyperspectral sensor mounted in a camera (Specim FX10, with 224 bands in the VNIR spectral range) [11] (Fig. 4).

Additionally, an IDS RGB camera has also been set up in the UAV with the goal of providing extra information to improve the spatial quality of the acquired hyperspectral data. These two sensors have been placed in a DJI Ronin MX gimbal for softening the vibrations and increasing the quality of the acquired images. The system also includes a Jetson TK1 NVIDIA embedded device used as on-board computer for autonomously controlling the drone and managing the data acquisition. This board also enables the possibility of carrying out the on-board processing of the acquired data, which highly increases the applicability of this acquisition system (Fig. 5).

The goal of this system is twofold: on the one hand, it generates a set of vegetation indices (VI) able to give information about the status of the crops [12, 13]. Among all of the VI, one of the most well-known is the Normalized Difference Vegetation Index (NDVI) [14], which quantifies vegetation by measuring the difference between near-infrared (which vegetation strongly reflects) and red light (which vegetation absorbs).

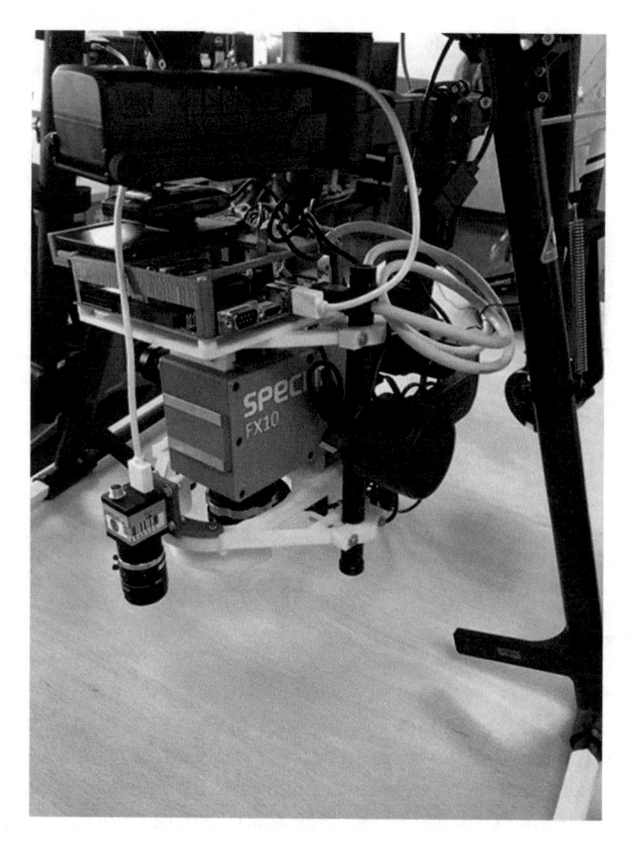

Fig. 5 Detail of the payload of the hyperspectral flying platform

The second goal of the flying platform equipped with the hyperspectral sensor is detecting persons or animals in front of an autonomous harvester to avoid fatal accidents. Once detected, the on-board system sends a signal to the harvester to stop its trajectory or avoid any kind of obstacles. The developed collision avoidance algorithm is based on Anomaly Detection (AD) [15], and is an increasingly important task when dealing with hyperspectral images in order to distinguish rare objects whose spectral characteristics substantially deviates from those of the neighboring materials.

2.1.2 Autonomous Farming Vehicles

In the context of the autonomous farming vehicles we focus on two aspects: Motion planning and perception. Methods for trajectory planning and vehicle control have been taken from automated driving [16] and successfully applied for automated farming.

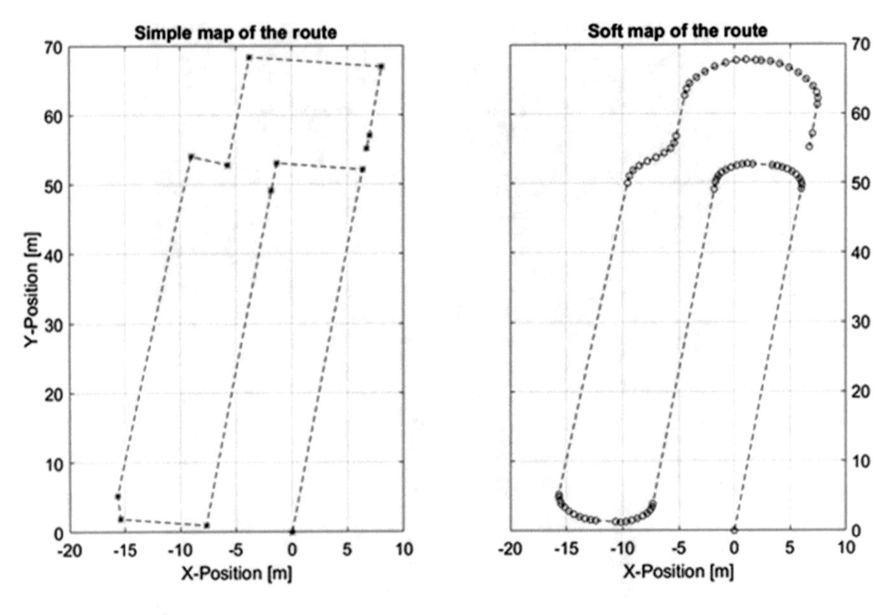

Fig. 6 Route map for a crop collecting process

For the path planning within the farming use case, a map based on simple points—global planner—was designed to completely cover the harvesting area (see Fig. 6, left side). Moreover, a soft and continuous route is generated to complete the scenario (Fig. 6, right side). This route then represents the process of collecting the crops.

The method to set the global points of the path is based on a static algorithm to minimize the non-cropped areas during the harvesting process. The planning system generates spiral shapes with straight lines ensuring a reduction of the non-cropped areas and a reduction in fuel consumption as well.

For the path generation, a simulation test case is presented in Fig. 7. The red dots represent the global map, the blue line is the local map generated using Bézier curves and the black line is the position of the harvester during a collecting process. The spiral shapes ensure the possibility of unloading using an automated tractor on the left side after the first complete process. When the system arrives at a closed turn, it opens the curvature radius to the vehicle minimum. In the figure, the black dashed line is the cropping area.

Figure 8 shows a sequence of the simulation tool from Creanex based on the results obtained from the path planning. These figures depict the areas left in the corner sides during the harvesting process. It is relevant to say that these figures are minimized during trajectory tracking with the trade-off of reducing fuel consumption. The automated driving architecture integrated with the agriculture simulator permits validation and verification of multidisciplinary algorithms (automated driving and automated harvesting).

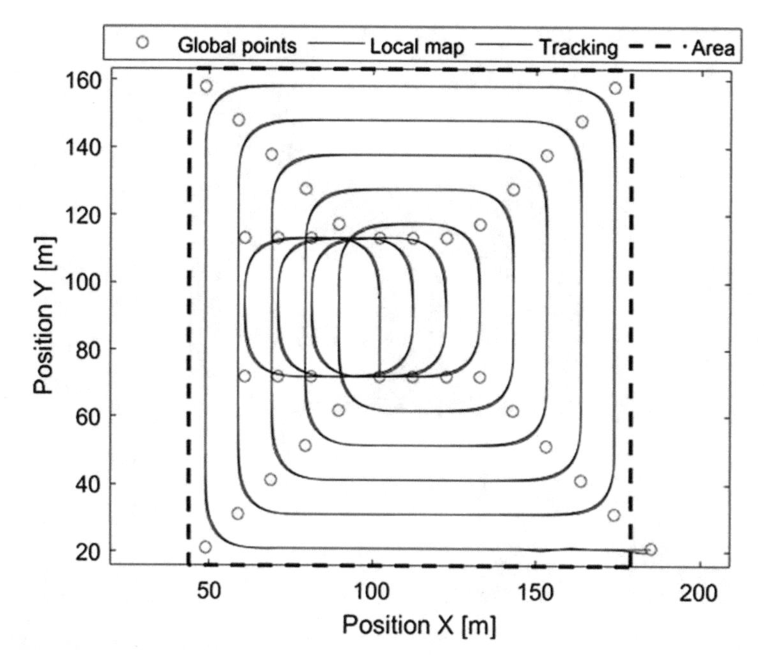

Fig. 7 Simulation test case for automated harvester

Fig. 8 Automated harvesting process in simulation

Sensors in the context of farming are applied in two categories. One category is related to smart farming capabilities such as soil moisture sensors or hyperspectral cameras that determine status of leaves or degree of ripeness of fruits. The other category contains sensors that enable localization and motion control of complete farming machines or machinery parts. Those are sensors such as camera or radar, but also sensors to monitor the status of a machine such as tire pressure. Of course, there are also hybrids such as hyperspectral cameras that can be used to generate detailed local maps and simultaneously determine the status of crops.

Although radar can be applied in many different ways for farming, such as determining soil moisture or crop density and height, within the defined scenario it is use as environment sensor. During drive-by unloading of the harvest from the harvester to the tractor it is intended to keep a constant distance between both vehicles. Due to the dusty environment featuring a high density of crop particles, radar is considered a feasible perception sensor.

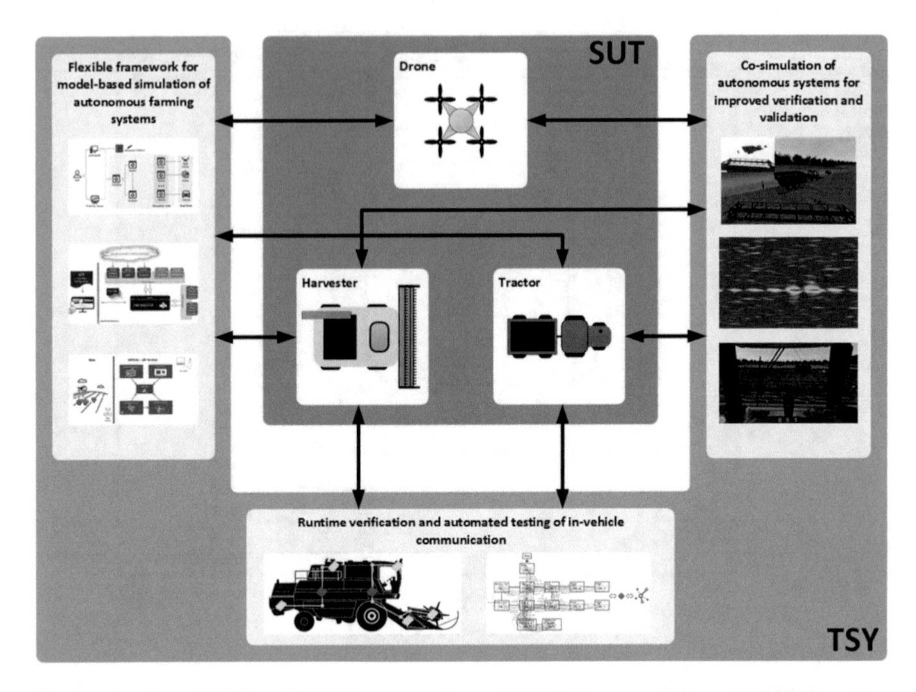

Fig. 9 Mapping of SUTs to TSY. The arrows show which TSYs are used for which SUTs

2.2 *Test Systems*

Within the farming use case, multiple test systems have been developed for testing different concepts in the farming application. As presented before, there are in principle three SUTs (Drone, Tractor and Harvester) where different concepts will be validated. In this section, we will present three test systems (TSY) that are targeting the different SUTs and the different aspects of the SUTs. Figure 9 shows which test systems are interacting with which SUT. Two TSYs are targeting all the SUTs, whereas the validation of the in-vehicle communication is only targeting either the harvester or the tractor.

More details regarding the test systems and what they are validating at the SUTs is presented in the upcoming subsections.

2.2.1 Co-simulation of Autonomous Systems for Improved Verification and Validation

The farming simulator from Creanex provides a holistic simulation consisting of a virtual farming environment, a drone and a combine harvester simulation. It presents a SiL simulation to control farming vehicles and drones and to collect sensor data. The main goal of the co-simulation is to provide system developers an improved

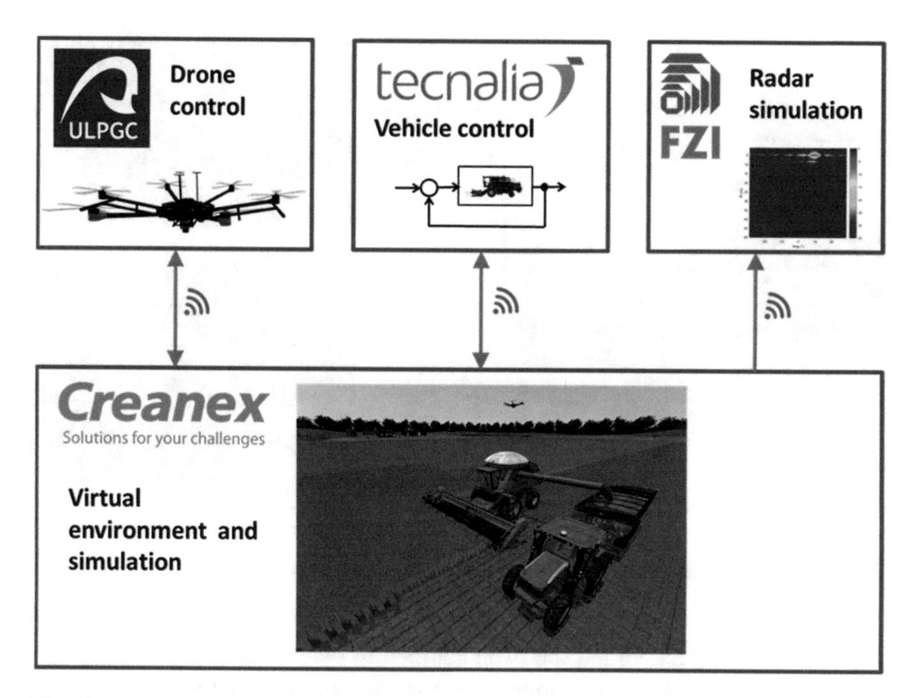

Fig. 10 Co-simulation of autonomous farming systems

testing, verification and validation environment for autonomous farming systems. The co-simulation system covers various farming scenarios, composed of farming vehicles (e.g. combine harvester, tractors), drones and a collection of virtual sensors (e.g. hyperspectral imaging and radar). As presented in Fig. 10, the co-simulation environment consists out of the virtual environment and simulation tool for the farming vehicles, combined with drone and vehicle control, using real control data to control the vehicles and drones in the simulation. Additionally, a radar simulation solution has been integrated in the co-simulation.

Additionally, the virtual environment gathers ground-truth data that is required for the sensor simulation. To collect ground truth data to generate artificial hyperspectral images or radar data, a general model of a virtual camera sensor has been developed in the farming simulator. Depending on the configuration of the virtual sensor, the output data can be a line of pixels having identification (material) color or a rendered image is converted to a list of objects detected in the field of view. In the case of the hyperspectral line scanner, the simulation detects different objects as unique colors, which simulate material differences detected by a virtual hyperspectral camera. Based on the material info in each pixel of the simulated line scanner, a hyperspectral image is reconstructed. In the simulation, material IDs can be set for specific objects. This unique ID allows to identify all the objects and their material info in the camera sensor view. From this image data, it is possible to create object lists including information of where objects are and how they are

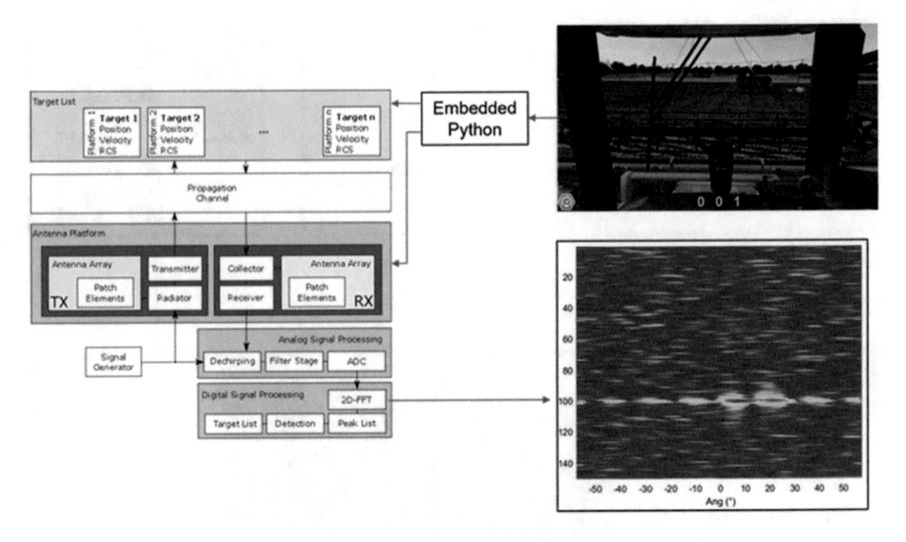

Fig. 11 Co-simulation between the farming simulation and radar sensor model

moving relative to the virtual sensor in addition to the material data. This is used as the environment simulation input to the connected radar sensor model. Based on the provided object lists, the resulting received radar backscatter is computed. Physical features of the radar sensor such as antenna element dimensions and array placement are taken into account. Virtual transmit signal are also provided by the radar simulation. The output of the radar simulation consists of the range-azimuth map that depicts peak levels obtained from the received backscatter depending on range distance and horizontal location. Afterwards, radar specific weather and environment simulations can be applied at this signal processing state. The co-simulation was implemented using TCP/UDP sockets in embedded Python for Matlab, since the radar sensor model exists in Matlab. In Fig. 11, the scheme of the demonstrator regarding the connection between the farming simulation and the radar sensor model is depicted.

The sensor services of the simulator together with a testing library allow reading of this virtual camera sensor data remotely. Figure 12 depicts a virtual sensor together with the simulated world.

2.2.2 Flexible Framework for Model-Based Simulation of Autonomous Farming Systems

The art2kitekt© tool suite is intended to facilitate and assist in the design, modelling and V&V process of real time systems. This web-based toolchain provides features to help the engineers in the process, guiding the user through different views offered by the tools.

Fig. 12 View from a camera sensor mounted on the top of the combine harvester cabin

In the Farming Use Case, art2kitekt© contributions are twofold: on the one hand the extended functionalities are focused on simulation capabilities, providing a platform where the user may simulate scenarios of autonomous farming systems. On the other hand, a Quality Management (QM) tool has been implemented to give support to V&V tasks when simulating these environments. The result of this work is a framework able to create and customize different farming scenarios with a high level of flexibility allowing the user to automate the testing necessary to verify such systems.

Thus, the main objective within the use case is to improve the testing workflow for autonomous farming systems allowing high flexibility in their simulation and providing additional features for test automation.

Figure 13 depicts the solution implemented for the Farming Use Case. The user creates simulation units, inspired by the standard FMI,[1] and configures these units for later testing activities. Configurating the interconnections among these units and other external modules (Drone, Radar, Vehicle in the figure) are also an essential part of the platform and the key point in order to leverage the testing progress provided by art2kitekt©.

[1] https://fmi-standard.org/

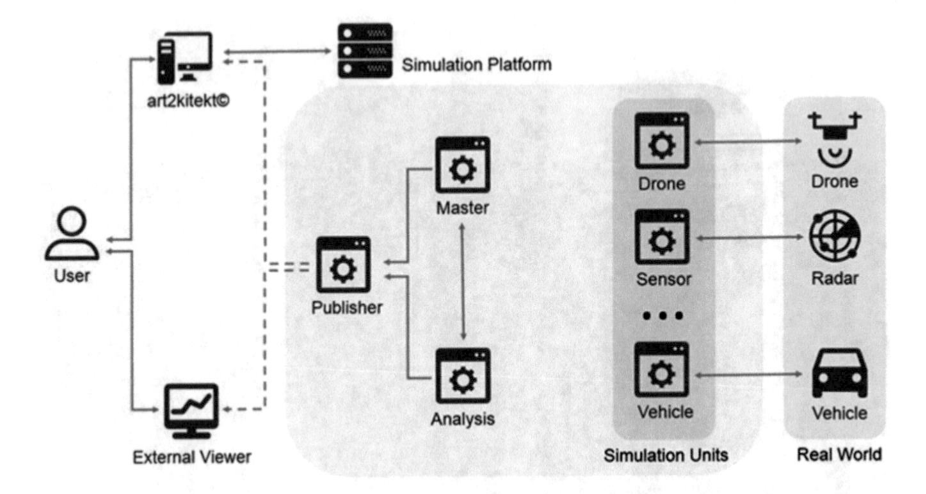

Fig. 13 Farming Use Case Solution with art2kitekt© (Icons made by Freepik, SimpleIcon, Google, Smashicons, Lucy G from www.flaticon.com)

art2kitekt© Simulation Platform

In the context of a co-simulation platform for farming environments where several tools work together to verify and validate autonomous farming systems, art2kitekt© simulation platform is focused on the model in the loop (MiL) part. The co-simulation services are intended to be used in early stages of the development of farming systems, where model-based analysis and refinement of algorithms help in the process of verify such systems.

The art2kitekt© simulation platform provides a set of components, called *simulation units* (see Fig. 13 above) as the building blocks to compound and design scenarios. These units can be both, physical objects and control algorithms. When designing farming scenarios, simulations units represent physical objects and are designed to simulate the behavior of farming vehicles, quadrotors and obstacles. Obstacles may be designed from a static perspective, such as farming buildings, or take a dynamic view, such as animals. See in Fig. 14 a representation of a static view of a scenario in art2kitekt©.

The simulation units that represent control algorithms can be pilot, guidance and path planning units. The goal of the guidance unit is to command the velocity and attitude of the vehicle to get at desirable waypoints. The goal of the pilot unit is the vehicle to reach the position and attitude commanded by the guidance. To do that, the pilot simulation unit makes use of vehicle actuators. Finally, the goal of the path planning is to optimize the route of the farming vehicle in the work zone.

As the platform also may simulate multi-domain dynamical systems, it is also possible to add more units to the platform such as sensors, avoidance algorithms, communication interfaces, etc. These units can be configured both by the user and by external components such as a path generator, etc.

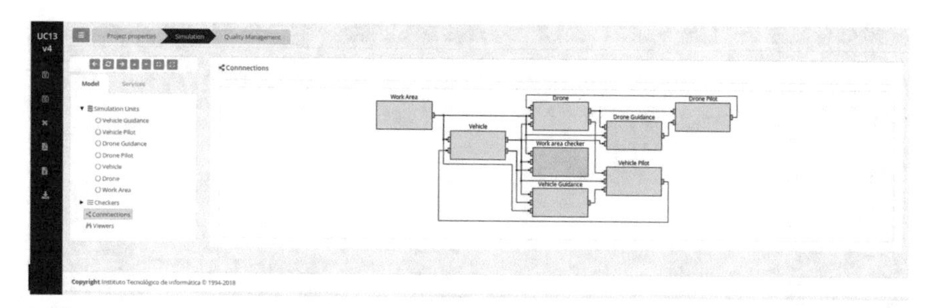

Fig. 14 art2kitekt© Simulation Platform—Static View

The simulation platform may also accept FMU (Functional Mock-up Unit) allowing the integration with different technologies and external connections to RT systems for monitoring.

Simulation units are configured via inputs and outputs to interconnect each other. art2kitekt© provides features to make these interconnections and allows the user to build more complex systems easily and monitoring their behavior in functional simulations.

When executing simulations, the correct behavior of the system or any of its components can be verified by checkers. Checkers test the behavior of simulation units individually or test specific parameters of the mission indicated by the user, such as the elapsed time of a mission or the worked area by the farming vehicle. Simulations can be carried out in two modes, the "live" mode where the engineer can observe the behavior of the different components of the simulation. This mode is necessary in cases of using external components, such as a real sensor or drones (or realistic models of these) and the "fast" mode where the simulation is executed at the maximum computing speed, accelerating the results computation, and therefore reducing the testing time. This mode is essential in many simulations for test cases that could last for hours. The simulation platform offers tight integration with art2kitekt© quality manager tools described below.

In comparison to the simulation tool, previously described, art2kitekt© enables the user to work with different components (drones, etc.) and scenarios at different levels of abstraction. This means that the user may design simple components and move on until configurations of advanced scenarios are closer to the real ones. art2kitekt© tool is then more focused on the initial stages of the design and analysis of farming systems rather than a realistic-virtual environment. Additionally, it is also integrated with the Quality Management components; thereby tests may be designed and configured in parallel with the configuration of the scenarios (components, etc.) and therefore implement checkers over these configurations.

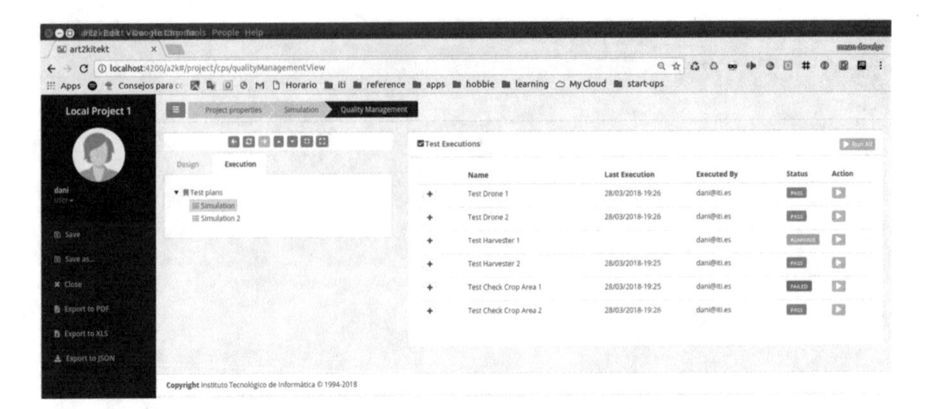

Fig. 15 art2kitekt© Quality Management—Design and Execution Tabs

art2kitekt© Quality Manager

art2kitekt© has been enhanced to support V&V activities, integrating a QM component that allows the engineer to generate and execute test plans and test cases easily and then show the results through specific viewers.

Test plans and test cases can be defined by the user, making use of wizards under the *design tab* (see Fig. 15). Moreover, the QM component provides features to generate test cases from configurations created in the simulation platform (see above section), allowing the engineer to make changes in the model, modifying parameters or checkers and save them as new test cases, making the creation of test plans fast and intuitive.

The execution of test plans and test cases is done under the Execution section and can be executed in batch mode (as a test plan) or launch individual test cases. The presentation of results is also shown to the user in the Execution section, where you can see the results of the last execution, as well as the history of the results of the execution of each test case.

2.2.3 Runtime Verification and Automated Testing of In-Vehicle Communication

In highly automated and autonomous vehicles, the real-time properties of the various components/sub-systems are fundamental to ensure full system safety. Real-time analysis and associated verification methods allow to mathematically define and analyze scheduling requirements and the sharing of computational resources in applications and communication as well as their interaction with the external environment.

Traditionally, based on classic real-time and communication analysis, tool and frameworks can be constructed in order to provide the generation of more realistic tests that ultimately can improve the overall verification process one may targeting.

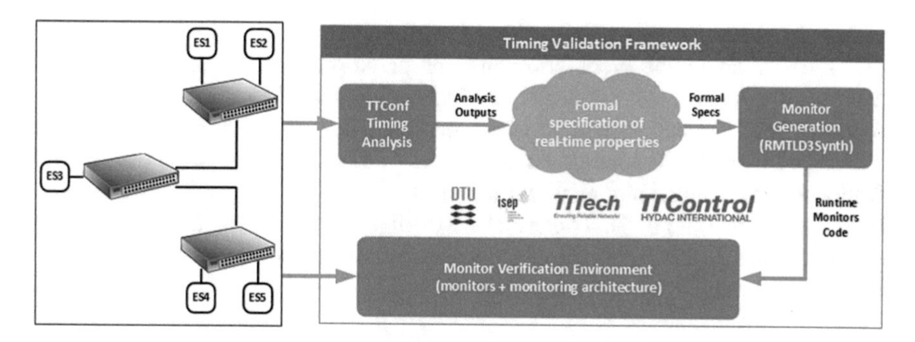

Fig. 16 Proposed runtime verification based in-vehicle communication testing framework

However, testing is in essence an incomplete process and a defined set of tests may fail to address aspects of the target system under test which can compromise its safety when deployed in the real-world. Therefore, more advanced and robust approaches are necessary, namely those based on formal verification methods.

Motivated by the limitations of testing, we propose a new approach, where test generation frameworks and rigorous mathematically grounded analysis and formal verification techniques are combined with the aim of providing a more complete and robust testing framework for intra-vehicle communication. The high-level architecture is presented in Fig. 16, where we depict the interaction between the main involved components: (1) an input network topology consisting of switched and end-systems and a set of communication traces simulated for that topology; (2) the TTConf Timing analysis which is responsible for calculating worst-case delay times for the transmission of network packets; (3) and the RMTLD3Synth tool which is responsible for generating monitors that are derived from formal specifications that take into consideration the properties of the network topology and the extra timing constrains calculated by the TTConf timing analysis tool.

In what follows, we briefly describe the role of each tool in the overall approach.

The TTConf Tool and the RMTLD3Synth Framework

TTConf

In terms of network communication analysis, the TTConf tool can perform both timing analysis and configuration optimization for Deterministic Ethernet. The schedulability of the Time-Triggered (TT) traffic can be guaranteed during design phase, by synthesizing of the Gate-Control-Lists (GCLs). However, an Audio/Video Bridging (AVB) flow is schedulable only if its worst-case end-to-end delay (WCD) is smaller than its deadline. Although latency analysis methods have been applied to AVB traffic in AVB networks [17], they do not consider the effect of TT traffic on the latency of AVB traffic. In ENABLE-S3, the TTConf tool was extended with a Network Calculus-based approach to determine the worst-case delays of AVB traffic in a Time-Sensitive Networking (TSN) network by considering the effects of TT traffic, including preemption and non-preemption modes, and supplements a proof of non-overflow condition for AVB credit [18]. Regarding configuration optimization, TTConf determines the worst-case end-to-end delays of messages exchanged over "Deterministic Ethernet", i.e., TTEthernet and IEEE 802.1 TSN. It supports the analysis and configuration of several traffic classes, Time-Triggered, Rate Constrained and AVB. It determines a configuration such that the dependability requirements are guaranteed (timing, but also redundancy levels for streams, see the 802.1QCB, Frame Replication for Reliability[2]).

RMTLD3Synth

The RMTLD logic [19, 20] is a restricted fragment of the *Metric Temporal Logic with Durations (MTLD)* [21] and was designed for specifying real-time properties based on relevant timed events, whereas the RMTLD3Synth framework [22] is an instantiation of RMTLD which allows to automatically generate monitor code that embodies the semantic of the formulas specified in RMTLD formulas. Both the logic and the toolbox were designed to be cross-domain applicable, which enables them to be coupled with auxiliary frameworks (such as the ones considered in this use case). This coupling requires an instrumentation of monitored application for it to capture the events of interest for the specification from which the monitors have been derived. In our developments, we will investigate how RMTLD3Synth can be integrated with more real-time analysis of intra-vehicle communications, namely the TTConf tool, to ensure the required safety levels, and guarantee that the generated monitors will not affect the overall real-time requirements of the applications being monitored and verified during execution time. The generated monitors can be used within/connected to virtual testing components, and if working according to what they were specified to verify, they can be deployed into the final system without needs of modification and contribute to its verification and validation. In terms of practical applications, the framework was previously used to verify at runtime a prototype auto-pilot, as reported in [23].

3 Verification and Validation Approaches and Results

First verification and validation tests have been performed using the systems under test and the test systems. The following subsections will provide first results of the tests performed within the farming use case.

[2]TSN Task Group, "IEEE 802.1CB/D2.4: Frame Replication and Elimination for Reliability," 2016.

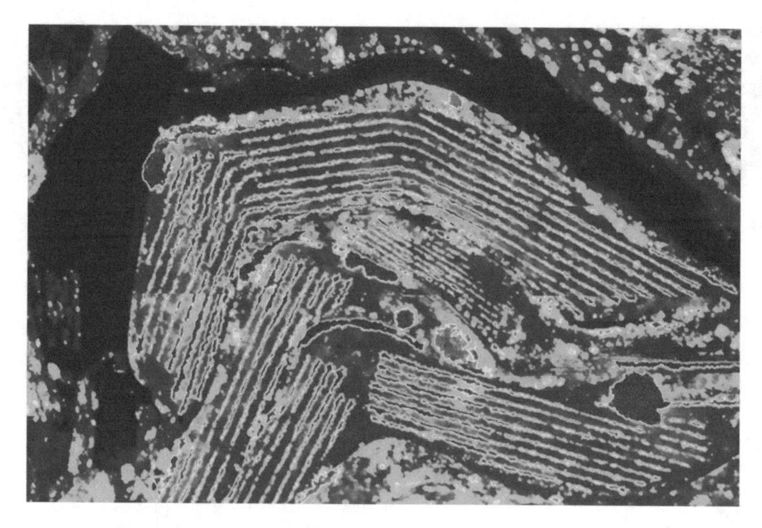

Fig. 17 Vineyard Normalized Different Vegetation Index (NDVI) map obtained with the ENABLE-S3 flying platform

3.1 Hyperspectral Flying Platform

Part of the evaluation of the hyperspectral flying platform has been carried out in some vineyards located in Gran Canarias, Canary Islands, Spain (27°59′14.9″N 15°35′51.9″W). During a flight of 4 min, data were acquired for 224 bands from 400 to 1000 nm (VNIR), and different vegetation indices were obtained and represented in a false-colored map. As an example, the NDVI map in Fig. 17 was generated showing in *red* color those areas with healthy vegetation while *blue* color represents soil, roads, houses and dead vegetation.

Once the maps are generated, coordinates of regions of interest (ROI) can be obtained in order to be transferred to an autonomous farming vehicle or to farmers.

3.2 Autonomous Driving

For the autonomous driving validation, the algorithms are tested first in the simulator, and afterwards in the real platform. It is therefore mandatory to model the scenarios as close as possible to the reality. As shown in Fig. 18, some of the work regarding the autonomous driving has been performed on the simulation environment (vehicle model). The left side of Fig. 18 shows the environment recognition. Then, this information is used to describe the scenario. A simulation model (middle picture) has been designed based on this information that permits to know the real space and conditions, including visual models. For the farming use case, a new scenario has been designed (picture on the right side). It is used as a

Fig. 18 Testing place (left side), test scenario in simulation (middle) and farming layer (right side)

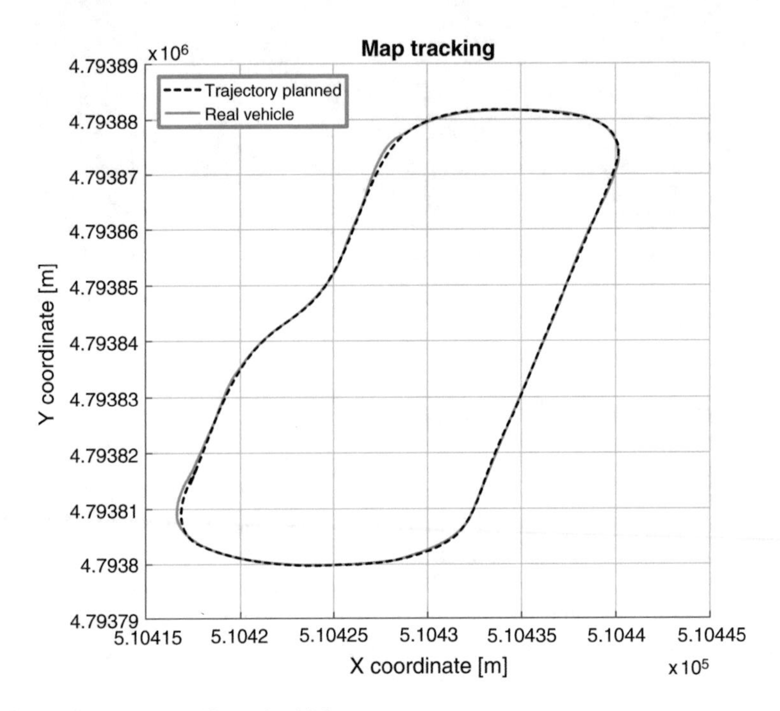

Fig. 19 Trajectory tracked for real vehicle

layer of farming models, to give a better understanding of the correct execution of the developed algorithms.

For the planning of the trajectory, a map based on simple points—global planner—has been designed to completely cover the whole harvesting area. To handle the curves of the trajectory, a soft and continuous route is generated to complete the scenario.

The verification tests for the trajectory are related with the tracking of the complete field to verify the correct application of the maps, receiving correctly the information from the DGPS and the dimensions of the circuit to recreate a scenario of an automated farming vehicle. In this sense, Fig. 19 shows a path tracked by the real platform covering the total area for the scenario (private track at location). This experiment shows a closed circuit, obtaining results with tracking errors under 40 cm of magnitude, and using differential GPS information.

The first results, in simulation and then in the real platform, show promising results for the upcoming integration with the drone in the final demonstration.

3.3 *Virtual Sensor Verification and Validation*

For virtual verification and validation, virtual sensor stimulation needs to be provided that is as realistic as possible or sufficient to a specific testing aim. For that reason, data-driven models for adverse weather and environmental conditions that can be applied to both, real and virtual radar are developed. Previously, a rain and particle simulation was realized for on-road video sensors at the University of Tuebingen [24]. In Fig. 20, the scheme is depicted. After real measurements are used to extract backscatter features introduced by adverse weather and environment conditions, data-driven models are generated that can be applied to the same interface for real and virtual radar. Beforehand, the radar sensor model was validated against the real-world radar. The simulated environment corresponds to any source of object lists with diverse abstraction levels. The input from the simulated environments determines how close to reality the radar output will be. The interface of the radar sensor model (see chapter "Radar Signal Processing Chain for Sensor Model Development") is chosen at the lowest level where the signal is accessible in the original real-world radar sensor. At that interface, which is the analog to digital converter (ADC) output, measurements are recorded in various conditions reflecting adverse weather and environment models.

Rain has only little effect on radar within the relevant ranges. Nevertheless, conditions that influence the radome such as splash water or spraying, attenuate the receive signal. Splash water or spraying can either be caused by mud or puddles

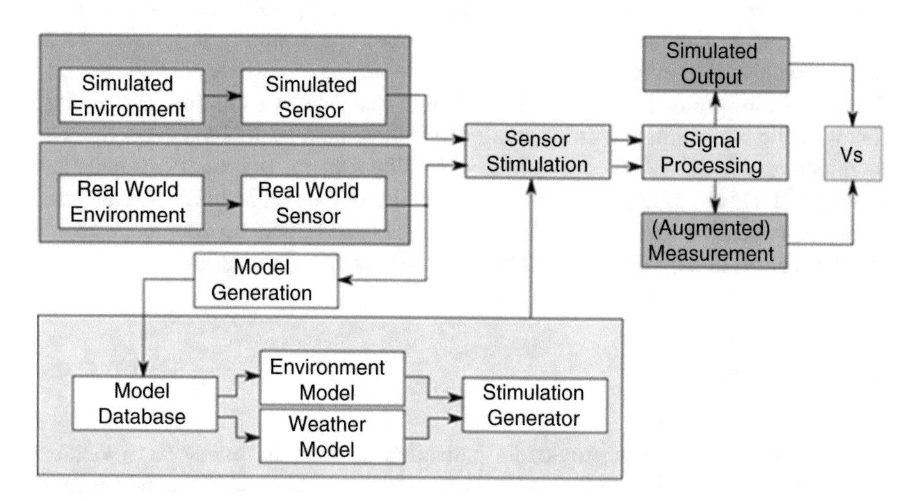

Fig. 20 Application and validation of adverse conditions on radar

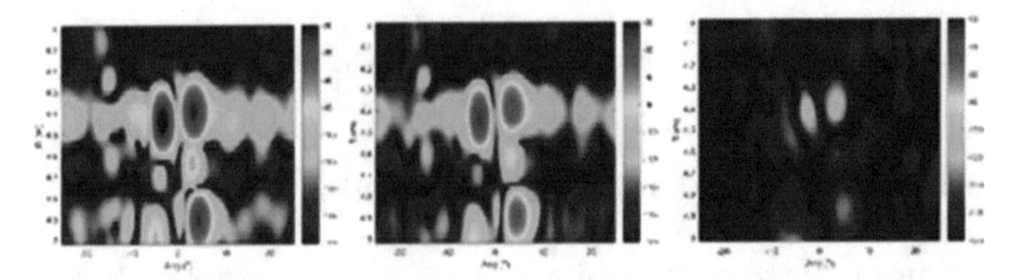

Fig. 21 Measurement scenario from radar perspective with different spray intensities. Left: none. Middle: light. Right: constant

in the trajectory of the vehicles or spraying of pesticides can land on the sensors. Additionally, it is also possible that during the farming scenario, weather conditions change, or it starts to rain.

With static measurement setups in a highly controlled measurement environment such conditions can be isolated and extracted. In Fig. 21 range-azimuth maps show how the additional radome attenuation impacts the receive signal during the sequence of a dry, waterfilm covered and drying radome. The measurements were taken while the radar was mounted on a railing and two targets, one of cuboid and one of spherical shape, were placed at a distance of 4.5 m.

3.4 Runtime Verification Results

Regarding the Runtime Verification and Automated Testing of In-Vehicle Communication, the current status is an initial prototype of the architecture and interactions of the three main components involved. In this sense, it is possible to establish useful specifications in the RMTLD logic that embedded the results from the worst-case communication delays produced by the TTConf timing analysis tool. The next step for this demonstrator is to provide a concrete instance of a runtime architecture that will allow to connect with simulation data generated by TTTech tools. Such an architecture will be based on a publish-subscribe system, where end-systems and switches of a given topology are encoded as nodes of the architecture and simulated traces of execution are feed to those nodes. The node will be coupled with the monitors generated by the RMTLD3Synth tool to enable verification.

4 Future Work

The farming use case has provided a collection of tools and methods to validate technologies that can be applied in the farming domain. Nevertheless, the concepts presented so far have mainly been targeting simulation tools. One of the most important next steps for the continuation of this work is the integration of real hardware (i.e. real vehicles, drones, sensors, etc.) into the simulation platforms. First tests on the real hardware have been performed, but those were currently completely stand-alone tests. The integration of hardware and the development of Hardware-in-the-Loop (HiL) test systems will be the next important step towards improving and reducing the validation and verification effort required.

Even if the real hardware is not even available, data coming from the real environment, collected by the hardware out in the field provides the V&V processes more realistic input and as a result, more realistic evaluation of the systems. As an example, integrating information on weather and environment conditions in a common simulation platform will allow to benchmark a sensor equipped SuT regarding robustness towards adverse conditions. This supports the evaluation of complementary sensors in order to improve the robustness of the whole system. As a result, the applied simulators in the farming use case (i.e. the farming simulation from Creanex and art2kitekt© from ITI) will be extended with additional features to include data from real hardware, like sensors. Creanex and ITI are already working on integrating sensor data in their simulation tools.

As mentioned before, currently most of the SUTs are still stand-alone. Part of the future work will be related to integrating the vehicles and drone into a single, but larger demonstrator showing the overall farming scenario. New features need to be introduced, especially around data communication. Wireless communication concepts have already been evaluated (e.g. LoRa, 4G, etc.) for their applicability in the field of farming. First tests showed good results, but further testing and evaluations are required to apply the demonstrators deployed in the farming use case.

5 Conclusions

The agriculture domain offers a great number of opportunities to develop new technologies for improving human life and safety, but at the same time many challenges arise to validate and verify these technologies and to finally integrate them into real farming machines. Within the ENABLE-S3 project, a first attempt has been made to validate varying technologies, among other some coming from the automotive domain, for their applicability to the farming domain. Within the project, the project team has been encouraged to transfer technologies from different domains and to create synergies between automotive and farming. Nevertheless, many obstacles have been identified, as many technologies are not directly applicable and need to be validated in different ways and with separate input than e.g. the input from automotive.

Within the farming use case, the project team has validated mature technologies combined with new sophisticated ones to create solutions that target many challenges from the farming domain, like e.g. different sensor values, environment parameters, communication, vehicle configuration, etc. The usage of drones, hyperspectral and radar sensors, autonomous vehicles, advanced integrated co-simulation frameworks with different weather models and simulated stimuli and vehicle communication are some of the identified technologies that have been intensively validated for their applicability.

The aim of the project team was to create an extensive validation framework targeting automated and autonomous farming vehicles to improve safety and improve the life of the farmer of the future. The results presented in this chapter are a first step towards better and faster validation of farming systems, enabling continuous validation without having accessibility to the actual farming vehicles or the limited natural resources, like farming fields. The solutions presented here will open new opportunities in the farming domain with a large impact on social, economic and environmental aspects.

References

1. Bell, T., Elkaim, G., Parkinson, B.: Automatic steering of farm vehicles using GPS. In: FAO International Conference on Precision Agriculture (1996)
2. Lenain, R., Thuilot, B., Cariou, C., Martinet, P.: Adaptive and predictive non-linear control for sliding vehicle guidance: application to trajectory tracking of farm vehicles relying on a single RTK GPS. In: IEEE International Conference on Intelligent Robots and Systems, pp. 455–460 (2004)
3. Cariou, C., Lenain, R., Thuilot, B., Berducat, M.: Automatic guidance of a four-wheel-steering mobile robot for accurate field operations. J. Field Rob. **26**, 504–518 (2009)
4. Lattarulo, R., Pérez, J., Dendaluce, M.: A complete framework for developing and testing automated driving controllers. In: IFAC World Congress 2017, pp. 258–263 (2017)
5. Gonzalo, J., López, D., Domínguez, D., Gracía, A., Escapa, A.: On the capabilities and limitations of high altitude pseudo-satellites. Prog. Aerosp. Sci. **98**, 37–56 (2018)
6. Chang, C.-I.: Hyperspectral Imaging. Springer, New York (2003)
7. Lopez, S., Vladimirova, T., Gonzalez, C., Resano, J., Mozos, D., Plaza, A.: The promise of reconfigurable computing for hyperspectral imaging onboard systems: a review and trends. Proc. IEEE. **101**(3), 698–728 (2013)
8. Teke, M., Seda-Deveci, H., Haliloglu, O., Zubeyde-Gurbuz, S., Sakarya, U.: A short survey of hyperspectral remote sensing applications in agriculture. In: Proceedings of the 6th International Conference on Recent Advances in Space Technologies (RAST), Istanbul, Turkey, 12–14 June 2013
9. Sahoo, R.N., Ray, S.S., Manjunath, K.R.: Hyperspectral remote sensing of agriculture. Curr. Sci. **108**(5), 848–859 (2015)
10. Adao, T., Hruska, J., Padua, L., Bessa, J., Peres, E., Morais, R., Joao-Sousa, J.: Hyperspectral imaging: a review on UAV-based sensors, data processing and applications for agriculture and forestry. Remote Sens. **9**(11), 1110 (2017)
11. Rodríguez, A.S., Horstrand, P., López, J.F., López, S.: Setting up an autonomous hyperspectral flying platform for precision agriculture. In: SPIE Remote Sensing, Berlin, Germany, 10–13 September 2018
12. Xue, J., Su, B.: Significant remote sensing vegetation indices: a review of development and applications. J. Sensors. **2017**, 1353691 (2017)
13. Thenkabail, P.S., Smith, R.B., De Pauw, E.: Hyperspectral vegetation indices and their relationships with agricultural crop characteristics. Remote Sens. Environ. **71**(2), 158–182 (2000)

14. Huang, J., Wang, H., Dai, Q., Han, D.: Analysis of NDVI data for crop identification and yield estimation. IEEE J. Sel. Top. Appl. Earth Obs. Remote Sens. **7**(11), 4374–4384 (2014)
15. Horstrand, P., López, S., López, J.F.: A novel implementation of a hyperspectral anomaly detection algorithm for real time applications with pushbroom sensors. In: IEEE 9th Workshop on Hyperspectral image and Signal Processing: Evolution in Remote Sensing, Amsterdam, The Netherlands, 23–26 September 2018
16. Lattarulo, R., González, L., Martí, E., Matute, J., Marcano, M., Pérez, J.: Urban motion planning framework based on N-Bézier curves considering comfort and safety. J. Adv. Transp. **2018**, 6060924 (2018)
17. Azua, J.A.R., Boyer, M.: Complete modelling of AVB in network calculus framework. In: Proceedings of the 22nd International Conference on Real-Time Networks and Systems (2014)
18. Zhao, L., Pop, P., Zheng, Z., Li, Q.: Timing analysis of AVB traffic in TSN networks using network calculus. In: Proceedings of the IEEE Real-Time and Embedded Technology and Applications Symposium (RTAS), pp. 25–36 (2018)
19. Pedro, A.: Dynamics contracts for verification and enforcement of real-time systems properties. PhD Thesis, Braga, Portugal, 10 April 2018
20. Pedro, A., Pereira, D., Pinto, J.S., Pinho, L.M.: Monitoring for a decidable fragment of MTLD. In: The 15th International Conference on Runtime Verification (RV'15), Vienna, Austria, 22–25 September 2015
21. Lakhneche, Y., Hooman, J.: Metric temporal logic with durations. Theor. Comput. Sci. **138**(1), 169–199 (1995)
22. Pedro, A., Pereira, D., Pinto, J.S., Pinho, L.M.: RMTLD3Synth: runtime verification toolchain for generation of monitors based on the restricted metric temporal logic with durations. https://github.com/cistergit/rmtld3synth. Accessed 23 Nov 2018
23. Pedro, A., Pereira, D., Pinto, J.S., Pinho, L.M.: Runtime verification of autopilot systems using a fragment of MTL-$\int$. Int. J. Softw. Tools Technol. Transfers (STTT). **20**(4), 37–395 (2018)
24. Hospach, D., Müller, S., Rosenstiel, W., Bringmann, O.: Simulation of falling rain for robustness testing of video-based surround sensing systems. In: Proceedings of the 2016 Design, Automation & Test in Europe Conference & Exhibition (DATE) (2016)

A Virtual Test Platform for the Health Domain

Teun Hendriks, Kostas Triantafyllidis, Roland Mathijssen, Jacco Wesselius, and Piërre van de Laar

1 Introduction

Innovation in the health domain revolves around improving the quality and efficiency of healthcare through a focus on care cycles [1]. Central to care cycle thinking is a patient-centric approach that optimises healthcare delivery for all major diseases. Combining expertise in medical technology with clinical know-how of hospitals produces innovative solutions that meet not just the needs of individual patients, but which also enable healthcare professionals to work faster, more easily and more cost-effectively.

To achieve these innovations, a significant percentage of sales is invested in R&D of complex medical equipment such as Philips' Azurion image-guided therapy systems (see Fig. 1; [2]). Such equipment is typically provided in many variant configurations, tailored to the needs of a hospital and its healthcare professionals. It must be ensured that all these variant configurations work properly and are safe for the patient, the surgeon and the hospital staff.

These systems now have the ability for semi-autonomous robotic movements. These are intended to assist the surgeon to manoeuvre the robot arm with the X-ray tube and the image detector around the patient in a user-friendly way to obtain optimal X-ray images, while avoiding collisions with patient, staff or devices in the operating room. Verification and validation of such advanced features for many configurations and workflows requires a significant increase of test effort. At the same time, business competitiveness demands the time to market to be shortened. Still, compliance with standards and regulations must be proven.

T. Hendriks · K. Triantafyllidis · R. Mathijssen (✉) · J. Wesselius · P. van de Laar
ESI (TNO), Eindhoven, The Netherlands
e-mail: teun.hendriks@tno.nl; kostas.triantafyllidis@tno.nl; roland.mathijssen@tno.nl; jacco.wesselius@tno.nl; pierre.vandelaar@tno.nl

A. Leitner et al. (eds.), *Validation and Verification of Automated Systems*,
https://doi.org/10.1007/978-3-030-14628-3_21

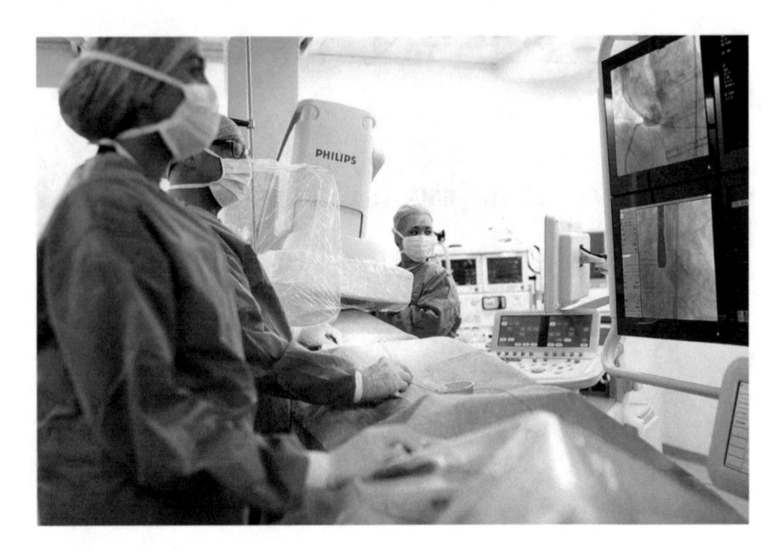

Fig. 1 A Philips image-guided therapy system

Within the Enable-S3 project [3], the Dutch project partners, Philips, Reden, TU/e, and TNO, investigated the use of virtualisation and virtual or mixed virtual/real test systems to improve the verification and validation (V&V) of complex medical equipment with semi-autonomous movement capabilities. Philips, in particular, investigated concrete improvements for its medical equipment product families, considering five main testing use cases as described next. ESI (TNO) investigated the concepts, general applicability, and conceptual frameworks needed to sustain virtual testing in the health domain.

This chapter reports on the progress towards a virtual test platform for the health domain, its concepts, and the underlying conceptual framework. First the considered testing use cases are described, then the opportunities and challenges of virtualisation are outlined for V&V in the health domain. Next the concept of a virtual test platform is described, and the conceptual framework for a virtual test platform is presented. Finally, demands of the considered testing use cases on this virtual test platform are listed and conclusions are summarised.

2 Use Cases of the Health Domain in Enable-S3

Given the anticipated benefits of mixed-reality and virtual prototypes, the health domain decided to experiment within the Enable-S3 project with mixed real and virtual and fully virtual prototypes in the following use cases:

- *Early Validation Test* is about ensuring that the needs of clinical users are properly captured before the system is being realised.
- *Informal Feature Test* refers to confidence testing a specific system feature.

- *Software Unit Test* denotes testing a specific software component in a relatively isolated context.
- *Informal Software Integration Test* relates to confidence testing of software components in the context of a complete system.
- *Usability Test* denotes evaluating the usability of a system by clinical users, with particular focus on specific system features.

2.1 Early Validation Test

In this use case a mixed prototype has been developed to ensure that the needs of clinical users are properly captured and thus the right system will be developed. This prototype is called the Virtual Cathlab [4] and is shown in Fig. 2. This use case has revealed that clinical users appreciate the use of real UI components since these UI components are part of their frame of reference. Furthermore, a mixed-reality prototype can be realised relatively quickly by using real product components. Real components may, however, complicate the realisation of new behaviour due to their complexity, and the built-in safety precautions.

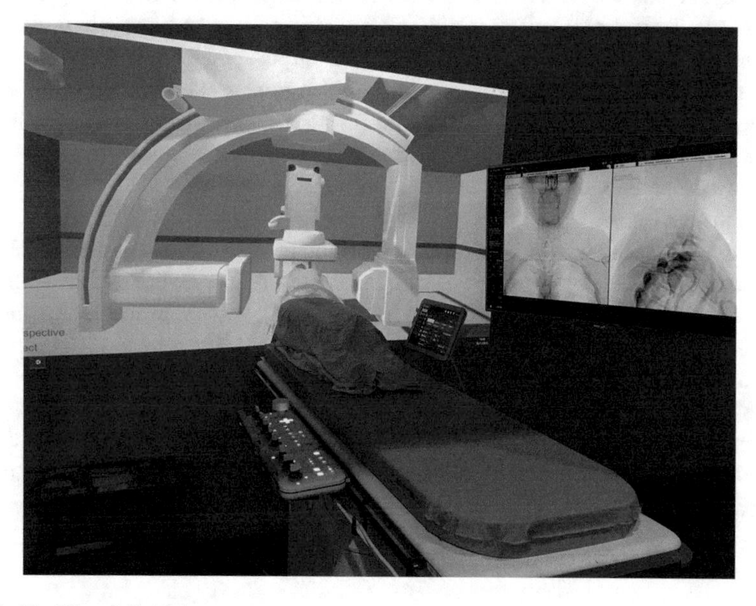

Fig. 2 The Virtual Cathlab

2.2 Informal Feature Test

Miscommunication and misunderstanding may easily occur between clinical users and developers when developing a new feature based on specifications written down in a document. To avoid this, informal feature tests with simple prototypes are needed to increase the confidence that the right feature will be developed and delivered. In this use case, mixed and virtual prototypes, such as the already mentioned Virtual Cathlab, have been used to perform the informal feature tests. This use case showed that mixed and virtual prototypes can prevent issues caused by miscommunication in early phases of development. Informal feature tests provide a strong link to the natural frame of reference of the various stakeholders, e.g. clinical experts 'see' the working of new a feature by trying out its concept in a prototype. Similar observations were also made in other domains, such as the lighting domain [5, 6].

2.3 Software Unit Test

Normally, software units are tested in relative isolation before integrating them into the system. To isolate a software unit, all function calls over its required interfaces must be handled. Test stubs are created for this purpose; they typically return a fixed response, i.e. independent of the function call's arguments and of their state. Virtual components can provide a more realistic response, which provide more variation than a test stub. A virtual component can more realistically reflect the state of the stubbed component, and it can tailor its response to the function call's arguments and its internal state.

Within the health domain, the way of working has been changed, see also Fig. 3, to include virtual components for testing the positioning software (software units to manoeuvre the system's robot arm and patient table). Instead of testing positioning software with a single physical system configuration (Fig. 3 left), the positioning software can now be tested with many virtual prototypes too (Fig. 3 right). In this use case, more checks can be performed before integration; this way more errors can be isolated. As a result, the quality of the software delivered to integration testing is much higher, which significantly reduces the number of integration issues and total integration time.

2.4 Informal Software Integration Test

Regulatory bodies, such as the U.S. Food and Drug Administration (FDA), require test evidence when releasing a medical device. Before creating test evidence, medical companies already test their products to increase their confidence in the device under development. In this use case, integration in mixed and virtual prototypes is

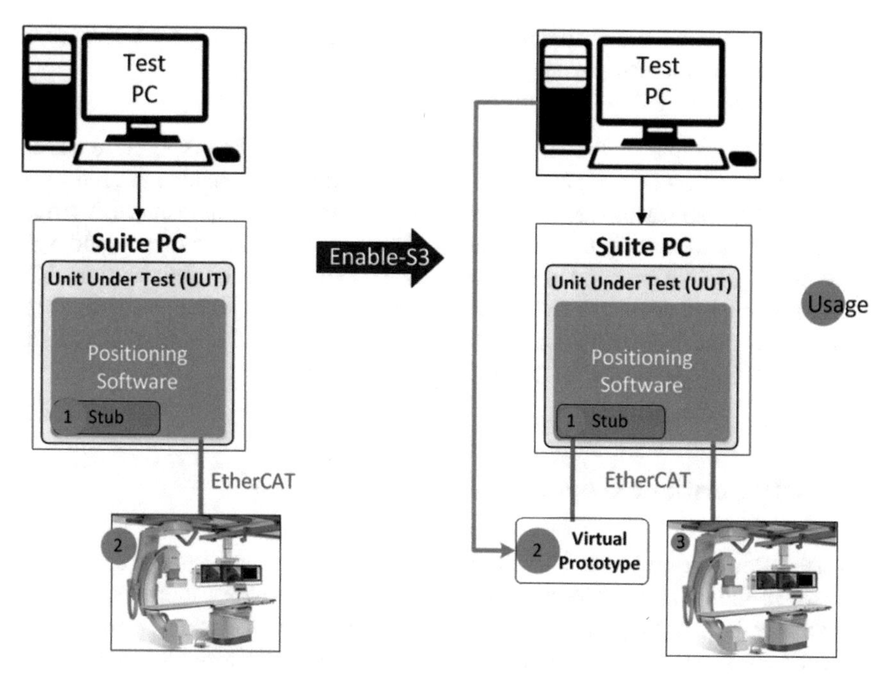

Fig. 3 Changes in software unit testing

used to increase the confidence level. In particular, software components related to manoeuvring the robot arm and acquisition and processing of X-ray images are integrated into mixed-reality and virtual prototypes, that represent a complete system. This use case clearly shows the opportunities, but it also indicates that mixed and virtual prototypes will only be accepted when a large percentage of the system errors are caught and only a small number of additional errors are introduced by virtualisation.

2.5 *Usability Test*

For the development of medical devices, proof of good usability is required to release a product into the market. So far, usability tests are performed with actual users on physical prototypes. In this use case, the health domain investigated under which conditions usability test results on mixed and virtual prototypes are comparable to those on physical prototypes.

Observed challenges of using mixed and virtual prototypes are the following:

- to prevent that the usability test becomes biased to evaluation of the mixed or virtual reality instead of evaluation of the system's usability in its environment;
- to ensure consistent and complete sensory inputs, e.g. consistent visual and audio feedback of robot arm movements;

- to prevent users from changing their behaviour, e.g. by taking more risks since mistakes will not result in any real damage;
- to ensure observability of the users.

The investigations within the health domain confirm the importance of the completeness of sensory input. Mixing real and virtual components in prototypes eases the observability of users over purely virtual prototypes. For instance, if virtual reality goggles were to be used in a purely virtual prototype, then they would need to be equipped with cameras to expose the facial expressions of their wearers. The investigations also showed that the above mentioned challenges can be sufficiently addressed for the health domain, and they demonstrate the suitability of mixed and virtual prototypes for usability testing.

3 Virtualisation: Opportunities and Challenges

Virtualisation can be found in many different forms, e.g. augmented reality with humans-in-the-loop, simulations of hardware, simulation of user inputs, simulation of peripherals such as picture archiving and communication systems, or electronic patient record systems. The use cases in the previous section already illustrated some of the advantages of virtualisation. In this section we describe the opportunities and the challenges of virtualisation for the development of complex medical equipment.

3.1 Opportunities with Virtualisation

Virtualisation in the health domain has been a long standing research topic. Early applications considered surgical procedures such as remote surgery, planning and simulation of procedures before surgery, medical education and training, and also the architectural design of healthcare facilities [7].

More recently, virtualisation to improve the development and testing of medical equipment itself has become an active research topic. For this purpose, the availability of virtual systems can be exploited in several ways: by strengthening communication between stakeholders, by providing earlier access to systems (under development), by reducing cost, by providing opportunities to handle variability better, by lifting real-world restrictions, and by increasing the level of flexibility and control for integration and test engineers. These opportunities are described in the next sections.

3.1.1 Strengthen Communication Between Stakeholders

Using virtualisation, an experience can be delivered that closely approximates the experience one would get when interacting with the real product or with an envisioned product by means of a virtual mock-up. The chance of miscommunication is minimised, since experiences and discussions are directly linked to the natural frame of reference of the various stakeholders. For example, virtual X-ray enables evaluation of the imaging chain to be based on computed images instead of abstract key performance indicators (KPIs).

Another possible application of virtualisation to improve communication about the expected system behavior is to visualise areas in the virtual world, that are invisible in the real world, like safety zones around the patient. This helps to clarify the change in behaviour of the system when entering or leaving those areas. Virtualisation thus improves the quality of the communication between stakeholders and assures that the right decisions are made, thereby improving the success rate of an organisation.

Interaction with a virtual system is worth a thousand movies: one movie shows only one fixed scenario, while many more questions of stakeholders about a system's behaviour can be answered by interacting with its virtual equivalent.

3.1.2 Earlier System Access

Having a system virtually available prior to its construction, enables earlier usability testing. This does not only ensure that the right system will be built but it also results in recorded runs of user interactions for analysis, development, and test purposes. Furthermore, a virtual system can be used for design space exploration to optimise the design for desired KPIs. Finally, by virtualising a component, one can check the behaviour of related components, e.g. check the software that will control hardware still in development.

3.1.3 Cost Reduction

Various costs can be reduced by virtualisation. The bill of materials of test facilities can be reduced by virtualising the most expensive parts of the system. Footprint of virtual test facilities is typically smaller, reducing required building space. Operational costs can be reduced by virtualising test subjects, the associated consumables such as contrast fluid needed for X-ray imaging, and the parts with the costliest wear and tear. The transportation costs, associated with demonstrating a system at a customer site, fair, conference, or job market, can be reduced by virtualising the bulkiest parts. The add-on costs for virtualisation can be kept modest by standardising on COTS components for virtualisation, e.g. by running a simulated robotic arm on a standard PC.

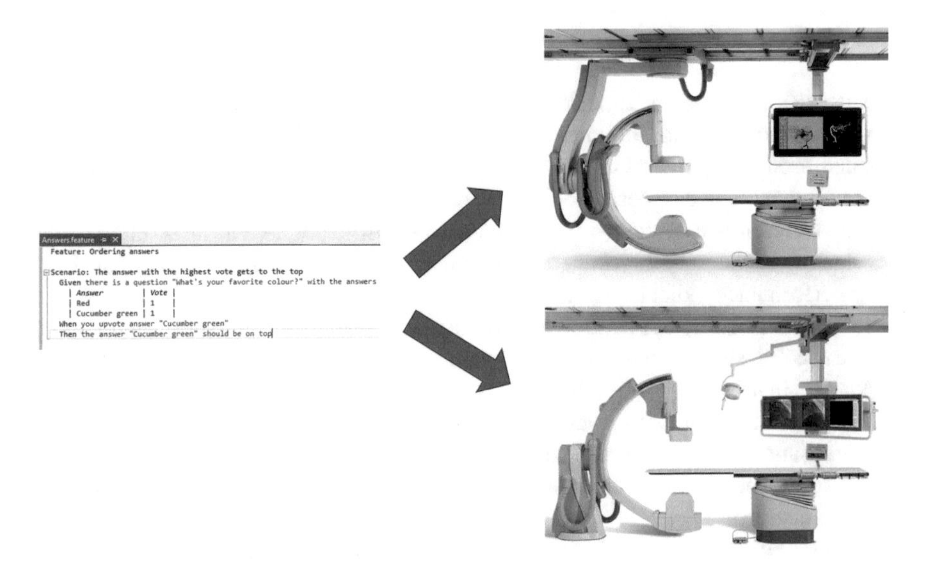

Fig. 4 Executing the same test on multiple configurations

3.1.4 Handle Variability Better by Changing Configurations Quickly

Development in the health domain must handle a huge variability. A product platform of image-guided therapy systems typically contains multiple products. Those products are combined with a variety of third-party equipment in the operating room, such as ECG monitors, patient tables, and contrast injectors, to perform a wide range of medical procedures, including cardio and brain surgery, on a variety of patients, i.e. from babies to adults, males and females, and of different lengths and weights. On top of this, it is important to note that these elements evolve over time.

Using a virtual system mitigates the impact of the growth in this variability. Where changing the complete physical configuration of a system takes a few weeks of manual work, changing a system virtually can be done almost instantaneously and automatically. Thus, with virtualisation, a component that is used in multiple configurations of a product family can also be quickly and automatically tested for multiple configurations, as shown in Fig. 4, *without* needing all those configurations as physical test systems.

3.1.5 Lifting Real-World Restrictions

Using and testing of real-world systems typically imposes restrictions on the use of the system, e.g. for safety reasons. Such restrictions can be lifted by virtualising the components for which these limitations are needed. For instance, humans can interact more freely with the system when the 'dangerous' parts are virtualised, thus preventing damage from collisions and exposure to X-rays. Endurance tests profit from virtualising consumables, such as contrast fluid, making them infinitely available.

Many tests can be executed faster when changes in system state are no longer limited by real-world restrictions, such as physical laws and safety regulations. X-ray imaging speed is restricted to avoid overheating of the X-ray tube. Changing the position and orientation of the robot arm of an image-guided therapy system is limited by its inertia, and by the system itself to prevent severe damage. Due to these limitations some changes in position and orientation of a robot arm take multiple seconds to be executed. Virtually such robot arm movements could be done instantly, hence reducing test time significantly.

3.1.6 Increasing Flexibility and Control for Integration and Test

By using virtual components, one also enables new partial integrations by including virtual components, instead of real, product components, possibly still in development. Hence, integrators and testers have more options to choose those integrations that maximally reduce risk and simplify root cause analysis.

Virtual components can support test interfaces that increase the options to control their behaviour. Examples include testing the system's behaviour while controlling the virtual component's timing behaviour to respond as slowly as allowed or to generate bursts of activity, and erroneous behaviour by injecting non-standard events, such as an X-ray tube overheating, or paper jamming inside a printer.

3.2 Challenges for Virtualisation

To achieve the benefits of virtualisation, several challenges need to be addressed, to ensure that key capabilities are in place and anchored in the product creation process. At a high level, the most important challenge to sustain the benefits of virtualisation throughout the product creation process is to *manage consistency between real, mixed, and virtual systems*, as illustrated in Fig. 5.

To achieve a managed consistency between real, mixed, and virtual systems, three key capabilities need to be realised as follows:

- ability to mix and match virtual and real components;
- consistent behaviour of virtual/mixed/real systems over test scenarios;
- co-governance of the development of real and virtual components.

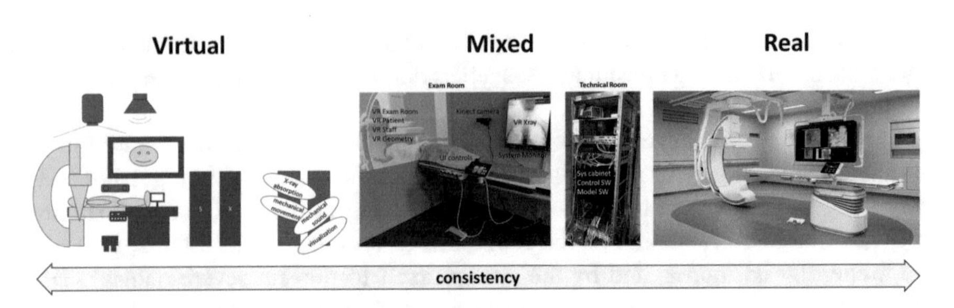

Fig. 5 Need to manage consistency between real, mixed, and virtual systems

3.2.1 Ability to Mix and Match Virtual and Real Components

The ability to mix and match virtual and real components is crucial for the success of a virtual test platform. This ability requires a well-thought-out technical strategy, which is usually captured in a reference architecture [8, 9]. Before the advent of virtualisation, reference architectures typically supported product families and their evolution. Virtualisation puts new demands: a reference architecture should also support mixing and matching of virtual and real components.

Such a reference architecture should enable the quick instantiation and change-over of virtual and mixed reality system configurations, in a possibly virtualised environment and test context. This places new requirements on: interface management; linkage to the test infrastructure; system start-up and shutdown; and configuration change-over facilities, to replace real with virtual components and vice-versa. The reference architecture should support the necessary virtualised/mixed test configurations and support testing needs. A conceptual framework defining the basic virtual/mixed/real system logical decomposition structure and testing context is a key element of such a reference architecture.

3.2.2 Consistent Behaviour of Virtual/Mixed/Real Systems

To obtain the benefits of virtualisation, the behaviour of virtual and mixed virtual/real systems should be consistent with the behaviour of the real system being virtualised as needed for the considered test scenarios. Each virtual component should comply with the interfaces specified for the corresponding real component in the product architecture. Virtual components should use appropriate models to support the considered test scenarios. The models, virtual components, and virtualisation infrastructure need to meet requirements with respect to those scenarios, including test cases, working ranges, and which properties are reflected in the virtual components and to what fidelity.

Virtual components may not need to replicate all properties exactly, e.g. for functional testing, requirements on timing equivalence can be relaxed. To reproduce anomalies as occurring in a real system in a real hospital environment, virtual components and the virtualised context need to have means to detect these anomalies, or they should at least provide support to test engineers to detect and diagnose anomalies. Such anomolies could include failures to disable functionality when specific exception conditions occur.

3.2.3 Co-governance of the Development of Real and Virtual Components

Systems and their constituent components are evolving constantly [10, 11]. The virtual counterparts need to co-evolve to stay in sync. Users of a virtual system and virtual components must be aware of their equivalence relation with the real system: under which conditions may the behaviour of the different versions of both the virtual and real system be considered equivalent and thus replaceable in a test scenario.

To manage this co-evolution, virtual components should be governed like real components are governed. Virtual components, test and virtualisation infrastructure and configurations have to be managed, and have to stay in sync with the evolution of the product and its components. Correspondence has to be maintained between virtual component versions and variants to real product component versions and variants for intended test activities.

Virtual components need to have owners assigned. A logical choice is to have the same owner for a real component and its virtual equivalents, to assure that changes in a real component are always propagated to its virtual equivalent. These owners have to verify that virtual component implementations indeed duly replicate their real-world counterpart behaviour and meet test requirements.

4 Towards a Virtual Test Platform

A virtual test platform aims at linking and unifying all virtualisation initiatives for testing. The objective is to achieve one single, unified approach and a consolidated, maintainable platform for virtual testing. A virtual test platform aims at reducing testing costs by exploiting the commonality while supporting the variability between virtual test configurations. In this section, we describe what a virtual test platform is, what opportunities it provides, and what is needed to create and maintain it.

To understand what a virtual test platform is, we start with describing product platform, configuration, and instance. Next, we describe the concepts of a virtual test platform, configuration, facility, and instance. Then we describe the conceptual framework for a virtual test platform. Finally, we look at the demands of testing use cases on a virtual test platform.

4.1 Product Platform, Configuration, and Instance

An organisation's *product platform* enables series-of-one products by exploiting the commonality while supporting the variability between customer system configurations. The architecture of a product platform defines a structure consisting of a hierarchy of components, interfaces, and constraints. This structure has variation points to support the variability needed in the organisation's range of products, while shared functionality in the range of products is localised within components to maximize reuse. The method to define the architecture of a product platform is called either product family, product line, or domain engineering [12, 13]. For an example in the domain of consumer electronics see [14].

A *product configuration* is created by selecting a variant for each variation point in the product platform. Hence, a product configuration has no variation points left; all instances of a product configuration exhibit the same behaviour. Many companies limit their product configurations to predefined catalogue items. Then, the customer has a limited set of choices. Companies selling one-of-a-kind products offer more freedom to their customers and allow them to create their own product configuration using the components defined in the product platform. Of course, the resulting product configurations must adhere to the product platform constraints.

A *product instance* is a physical realisation of a product configuration. Some hospitals purchase multiple instances of the same product configuration to increase their capacity.

Figure 6 illustrates the relation between product platform, configuration, and instance. This figure shows that the product instance in the LUMC hospital, room 4 is realised by deploying product configuration 3F15, which is part of the Azurion product platform.

4.2 Virtual Test Platform, Configuration, Facility, and Instance

An organisation's *virtual test platform* aims at reducing testing costs by exploiting commonality, while supporting the variability between virtual test configurations. A virtual test platform enables virtual testing of a product instance, or part thereof, by replacing some real components with their virtual equivalents. A virtual test platform enables more options for partial integration by introducing variation points for virtualisation: either the equivalent real set of components or their virtual equivalents are contained.

The architecture of a virtual test platform extends that of a product platform with environmental elements (e.g. the patient), virtual components, virtualisation infrastructure, and testing infrastructure. In addition to these, it also brings new constraints. These constraints are not only among environmental elements, virtual components, and virtualisation infrastructure, but also between these and the

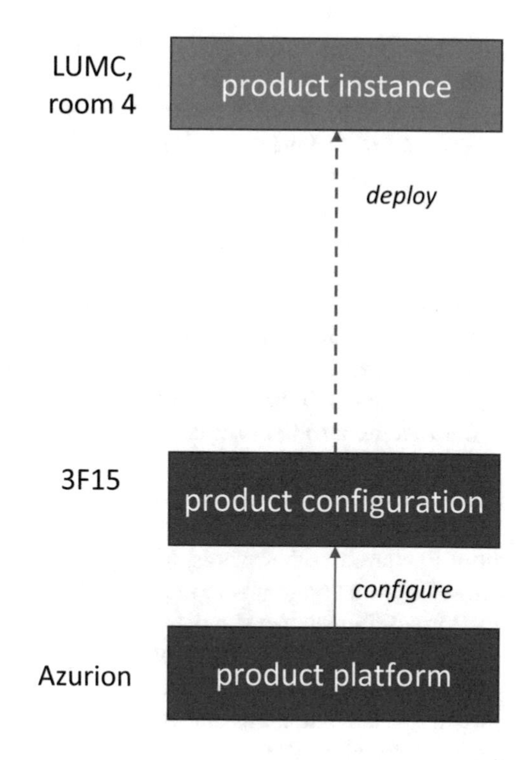

Fig. 6 Relations between product platform, configuration, and instance, with an example from Philips Healthcare

product platform itself. For example, a virtual component for image processing may only be available on Linux.

The environmental elements are needed because a product interacts with its environment using sensors and actuators. When virtualising the components involved in the interaction with the environment, the interacting environmental elements have to be virtualized too. To virtualize the X-ray imaging chain, for instance, including the X-ray tube and the X-ray detector, a model of the patient's anatomy is needed. Environmental elements can have diversity interfaces for testers to configure them. A model of the patient's anatomy could provide a diversity interface to configure the patient's sex, weight, and height.

A virtual component is the virtual equivalent of one or a group of real components. A virtual component does not have to be identical to a real component; it only has to provide an adequate approximation for the purposes considered. A virtual test platform may contain multiple virtual components for the same set of real components. The virtual component variants may differ in some key performance indicators. An imaging component may have two virtual component variants for example, one with a low latency and average image quality; another one with high image quality but no guarantees for latency.

It is useful when virtual components interact with virtualised environmental elements via standardised interfaces. This way, a virtual component can be used with a variety of virtualised environmental elements. For example, a virtual component of an X-ray detector could be connected via the same interface to simulators of either static patients, phantoms, or beating hearts. Having such standardised interfaces eases the development of a variety of virtual components with different working ranges and different levels of fidelity of system properties as needed to support test scenarios (see also Sect. 3.2.2). Having such standardised interfaces also eases the mixing and matching of real and virtual components, and can contribute to achieving a quick instantiation and change over of virtual and mixed reality system configurations (see also Sect. 3.2.1).

Infrastructure for virtualisation is needed for human interactions. Virtualisation includes localisation, audio and vision. This infrastructure must provide standardised interfaces not only to interact with humans but also to create events for the system. When both the patient and robot arm are virtualised with the virtualisation infrastructure, also the (virtual) collisions need to be detected and fed back to the virtualised product.

A *virtual test configuration* is created by selecting a set of compatible virtual components, and the necessary environmental elements and virtualisation infrastructure. In other words, all decisions necessary for virtualisation are made. However, variation points in the product platform may still be open. For example, the Virtual Cathlab can simulate both single and double robot arm product configurations using an X-ray image processing system that can support two streams of images. As with product configurations, a virtual test platform can offer predefined virtual test configurations that act as catalogue item that a tester can select, or the platform can allow the tester to define its own unique configuration while satisfying the constraints of the virtual test platform.

A *virtual test facility* is a physical instance of a virtual test configuration. A company can deploy multiple physical instances of the same virtual test configuration to increase their test capacity, and gain also extra flexibility in planning test facilities. On a virtual test facility, one or more (partial) product configurations can be virtualised.

A *virtual test instance* is a virtual, or mixed real/virtual, realisation of a (partial) product configuration together with environmental elements and test harness as needed. Using a virtual test facility, a tester can test a component that is shared within a product platform by rapidly creating successive test instances for relevant product configurations as needed, using applicable virtual components.

Figure 7 illustrates the relation between a virtual test platform, configuration, facility, and instance. The virtual test instance for the LUMC hospital, room 4, system is realised by deploying and configuring it on the Virtual Cathlab test facility located in room QP1.441. This Virtual Cathlab test facility is a realisation of the Virtual Cathlab virtual test configuration, which is part of the Azurion virtual test platform.

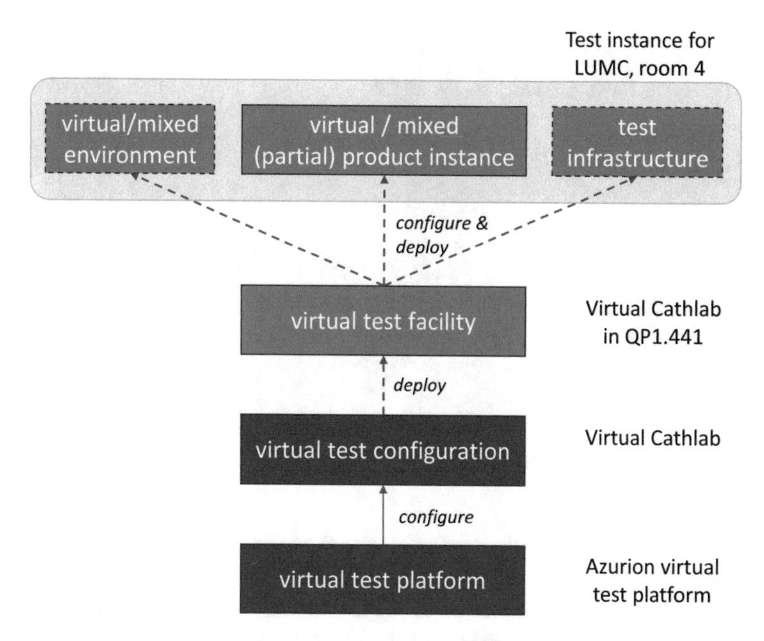

Fig. 7 Relations between virtual test platform, configuration, facility and instance

4.3 *A Conceptual Framework for a Virtual Test Platform*

In the previous sections, we considered the relations between product platforms, configurations, instances, and between virtual test platforms, configurations, facilities and instances. In this section we first describe a logical view on a system in its context and then we introduce the conceptual framework for a virtual test platform, followed by the necessary aspects to consider. We argue that the product platform and virtual test platform must be closely related for such a virtual test platform to be sustainable throughout the R&D development cycles and the product's life cycle.

4.3.1 Logical View on a System in Context

Before describing a conceptual framework for a virtual test platform, we first consider a logical view of a system in its context. This logical view already addresses some of the requirements for a virtual test platform. We relate this system (product) logical view to the logical view of the virtual test platform in the next two sections and explain our choices and considerations there.

The logical view for a (partial) system in its context is depicted in Fig. 8. The system's context consists of operators, i.e. the persons controlling the system, and a physical world in which the system is placed. In case of an Image-Guided Therapy System, the context typically includes the patient, the hospital staff, and all the paraphernalia typically present in an operating room.

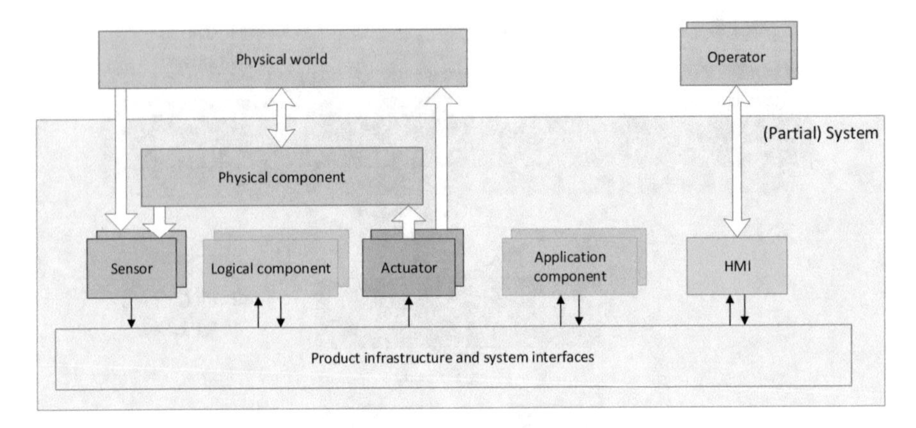

Fig. 8 System in context; logical view

Inside the system, we distinguish a small variety of component types. The Human-Machine Interaction (HMI) unit interacts with the operators through a variety of options such as (touch) displays, knobs, buttons, and pedals. Sensors either observe/measure the physical world, e.g. a proximity sensor, or some aspects of the system itself, e.g. encoders measuring an angle of a motion axis. Actuators similarly act on the physical world, e.g. the X-ray tube generates an X-ray beam, or on some part of the system itself, e.g. a motor driving a motorised patient table. Physical components consist mostly of the system's hardware, such as robot arms, displays, patient tables, and computing hardware. Furthermore, some logical components are present to control the physical hardware, e.g. motion controllers. Further application components may be present to post-process/enhance data or prepare/monitor work flows etc. Finally, a product infrastructure with system interfaces connects all components, and ensures communication between those components.

4.3.2 A Conceptual Framework for a Virtual Test Platform

A virtual test platform needs to include the system under test (partially virtualised) but it also needs to connect to the test infrastructure. A virtualised system may need additional support with respect to a real system; e.g. to virtualise X-ray imaging also a virtual patient (patient model) is needed. Key needs for a virtual test platform are that

- it has to be able to easily connect test infrastructure to a system under test;
- it has to be possible to switch system under test configurations quickly;
- it has to maintain consistency between the product evolutions and virtual test platform over development cycles and over the product's life cycle.

In Fig. 9, the conceptual framework for a virtual test platform is shown. The logical view of the system in its context can be found back in the top half of the schematic. In addition to the conceptual model of the system in its context, it contains three significant extensions:

- *Left*—Test infrastructure (test control, record/replay);
- *Right*—System and virtualisation deployment and configuration control;
- *Bottom*—Virtual system and virtualisation infrastructure.

The *product infrastructure* and system interfaces (should) have a central role in also hooking up these extensions. This will not happen by chance. It has to be deliberately architected in and is one of the key aspects to consider in deploying the conceptual framework.

The *test infrastructure* needs to be present also when testing completely real systems. When used for testing with partly virtual systems additional requirements apply: with a virtual test platform, the test infrastructure needs to be oblivious of whether real or virtual components are being tested, and whether real or virtual sensors/actuators are being used. Test infrastructure may consist of both test control infrastructure, and further components such as record & replay infrastructure.

Deploying a virtual test platform, especially to test multiple system configurations quickly, places special demands on the *deployment configuration and control*, but it may also put demands on the product itself. For new product deployments in a hospital, typically the hardware installation time is the major time factor. In a virtual test facility setting, however, this aspect disappears. Then product software and configuration change-over time quickly becomes dominant and limiting for the change-over time from one configuration to the next.

Next to these two aspects the virtualisation of both the system and the environment are considered too. In the conceptual model, the virtual components are meant to mirror real components. However, this also requires models of physical components to represent shape and dynamics of the robot arm in motion. Furthermore, it may also require a virtual world (a virtual patient, virtual staff) and additional models representing the real world, such as add-on equipment mounted on the table, to determine whether a virtual robot arm would virtually collide with this add-on equipment present.

These models of the physical world may need input from the system. The location of the add-on equipment is in many cases determined in part by the location of the physical table it is mounted on. Such dependencies need to be managed.

Underlying the virtual system, virtual world, and the model of the physical world, is a virtualisation infrastructure. This virtualisation infrastructure, much like an operating system for a computer, needs to tie together these models of the physical components, the virtual world, and the model of the physical world. These should drive the visualisation with all senses, not limited to vision only: the sounds of the robot arm motors and brakes are very important sources of information to physicians, as they provide audible feedback on the actions of the system that may be out of view as the physician, who, while controlling the robot arm, is watching the patient or the X-ray image instead of the robot arm. The virtualisation

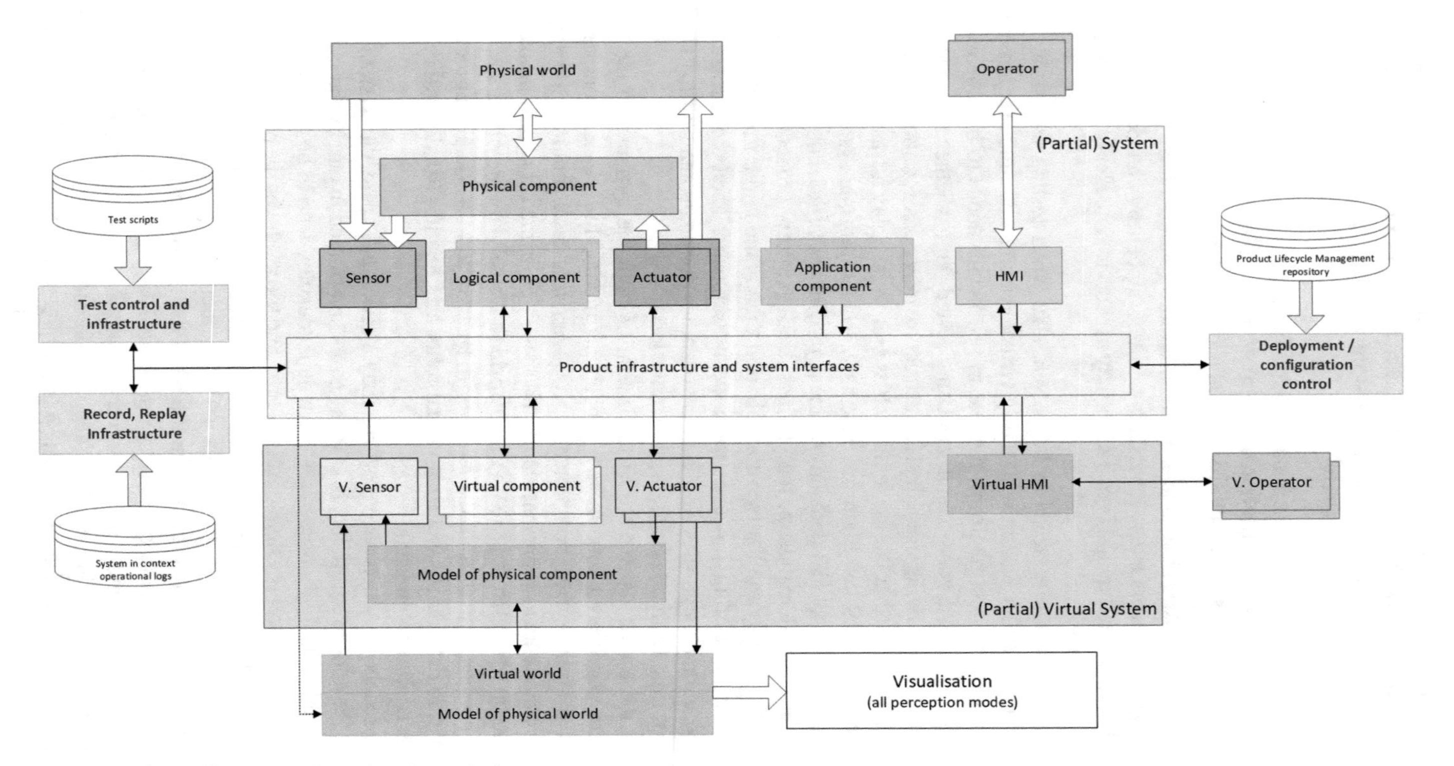

Fig. 9 A conceptual framework for a virtual test platform

infrastructure also takes in virtual actuator outputs and generates virtual sensor signals, by stimulating and coordinating the various models as appropriate.

Figure 10 shows a specialisation of the virtual test platform for a "model-in-the-loop" constellation. In this constellation, the complete system is virtualised, except for the application components and HMI. Virtual components model the logical components, receive input from virtual sensors, and drive virtual actuators. An operator can still control the system through an HMI. This constellation can be used to perform e.g. informal feature tests, or virtual testing of application components.

4.3.3 Key Aspects for a Successful Deployment of a Virtual Test Platform

Successful deployment of a virtual test platform delivering sustainable benefits requires a number of considerations and choices to be made. We mention here a few key aspects (besides ensuring usability) to help guide choices, as follows:

- managed units of virtualisation, along managed system interfaces;
- a product infrastructure handling both real and virtual components;
- managed specifications for virtual components;
- virtual components are tested against their specifications;
- virtual components are first class members of the product family.

The proper determination and selection of *managed units of virtualisation, along managed system interfaces* is a first key aspect influencing the success of a virtual test platform. Whereas prior to virtualisation, components were selected as unit of function/abstraction, or unit of variation, now components and units also need to be selected such that they can be virtualised properly. This implies that the real and virtual components should have the same interfaces (managed system interfaces), and that all relevant aspects of the component interfaces have to be covered in the interface specification. When a real component would need one or more physical interfaces, these should be carefully selected such that they do not hinder virtualisation. A key pitfall is to try to virtualise along physical interfaces with external vendor proprietary protocols, as this may greatly complicate or even hinder virtualisation.

A second key aspect is the *product infrastructure handling both real and virtual components*. The product infrastructure should provide facilities for installing components that implement the specified interfaces. It must provide transparent communication between virtual and real components, and allow quick and easy installation/switching of various system configurations, i.e. a set of real and virtual components. The same infrastructure should be usable both for virtual systems and for real systems (products).

A third key aspect is to have *managed specifications for virtual components*. Also virtual components need to have a specification, i.e. to specify for which use cases they can be used, which properties they emulate well, and which not, and how close the virtual components are to reality, i.e. for what test cases they can be considered equivalent to the real component.

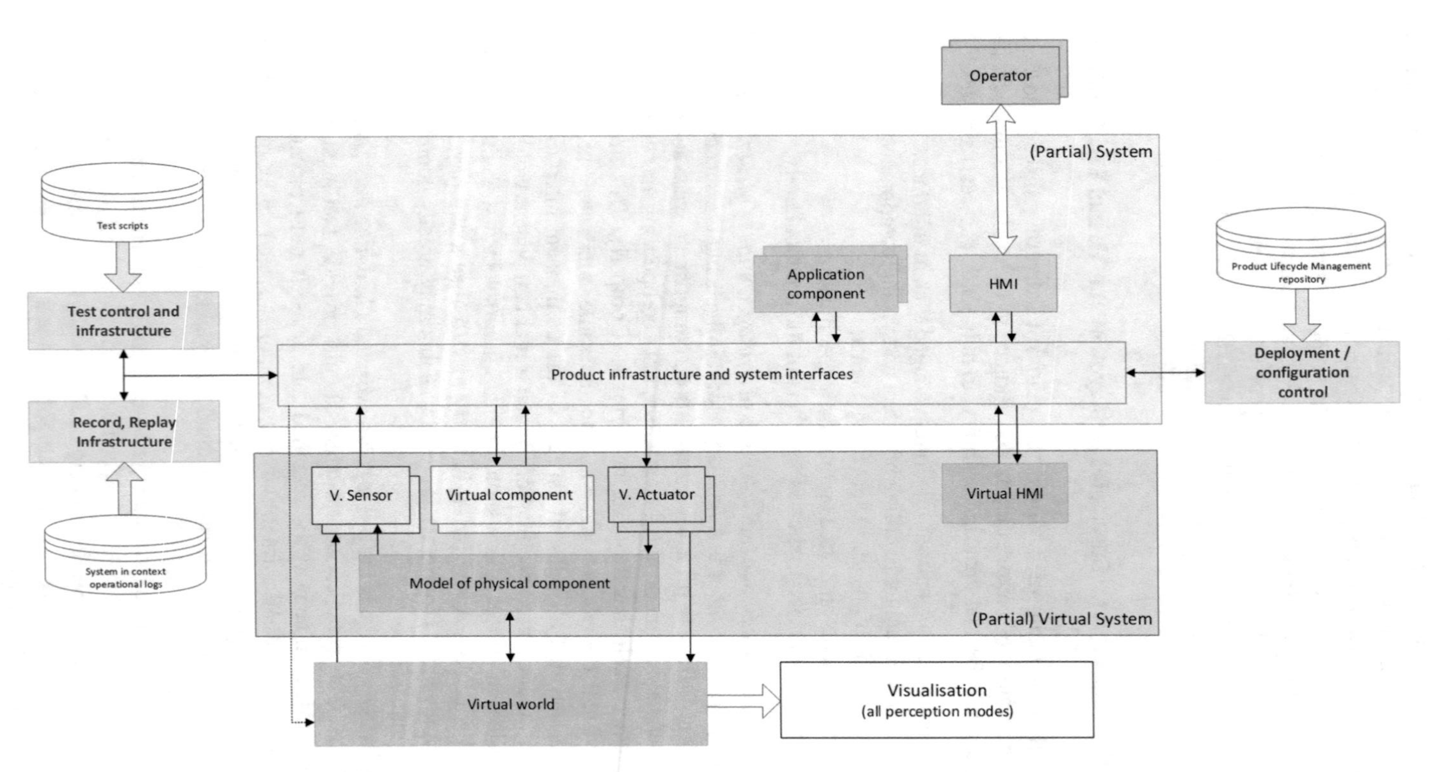

Fig. 10 Virtual test platform specialised for model-in-the-loop testing

A fourth key aspect is to ensure that *virtual components are tested against their specifications*. In the heat of development, often the attention is fully focused on the real component. However, the virtual components need to be qualified as well before they are used. When virtual components and systems are used in system verification tests, their full qualification against specificiations is crucial. Otherwise results of verification tests are questionable.

The fifth key aspect is to ensure that *virtual components are first class members of a product family*. A catalogue of released virtual components, and virtual environmental models, e.g. patient models, should be available. As the real components evolve, the virtual components should co-evolve. A product should not be ready for release when its virtual components are not up-to-date.

Proper consideration of these aspects and their effort involved, is needed to evaluate the business case for virtualisation. Frequently virtualisation efforts start as individual initiatives. Consolidation, embedding in the product creation process, and maintenance of the virtual test platform require an organisation's long-term commitment and stamina even during stressful times in product development.

4.4 Testing Use Cases and Demands on a Virtual Test Platform

A further consideration for the design of a virtual test platform is the types of testing use cases and the corresponding paradigms to be supported by the virtual test platform. Here we consider the Enable-S3 identified use cases for the health domain (see Sect. 2) plus two additional uses cases inspired by other domains. These two additional cases are the following:

- *Record/Replay Test* relates to reproducing a problem—including those reported by a clinical user—in a 'laboratory setting' for diagnosis purposes, such as root cause analysis;
- *KPI-based Performance Test* is about calibration, tuning or optimisation of components or systems with complex physical behaviour over their working range (in automotive, e.g. performance and emission optimisation of an diesel engine, for health e.g. X-ray skin dose w.r.t. to image quality).

Table 1 provides an overview of the various testing use cases and their demands on a virtual test platform and derived test instances. Some of the demands are on the opposite ends of the spectrum, e.g. fast test execution on a minimal test instance for software unit testing versus realism of the product configuration for usability testing. A common theme is the flexibility for integration of (new) elements and ease of change of configurations.

Satisfying all demands of all use cases requires an extensive virtual test platform, cf. the DriveCube concept in the automotive domain [15]. Prioritisation of test cases and their demands towards an incremental realisation of a virtual test platform is needed.

Table 1 Testing use cases and demands on a virtual platform

Testing use case	Demands on a virtual test platform
Early validation test	Ease of modification of virtual test platform to include prototype elements; ease of creation of validation tests; ease of use for non-technical personnel during tests.
Informal feature test	Ease of change of system under test; ease of modification of virtual test platform to integrate new elements of the feature; control interfacing for error injection; flexibility to test with diverse workflows and scenarios.
Software unit test	Test execution at or beyond real-time; fast integration of unit under test in test instance; control interfacing for error injection; fast change of configurations; automation of testing.
Informal software integration test	Ability to integrate before real components are available; ability to execute tests while relaxing some system properties (e.g. timing); down-scoping of virtual test platform to create minimum needed test instances.
Usability test	Close to realistic approximation of real product and product environment; ease of use for (non-technical) usability experts; fast change of configurations for delta testing and A/B testing.
Record/replay test	Fast build-up of customer specific product configurations as replica; infrastructure to replay captured configurations, workflow and scenarios, pause on error detection; system monitoring for root cause analysis.
KPI-based performance test	Possibility for integration of extra instrumentation and monitoring equipment; infrastructure to replay standardised scenarios and workflows; ease of change of configurations and system settings; ease of access to system-internal measurements for monitoring and optimisation.

5 Conclusions

In the health domain, many opportunities exist to apply a virtual test platform. First, low hanging fruit opportunities can address single aspects with few components, such as software unit tests against virtual components. These are applicable when simplified system behaviour, with relaxation of some system properties, is acceptable. Further opportunities require a more elaborate virtual test platform and should be driven by business cases/project needs.

The overall business proposition looks unmistakeably positive however. In the health domain, systems are sold in many variant configurations. Limiting the number of physical variations of test systems needed through virtualisation can reduce test costs considerably. Virtualisation of a product platform can make switching between configurations almost instantaneous, enabling overnight testing of a product platform. Virtualisation also improves the aspect of access to product configurations in development. Tests can be done before actual hardware is available, enabling earlier feedback to development, and resolving problems in an early stage of the project. Testing restrictions can be lifted by virtualisation, in particular when 'dangerous' parts are virtualised, such as the X-ray generation.

The more advanced opportunities will have impact on the product development, architecture, and infrastructure. The product architecture should be (made) composable with identified units of virtualisation. A (standardised) product infrastructure has to support the needs of virtualisation and virtualised testing. To sustain a virtual test platform in an organisation's product creation process, ambition levels should be defined and a corresponding roadmap should be created. Besides the technology development, also the co-governance of real and virtual development needs to be embedded.

Based on the success of the Enable-S3 project, our Enable-S3 project partner Philips has initiated the productization of their Virtual Cathlab towards a virtual test platform. Their activity addresses the following items:

- inclusion of virtualisation in the product architecture, in particular capture the interfaces and stubs necessary for virtualisation;
- assignment of owners for virtualisation artefacts;
- requesting the upcoming development projects to define further use cases for virtualisation and to provide requirements for the necessary models for virtual components;
- standardisation of the installation and initialisation of partial virtual configurations;
- documentation of the current design for virtualisation, partly in the product design history file, and partly in separate design documents for the test tools;
- upgrading the quality and performance of the models, e.g. the X-ray models.

Finally, for the virtual test platform to be sustainable throughout the product creation process, it has to remain aligned to the product platform. And similar to a product platform, also for a virtual test platform commonalities and variations should be governed to sustain the benefits of virtual testing.

Outlook Current work in the Enable-S3 project has focused on the sustainable use of a virtual test platform relatively early in the verification and validation process; to build confidence in the system under development, and to find and resolve problems early on. A key open question is whether, and if so, for what part, a virtual test platform and virtual testing can play a role in creating the required formal test evidence to be submitted to regulatory bodies for release and acceptance of medical devices into the market. This question still requires much research into topics such as: how can virtual components be validated; how can their equivalence to real components be shown, and what type of test evidence is appropriate with virtual testing as opposed to testing with real systems.

Acknowledgements We thank our Enable-S3 project partner Philips for the collaboration in this project, and specifically Jacco van de Laar, Martijn Opheij, Jaap van der Voet, and Peter Benschop for their support, know-how of the health domain, and review of this chapter.

The research is carried out as part of the Enable-S3 program under the responsibility of ESI (TNO) with Philips as the carrying industrial partner. The Enable-S3 research is supported by the Netherlands Ministry of Economic Affairs (Toeslag voor Topconsortia voor Kennis en Innovatie).

References

1. Verbeek, X., Lord, W.: The care cycle: an overview. Medica Mundi. **51**(1), 40–47 (2007)
2. https://www.philips.co.uk/healthcare/solutions/interventional-xray. Accessed 10 Jan 2019
3. ENABLE-S3 Consortium: Generic Test Architecture. https://www.enable-s3.eu/media/publications/ (2017). Accessed 10 Jan 2019
4. https://en.wikipedia.org/wiki/Cath_lab. Accessed 10 Jan 2019
5. Doornbos, R., Huijbrechts, B., Sleuters, J., Verriet, J., Ševo, K., Verberkt, M.: A domain model-centric approach for the development of large-scale office lighting systems. In: Bonjour, E., Krob, D., Palladino, L., Stephan, F. (eds.) Complex Systems Design & Management. CSD&M 2018. Springer, Cham (2019)
6. Verriet, J., Buit, L., Doornbos, R., Huijbrechts, B., Ševo, K., Sleuters, J., Verberkt, M.: Virtual prototyping of large-scale IoT control systems using domain-specific languages. Proceedings of MODELSWARD2019 (2019)
7. Moline, J.: Virtual reality for health care: a survey. Stud. Health Technol. Inform. **44**, 3–34 (1997)
8. Cloutier, R., Muller, G., Verma, D., Nilchiani, R., Hole, E., Bone, M.: The concept of reference architectures. Syst. Eng. **13**(1), 14–27 (2010)
9. Muller, G., van de Laar, P.: Researching Reference Architectures and their relationship with frameworks, methods, techniques, and tools. In: Proceedings of the 7th Annual Conference on Systems Engineering Research (CSER 2009), Loughborough, England, 20–23 April (2009)
10. van de Laar, P., Punter, T. (eds.): Views on Evolvability of Embedded Systems. Springer, New York (2011)
11. van de Laar, P., Hendriks, T.: A retrospective analysis of Teletext: an interoperability standard evolving already over 30 years. Adv. Eng. Inform. **26**(3), 516–528 (2012)
12. Van der Linden, F., Schmid, K., Rommes, E.: The product line engineering approach. In: Software Product Lines in Action, pp. 3–20. Springer, Berlin (2007)
13. https://en.wikipedia.org/wiki/Product-family_engineering. Accessed 10 Jan 2019
14. van Ommering, R., van der Linden, F., Kramer, J., Magee, J.: The koala component model for consumer electronics software. Computer. **33**(3), 78–85 (2000)
15. Schyr, C., Brissard, A.: DrivingCube – a novel concept for validation of powertrain and steering systems with automated driving. In: Advanced Vehicle Control: Proceedings of the 13th International Symposium on Advanced Vehicle Control (AVEC'16), Munich, 13–16 September 2016, p. 79. CRC Press, Boca Raton (2016)

Printed in the United States
By Bookmasters